D1187155

Affinity
Chromatography

Affinity Chromatography

C. R. LOWE
and
P. D. G. DEAN

Department of Biochemistry,
Liverpool University

A Wiley–Interscience Publication

JOHN WILEY & SONS

London · New York · Sydney · Toronto

Library of Congress Cataloging in Publication Data:

Lowe, Christopher R.
Affinity chromatography.

"A Wiley–Interscience publication."
1. Affinity chromatography. 2. Enzymes—Analysis.
I. Dean, Peter D. G., joint author. II. Title.

QP601.L69   547′.758   73–17598
ISBN 0 471 54940 1

Printed in Great Britain by
William Clowes & Sons, Limited,
London, Beccles and Colchester

Preface

The technique of affinity chromatography has been applied to almost every area of biochemistry. The range of applications includes simple enzyme purifications, studies of enzyme mechanism, the isolation of specific receptor sites in cells and the development of complex multi-enzyme reactors. Apart from its intrinsic research value, this technique has potential uses in both industry (e.g. food and drug processing) and medicine (from routine clinical services to advances in medical technology).

As early as 1910, Starkenstein, in a preparation of amylase on insoluble starch, illustrated the inherent advantages of affinity chromatography. However, despite the early work of Lerman's and McCormick's groups, a considerable time-lapse occurred before the full potential of the technique was realized and the major advances in affinity chromatography undoubtedly stem from two papers in the late 1960's. In 1967, Porath and his group published a procedure for specific immobilization of amines onto inert polymers. The second paper, by Cuatrecasas, Wilchek and Anfinsen (1968), as well as introducing the term 'affinity chromatography', demonstrated the use of specific adsorbents in enzyme purification. The inherent advantages of this technique over classical procedures have resulted in an extremely diverse literature, the exponential growth of which must surely present a daunting task to any worker interested in this field.

Our objective in writing this book is to guide the student through the maze of literature. We have attempted to bring together some of the principles of affinity chromatography, drawing both on published work and on theoretical considerations. Some sections have been devoted to applications of affinity chromatography in molecular biology and biochemical research, while the final chapter provides details of the chemistry and methodology required to prepare affinity matrices. In writing the book we had in mind both students and research workers who are interested in using theoretical principles to help them through the design and analysis of possible experiments. The examples have been chosen, not to provide a comprehensive review of papers in the field, but rather to illustrate some of the parameters which seem to us to be important to the fundamentals of affinity separations.

Many of our ideas were developed whilst we were teaching a course in the Biochemistry Department at Liverpool University; the references used have been selected from the literature up to the middle of 1973.

C. R. LOWE
P. D. G. DEAN

Contents

Chapter I
Introduction

Techniques for the purification of proteins have advanced considerably since the days when good fortune and an element of practical expertise were the essential qualifications of the protein chemist. Today, it is possible to devise separative procedures based on such rational parameters as the size, shape or charge of the protein, and more recently on the nature of the ligand with which the protein interacts. Indeed, the evolution of the tactics of protein purification has been such that any competent protein chemist can reasonably be expected to purify a protein to currently acceptable standards of homogeneity. Many of these advances are a direct consequence of our greater understanding of protein structure and function.

A. PROTEIN STRUCTURE

Proteins are extremely complex macromolecules which have discrete physical, chemical and biological properties. They are elaborate polypeptides containing a non-systematic but quite specific sequence of some 20 different amino acid residues.[1] Amino acids are chemical compounds which have in common an acidic carboxyl (—COOH) and a basic amino (—NH_2) function but differ in the nature of a group (R) attached to the same α-carbon atom.

$$\begin{array}{c} \mathrm{H} \\ | \\ \mathrm{R{-}C^{\alpha}{-}COOH} \\ | \\ \mathrm{NH_2} \end{array}$$

Table I.1 lists the structures of some common amino acids and emphasizes the variation in the substituent R, which may be highly reactive in some cases, like the sulphydryl group of cysteine, or chemically inert, as in leucine.

Peptide bond formation occurs when the amino group attached to the α-carbon atom of one amino acid is condensed with the carboxyl group attached to the α-carbon atom of a second amino acid, with the concomitant elimination of a water molecule (Fig. I.1). When additional amino acid residues are subsequently attached, polypeptides are formed. The fixed amino acid sequence of the polypeptide chain, the so-called primary structure, is the basis for other levels of structural organization collectively termed the *conformation* of the protein. The polypeptide chain can be folded and bent into a secondary structure which is stabilized by the formation of linkages and bonds between adjacent amino acids. The planar character of the peptide bond and the rotational restrictions of the two single bonds (Fig. I.1) markedly limit the number of allowed conformations. If the substituents (R) on the amino acids permit, intra-chain hydrogen bonding can restrain the polypeptide chain into an α-helical form.[2] However, despite the internal hydrogen bonding, this helical structure is not sufficiently stable to exist in aqueous

Table I.1. The structures of some amino acids

General structure: $R{-}C(H)(NH_2){-}COOH$

Amino acid	R
Tryptophan	indol-3-yl—CH_2— (benzene ring fused to five-membered ring with N)
Lysine	$NH_2{-}(CH_2)_3{-}CH_2{-}$
Histidine	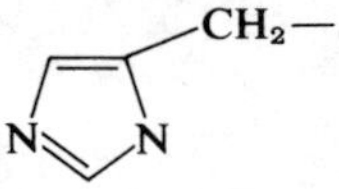 imidazolyl (ring with N, N)—CH_2—
Tyrosine	$HO{-}C_6H_4{-}CH_2{-}$
Glutamic acid	$HOOC{-}CH_2{-}CH_2{-}$
Serine	$HO{-}CH_2{-}$
Cysteine	$HS{-}CH_2{-}$
Leucine	$(CH_3)_2CH{-}CH_2{-}$

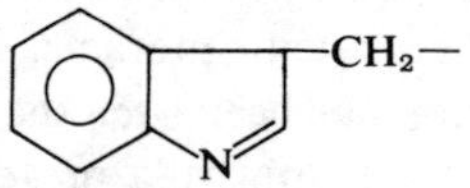

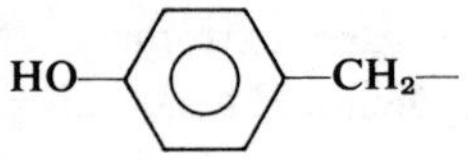

media without unfolding into a random disoriented strand. Environmental stability can often only be achieved if the helical chains are themselves folded and bent. This element of protein structure, the tertiary structure, is maintained by hydrogen bonding, electrostatic bonds, covalent linkages and hydrophobic interactions.[3] It is this level of organization of proteins that is particularly sensitive to environmental fluctuations and the one that confers on the protein the property of biological specificity. The tertiary level of organization is often augmented by a further level where polypeptide subunits are aggregated in various geometrical arrangements.[4] The resulting multi-subunit proteins are particularly sensitive to environment and the presence or absence of small ligands termed effectors. The majority of allosteric enzymes are multi-subunit proteins.[5]

The similarity in structure of most proteins is paralleled by a marked similarity in overall physicochemical properties, although small differences in amino acid content or distribution will be reflected in a slightly altered charge, size or shape of the protein. Fig. I.2 shows the structure of a relatively simple protein.[11]

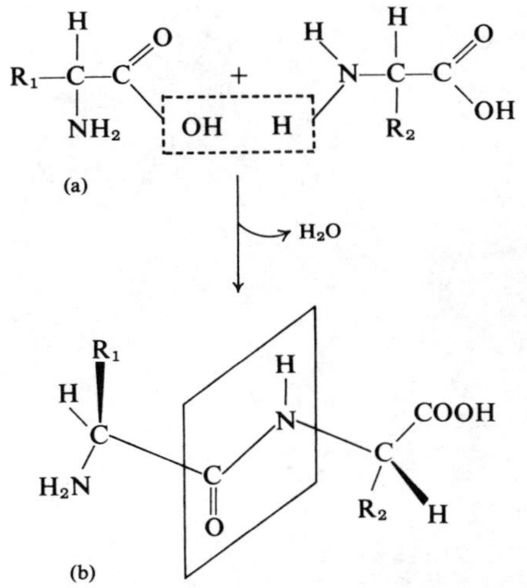

Figure I.1. The formation of a peptide bond

B. CLASSICAL PROTEIN PURIFICATION

Classical procedures of protein separation and purification are generally based on the relatively small differences in the physicochemical properties of proteins in the mixture. They are hence unselective, tedious and of poor resolution. The problems encountered by the protein chemist are thus subject to the following considerations.

The protein of interest may constitute only 0·1% or less of the dry weight of the starting material, and most of the remainder will consist of other proteins with closely allied properties. Furthermore, the classical methods adopted by the organic chemist, such as distillation and solvent extraction, are not applicable to the separation of proteins, owing to their size and instability. This instability is reflected in altered solubility properties occasioned either by changes in temperature, acidity or alkalinity or by a variety of chemical agents. The process of conversion of the 'native' protein to a

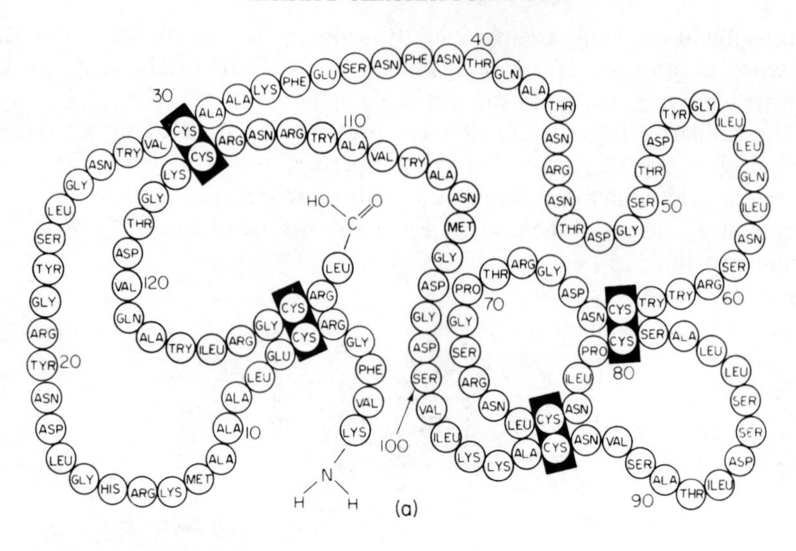

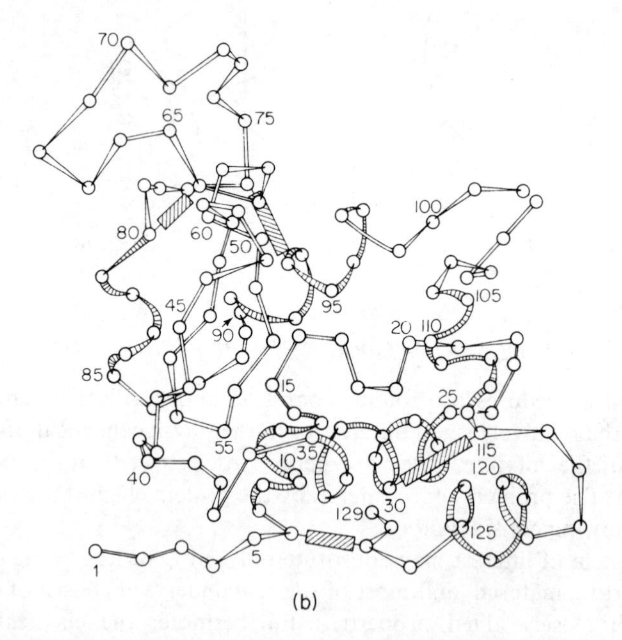

Figure I.2. The structure of egg white lysozyme. (a) The primary structure depicting the sequence of amino acid residues and the positions of the four disulphide bridges. (b) Schematic drawing of the main chain conformation. (a) Reproduced with permission from R. E. Canfield and A. K. Liu, *J. Biol. Chem.*, **240**, 1997 (1965). (b) Reproduced with permission from C. C. F. Blake *et al.*, *Nature* (*London*), **206**, 757 (1965)

product with altered properties is termed 'denaturation' and the product is a 'denatured' protein.

The first consideration in the isolation of proteins from natural materials is to find a source that is rich in the desired protein. This may frequently involve the assay of the discrete biological activity of the protein in a variety of potential sources to select the one that is the most favourable. Where the biological source is a micro-organism the amount of the desired enzyme or protein may be increased by genetic and environmental manipulation by harnessing such control mechanisms as induction, end product repression and catabolite repression. Since each control mechanism is influenced by environmental conditions, parameters such as temperature, pH, medium composition, aeration and stage of development are critical and optima must be determined empirically for the desired protein. The specific activity of enzymes can be altered several hundredfold by these methods.

Once the preliminary screening of natural sources has been performed, the next stage is to free the protein from the tissue. This is usually done by disrupting the cellular organization of the tissue by exposure either to a grinding or shearing action, to ultrasonic irradiation or to solutions of low ionic strength. If the desired protein is bound to a particulate portion of the cell, additional treatment with detergents or lipid solvents may be necessary to release it.

Since proteins are fragile molecules, several important precautions should be observed during the procedures necessary for extraction of the protein from the tissue and during the subsequent fractionation steps.

(i) The temperature should be maintained as near to the freezing point of the solvent as possible since exposure even to physiological temperatures can elicit slow denaturation *in vitro*.

(ii) The pH must be carefully controlled since the susceptibility of proteins to alterations in pH varies greatly.

(iii) The protein concentration should be kept as high as possible to prevent denaturation by dilution.

Many fractionation methods have been devised for the purification of proteins that depend on the small differences in stability, charge, size or shape of individual proteins. These will be considered briefly in turn.

(*a*) *Stability*. If the protein of interest happens to be unusually stable to heat or to extremes of pH, a brief heat treatment or exposure to high or low pH may serve to precipitate the bulk of contaminating material. This could effect a substantial purification. The solubility of proteins varies greatly with pH, and for each protein there is a pH value of minimum solubility, the *isoelectric point*.

(*b*) *Exploitation of charge proteins*. As already noted, proteins are large molecules with many positively and negatively charged groups. These can interact with each other, with small ions of opposite charge in the solvent, or with water. The force, F, between two charges of opposite sign (q_1 and q_2) is given by the expression:

$$F = \frac{q_1 q_2}{Dr^2}$$

where r is the distance between them and D is the dielectric constant of the medium. The latter, which is a measure of the influence of the medium of the interaction, can be altered by the addition of inorganic salts or organic solvents to the medium. The counter-ion concentration is increased until aggregation occurs and the protein is said to be '*salted out*'. Neutral salts such as ammonium sulphate, potassium sulphate or sodium sulphate are commonly used although the concentration and nature of the salt used are generally determined empirically. In general, selection of a pH value near the isoelectric point of the desired protein will lead to sharper precipitation.

The addition of organic solvents to the medium lowers the dielectric constant, reduces the activity of water, and hence favours protein–protein interaction and precipitation. Ethanol, methanol, acetone and 1,2-pentanediol are commonly used although the method suffers the disadvantage that enzyme activity can be lost if the temperature is not well regulated. The advantages of the method arise from the lower density of the resultant solution and hence lower centrifugation times. Neutral salts in general increase the solubility of proteins in organic solvents and hence careful optimization of temperature, pH, protein concentration, ionic strength and the dielectric constant are necessary in order to precipitate the desired protein. Under these conditions, individual proteins may be precipitated by neutral salts or by organic solvents within a comparatively narrow concentration range of the precipitants.

The selective adsorption and subsequent desorption of proteins on certain inert materials has been used as a purification procedure. Such adsorbents include calcium phosphate, starch and alumina gels, hydroxylapatite or diatomaceous earth (celite), and the process can be effected either on a column or in slurries of the materials.

A more universal approach to the purification of proteins employs column chromatography on ion-exchange resins. Resins with hydrophilic polysaccharide backbones that allow ready access of the protein to the ion-exchange group are most satisfactory for this purpose. The most commonly employed materials are an anion-exchange resin, diethylaminoethyl-cellulose (DEAE-cellulose) and a cation-exchange resin, carboxy-methyl-cellulose (CM-cellulose). The structures of these exchangers are shown in Fig. I.3. Such columns are developed by increasing the ionic strength or by changing the pH, temperature or dielectric constant of the eluting buffer. Providing that the correct methodology is employed, excellent resolution of complex mixtures of proteins and hence substantial purifications can be achieved.

Preparative electrophoretic techniques such as starch block, agarose electrophoresis or isoelectric focusing, may effect excellent resolution of some complex mixtures with almost quantitative recovery of the proteins. However, the low capacity of these procedures limits their general applicability.

(*c*) *Size and shape.* The use of molecular sieves, which fractionate proteins according to their size, has found considerable application in protein isolation. The most commonly employed sieves consist of cross-linked polysaccharide matrices that are available commercially in a variety of pore sizes. The exclusion limit is determined by the degree of cross-linking and can be employed for proteins with molecular weights ranging from less than 20,000 to several millions. Similar sieves based on cross-linked polyacrylamide have been utilized. More recently, DEAE- and CM-functions have been linked to these gels, such that simultaneous fractionation on the basis of size and charge may be effected.

A typical purification scheme will consist of a combination of these techniques to encompass the differences in stability, charge and molecular weight

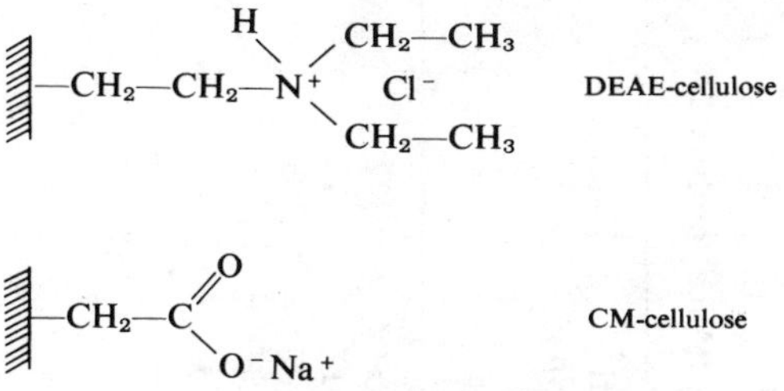

$$-CH_2-CH_2-N^{+}(H)(CH_2-CH_3)_2 \; Cl^{-} \qquad \text{DEAE-cellulose}$$

$$-CH_2-C(=O)-O^{-}Na^{+} \qquad \text{CM-cellulose}$$

Figure I.3. The structures of two common cellulosic ion-exchange adsorbents

of the desired protein over the contaminating materials. Provided that the use of these procedures has yielded a reasonably homogeneous protein preparation, crystallization may sometimes be achieved. Repeated recrystallizations can often significantly increase the purity of the final product.

The evaluation of a proposed step in an overall purification scheme rests on two criteria: the recovery and the degree of purification. These parameters are based on the specific activity of the protein. This is a convenient measure of the purity of the desired protein based on the enzymatic activity or some spectral property and related to the total amount of protein present. The yield represents the percentage of the total activity in the original extract that is recovered after each step, and the degree of purification the increase in specific activity at each stage. A typical purification scheme is presented in Table I.2, which demonstrates the relationship between, and usefulness of, the various quantities used to define the scheme.

Table I.2. Summary of a typical multi-stage protein purification scheme

Stage	Total protein (mg)	Volume (ml)	Total activity (units)	Specific activity (units/mg protein)	Yield (%)	Purification (fold)
1. Crude extract	19,200	372	192	0·01	100	1
2. Ammonium sulphate precipitation	5,400	74	120	0·02	62	2
3. DEAE-cellulose chromatography	340	103	100	0·29	52	29
4. Hydroxylapatite chromatography	50	20	80	1·6	41	160
5. Sephadex G-100 gel filtration	6	20	60	10·0	30	1,000

C. SPECIFIC MODIFICATION OF CLASSICAL PROCEDURES

The above considerations illustrate the empirical nature of conventional enzyme purification techniques and emphasize the need for a more rational approach. It is ironical that the very feature of protein structure that has hampered classical techniques for several decades should be the one that holds considerable promise for the future. The tertiary structure of proteins, primarily responsible for the physicochemical properties, also confers on them the property of biological specificity. Enzymes specifically and reversibly bind substrates, coenzymes, allosteric effectors and other ligands in such a way as to alter the overall physicochemical properties of the protein. A very low concentration of effector could produce a significant alteration in charge, size, shape or stability of the protein. Such effects can be used at any stage of the classical scheme as follows.

(*a*) *Stability*. The addition of a specific ligand to its complementary macromolecule can increase the resistance of the latter to the extremes of heat or pH. Contaminating proteins can thus be more readily removed.

(*b*) *Size and shape*. The binding of ligands to polymeric enzymes or proteins can lead to association or dissociation and hence to alterations in the apparent molecular weight. These changes can be detected in preparative ultracentrifugation or by gel filtration on molecular sieves.

(*c*) *Charge*. The selective binding of ligands to proteins can increase their solubility in organic solvents and alter the suceptibility to precipitation by inorganic salts. The alteration in overall charge of the protein occasioned by the binding of a specific ligand has been used to elute the desired protein from cellulosic ion exchangers and other adsorbents. Thus, yeast pyrophosphatase has been eluted from cγ alumina with dilute pyrophosphate[6] and fructose-1,6-diphosphatase and aldolase have been extensively purified from CM-cellulose columns by elution with low concentrations of fructose-1,6-diphosphate.[7] Furthermore, α-amylase has been purified and separated from β-amylase by adsorption on 'insoluble' starch and elution with 'soluble' starch.[8] Carminatti *et al.*[9] have eluted liver pyruvate kinase from CM-cellulose columns with dilute solutions of the positive allosteric effector, fructose-1,6-diphosphate.

In each of the foregoing examples increased selectivity was achieved by alteration of an overall property of the protein by specific binding of the ligand. In most cases the increased selectivity was achieved in the desorption phase of chromatography. The selectivity can also be enhanced at the adsorption stage of the chromatographic process and still leave the desorption stage free for further enhancement of specificity.

D. *AFFINITY CHROMATOGRAPHY*

The technique of affinity chromatography exploits the unique biological specificity of the protein–ligand interaction. This concept is realized by binding the ligand to an insoluble support and packing the support into a chromatographic column. In principle only enzymes or proteins with appreciable affinity for the ligand will be retained on such a column; others will pass through unretarded. Specifically adsorbed protein can then be eluted by altering the composition of the solvent to favour dissociation. The potential applications of this type of process for the purification and exploration of complex biological interactions are immediately apparent. In principle, affinity chromatography can be applied where any particular ligand interacts specifically with a biomolecule. For example, specific adsorbents can be used to purify enzymes, antibodies, nucleic acids and cofactors, vitamins, repressors, transport, drug or hormone binding receptor proteins. Furthermore, the technique can be employed for concentrating dilute protein solutions, separating denatured from biologically active forms of proteins and the resolution of protein components resulting from specific chemical modifications of purified proteins. These and other potential applications of affinity chromatography are summarized in Table I.3.

Table I.3. Some applications of affinity chromatography

1. Protein purification	Enzymes
	Antibodies and antigens
	Binding or receptor proteins
	Complementary proteins
	Repressor proteins
2. Separative procedures	Cells and viruses
	Denatured and chemically modified proteins from native proteins
	Nucleic acids and nucleotides
3. Concentration of dilute protein solutions	
4. Storage of otherwise unstable proteins in immobilized form	
5. Investigation of kinetic sequences and mechanisms	

Clearly, this method has several inherent advantages over the classical means of protein purification. The rapidity and ease of the separation as essentially a single-step procedure leads to ready resolution of the protein to be purified from inhibitors and destructive contaminants such as proteases. Furthermore, yields are often high, possibly because of protection from denaturation during isolation by stabilization of the tertiary structure.

At first sight the technique of affinity chromatography presents a panacea for enzyme purification. In practice, however, the restrictions and limitations of the technique are more numerous than the above elementary treatise would suppose. *Successful application of the method depends largely on how closely the experimental conditions chosen permit the ligand interaction characteristic of the components in free solution.* Clearly, we must try to simulate the enzyme–ligand interaction that would occur in free solution.

REFERENCES

1. Sanger, F. (1952): *Adv. Protein Chem.*, **7**, 1.
2. Pauling, L., Corey, R. B., and Branson, H. R. (1951): *Proc. Nat. Acad. Sci. U.S.A.*, **37**, 205.
3. Kendrew, J. C. (1963): *Science*, **139**, 1259.
4. Reithel, F. (1963): *Adv. Protein Chem.*, **18**, 124.
5. Monod, J., Wyman, J., and Changeux, J. P. (1965): *J. Mol. Biol.*, **12**, 88.
6. Heppel, L. A. (1955): in *Methods in Enzymology*, Vol. II (Eds. S. P. Colowick and N. O. Kaplan), Academic Press, New York, p. 576.
7. Pogell, B. M. (1962): *Biochem. Biophys. Res. Commun.*, **7**, 225.
8. Fischer, E., and Stein, E. A. (1960): in *The Enzymes*, Vol. IV (Eds. P. D. Boyer, H. Lardy and K. Myrback), Academic Press, New York, p. 315.
9. Carminatti, H., Rozengurt, E., and Jiménez de Asúa, L. (1969): *FEBS Lett.*, **4**, 307.
10. Canfield, R. E., and Liu, A. K. (1965): *J. Biol. Chem.*, **240**, 1997.
11. Blake, C. C. F., Koenig, D. F., Mair, G. A., North, A. C. T., Phillips, D. C., and Sarma, V. R. (1965): *Nature* (*London*), **206**, 757.

Chapter II

The Principles of Affinity Chromatography

Considerable interest has recently been shown in the development of water-soluble derivatives for the isolation and purification of biological macromolecules. This approach, affinity chromatography, is realized by covalently attaching a specific ligand that interacts with the desired macromolecule to an insoluble inert support. It is important that the experimental conditions chosen for the functional purification should reflect the interaction between the ligand and macromolecule that exists in free solution. Hence, careful consideration must be given to the nature of the inert matrix and to the steric problems associated with the immobilization of the ligand. Thus, the distance separating the ligand from the lattice backbone, the flexibility of the group interposed between the ligand and the matrix and the effects of immobilization on the stereochemistry of the ligand are important parameters which can influence the interaction. Furthermore, the nature, mode of attachment and concentration of the ligand can profoundly influence the adsorption of the complementary macromolecule and its subsequent elution. Consequently, the specific conditions selected for the purification of a particular macromolecule must be unique for each case and reflect the selectivity of the interaction.

A. THE SOLID MATRIX SUPPORT

1. General Principles

It has already been stated that the success of the method depends largely on mimicking the interaction that would occur if the components were in free solution. Careful consideration must therefore be given to the nature of the solid matrix. Clearly the matrix must have a number of favourable characteristics.[1] Bearing in mind the ideal circumstance, with no matrix at all, the insoluble support should form a loose, porous network which allows the uniform and unimpaired entry and exit of large macromolecules throughout the entire matrix. The gel particles should be uniform, spherical and rigid.

A high degree of porosity is a critical feature for ligand protein systems since it is the concentration of ligand freely available to the macromolecule that determines the behaviour of the system under chromatographic conditions. This consideration is more important for ligand–protein systems of relatively weak affinity (dissociation constant $\geqslant 10^{-4}$M) since a high ligand concentration is essential to permit interactions strong enough physically to retard the downward migration of the protein through the column. A high porosity of the matrix has two other desirable effects: firstly, it allows good flow properties and, secondly, it does not hinder the penetration of macromolecules of high molecular weight. It has been demonstrated that the exclusion effect of the matrix can be turned to advantage to exclude selectively the interactions of large macromolecules with affinity ligands;[2] an effect which will be discussed elsewhere in this book.

The matrix backbone must interact very weakly with proteins, in general, to minimize non-specific adsorption and, most important of all, it must possess chemical groups which can be activated or modified, under conditions which are not detrimental to its structure, to permit the covalent linkage of a variety of ligands. These chemical groups on the matrix backbone should be so abundant as to allow a high effective concentration of the coupled ligand and hence satisfactory retardation of proteins with ligand systems of low affinity. Furthermore, the solid support must be physically and chemically stable to the conditions selected for coupling, adsorption and elution. The dimensional rigidity, and hence the porosity, of the matrix should not be altered under these conditions. The properties of the ideal matrix are summarized in Table II.1.

Table II.1. Some qualities of the ideal matrix

1. The matrix must form a loose porous network which permits unimpaired movement of large macromolecules
2. The gel particles should be uniform, spherical and rigid with good flow properties
3. The matrix should not interact with proteins in general so that there is no non-specific adsorption
4. The inert supportant must have an abundant supply of chemical groups which can be activated or functionalized to allow the covalent attachment of a variety of ligands
5. The gel must be mechanically and chemically stable to the conditions of coupling, adsorption and elution

2. The Diversity of Insoluble Supports

The restrictions imposed on the choice of the water-insoluble carrier have seriously hampered the usefulness of the technique, since until recently nearly all solid supports have been derivatives of cellulose, polystyrene or synthetic poly-amino acids.

a. Cellulose

Hydrophilic cellulose derivatives have been used in the purification of antibodies and enzymes. In the pioneering work of Lerman a *p*-phenylazophenol was coupled to a *p*-aminobenzylcellulose by diazotization and utilized to purify mushroom tyrosinase,[3] whilst flavin derivatives of CM-cellulose and cellulose specifically adsorbed liver flavokinase and enriched the preparation several hundred times.[4,5] Cellulose derivatives were particularly effective in the field of nucleotide chemistry; thus, Sander *et al.*[6] separated nucleotides on thymidylate-cellulose and complementary strands of nucleic acids and certain species of transfer RNA were resolved on other derivatives.[7,8] Furthermore, the preparation of biotin-cellulose showed that even non-enzymatic proteins, such as avidin, could be enriched by using specific polymer matrices.[9]

Although cellulose derivatives may be advantageous in specific instances, their usefulness is limited by their fibrous and non-uniform character which

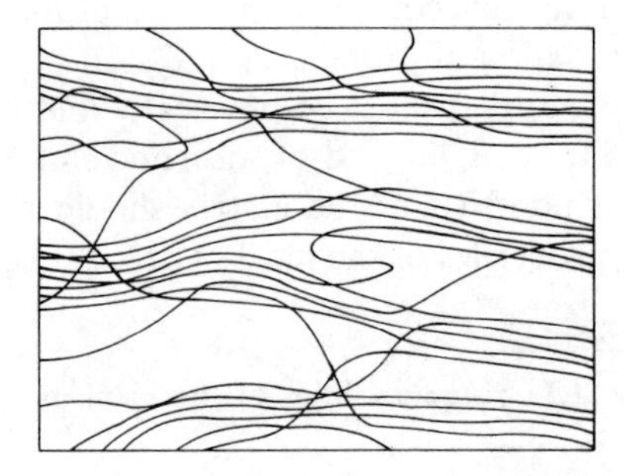

Figure II.1. Diagrammatic representation of the organization of molecules within the cellulose structure showing the areas of order and disorder. Reproduced from C. S. Knight, *Adv. in Chromatog.*, **4**, 61 (1967), by courtesy of Marcel Dekker, Inc.

impedes proper penetration of large macromolecules.[10] The cellulose fibre consists of an aggregation of glucosidic chains with a molecular weight $\geqslant 300{,}000$. The glucosidic molecules have a very high capacity for hydrogen bonding and consequently a high degree of molecular orientation. The hydrogen bonding is confined to a single plane at right angles to the chain axis, whilst van der Waals' forces exist between the molecules at right angles to this plane. The high degree of molecular structure leads to crystalline regions and disordered areas are generated where the glucosidic chains leave these regions (Fig. II.1). This molecular order–disorder relationship is very susceptible to chemical and physical attack and, under preparative conditions, preferential reaction at lower-order regions occurs. This micro-heterogeneity of ligand density within the cellulose matrix creates a spectrum of affinity for the complementary macromolecule and can lead to undesirable effects on the capacity of the adsorbent.

b. Polystyrene Gels

The deficiencies of cellulose also manifest themselves in the highly hydrophobic polymers, such as polystyrene, which display poor communication between the aqueous and solid phases.

Polystyrene gels, cross-linked with divinylbenzene, have formed the basis for most ion-exchange resins in the past. The application of these gels to gel chromatography has not yielded very promising results. Despite good swelling properties, a low porosity and predominance of adsorptive and partition effects have been demonstrated. Acrylic gels also display similar defects.

c. Cross-linked Dextrans

Various hydrophilic polysaccharide polymers have been used with some success. Of these, the cross-linked dextran derivatives possess most of the desirable characteristics detailed above, except for their low degree of porosity. This has severely restricted their use in enzyme purification since they are relatively ineffective as adsorbents even for enzymes of low molecular weight.

d. Polyacrylamide Gels

The synthetic polyacrylamide gels also possess many desirable features and are commercially available in beaded, spherical form, pregraded in sizes and porosities. Polyacrylamide beads are superior to many polymeric supports in that their polyethylene backbone endows chemical stability coupled with a statistically uniform physical state and porosity, permitting the penetration of macromolecules with molecular weights up to 500,000. The principal advantage of polyacrylamide is that it possesses a very abundant supply of modifiable groups (Fig. II.2) which, together with a versatility in derivatization techniques, allows the covalent attachment of a variety of ligands.[11] Thus, highly substituted derivatives may be prepared for use with interacting systems that display poor affinity for the attached ligand.

The use of polyacrylamide carriers for the affinity chromatography of enzymes has been limited. This appears to be principally the result of the low degree of porosity of the beads currently available and to the shrinkage observed during the chemical modifications required for attachment of ligands.[12] Furthermore, problems may arise in column chromatography with charged polyacrylamide derivatives with elution gradients that span ionic strengths somewhat above and below 0·01, since large bed volume changes can occur in the lower cross-linked forms.[11]

The deficiencies of polyacrylamide as an insoluble matrix for affinity chromatography are highlighted by the observation that derivatives containing a β-galactosidase inhibitor were totally ineffective in retarding migration of the *E. coli* enzyme.[12] An analogous agarose adsorbent which contained 50 times less ligand was effective. These differences are ascribed to the differences in the concentration of ligand freely available to the macromolecule and demonstrate that a capacity for chemical substitution for a solid support is not necessarily the best criterion for an operationally efficient adsorbent. It is

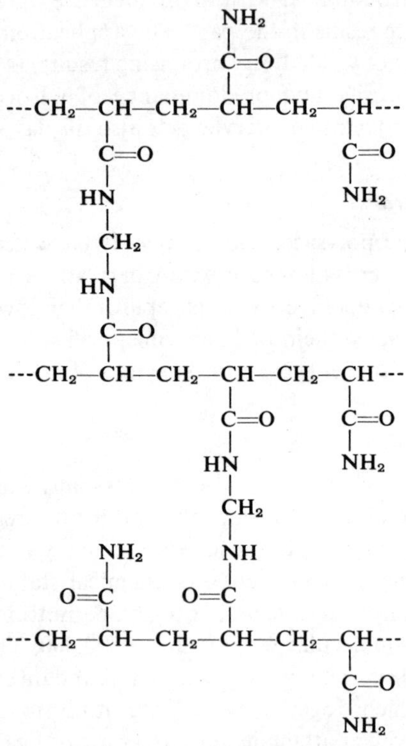

```
                        NH2
                        |
                        C=O
                        |
---CH2—CH—CH2—CH—CH2—CH---
       |                 |
       C=O               C=O
       |                 |
       HN                NH2
       |
       CH2
       |
       HN
       |
       C=O
       |
---CH2—CH—CH2—CH—CH2—CH---
                 |       |
                 C=O     C=O
                 |       |
                 HN      NH2
                 |
                 CH2
                 |
       NH2       NH
       |         |
     O=C       O=C
       |         |
---CH2—CH—CH2—CH—CH2—CH---
                         |
                         C=O
                         |
                         NH2
```

Figure II.2. The structure of part of a polyacrylamide matrix. Reproduced by permission of Bio-Rad Laboratories from *Gel Chromatography*

clear that accessibility to the ligand is a fundamental concept of affinity chromatography. The development of improved polyacrylamide beads for selective adsorbents of high porosity would certainly answer many of the present criticisms. Nevertheless, in some circumstances, such as intact cell separations, polyacrylamide gels may prove superior to the use of agarose. Furthermore, polyacrylamide has special advantages over dextran and cellulose as an inert matrix in terms of improved resolution, reduced background adsorption and chemical stability.[13,14]

e. Porous Glass

Weetall and Hersh[15] have recently described a technique for the immobilization of the organic ligands on glass surfaces. Glass matrices are available commercially as porous granules of high silica glass permeated by interconnecting pores of uniform and precisely controlled size. These column packings are rigid, insoluble and unaffected by changes in the eluant or solvent system, pressure, flow rates, pH or ionic strength. The controlled pore size generates sharp exclusion limits, good resolution and high reproducibility and, since there is no compaction, high flow rates are attainable and vary linearly with pressure. Consequently, separations are fast and the recoveries excellent. Furthermore, inorganic carriers such as glass have the inherent advantages of resistance to microbial attack and ease of sterilization by disinfectants or heat. This means that such columns can be re-used many times without significant deterioration in their capacity or resolution.

Like all silica glasses, however, the surface of porous glass beads consists of hydroxyl groups which exhibit a slight negative surface charge in aqueous solution. While most acidic proteins, viruses, polysaccharides or nucleic acids elute without any retardation or adsorption, all strongly basic proteins and some neutral proteins and viruses may be adsorbed. This non-specific adsorption of proteins to glass surfaces presents a serious hazard to their widespread use in selective functional purifications.

Glass beads may be employed to advantage with high-affinity systems and for the industrialization of current laboratory procedures. Unfortunately, prohibitively slow flow rates are often produced by the tight packing of the beads and lead to mechanical entrapment of particulate materials. These limitations in chromatographic technique can be overcome with batchwise techniques, provided that the system under study displays a high degree of affinity.

Despite these serious restrictions, glass beads coated with antigen have been moderately successful in the separation of immune lymphoid cells.[16,17] An enzymatically active derivative of NAD^+ has also been reported.[18]

f. Agarose

The beaded derivatives of agarose[19] have many of the properties of the ideal matrix and have now been used successfully in numerous purification procedures. They have a very loose structure which allows ready penetration by macromolecules with molecular weights in the order of several millions. The diffusion equilibria of substances of low diffusion constant should be attained most readily by fractionation in a gel of fine particles. The uniform spherical shape of the gel particles is thus of particular significance. Furthermore, these polysaccharides can readily undergo substitution reactions by activation with cyanogen halides, are stable and have a moderately high capacity for substitution.[20–22]

Agarose is a linear polysaccharide consisting of alternating residues of D-galactose and 3,6-anhydro-L-galactose[23] (Fig. II.3). The agarose gels, which are stabilized by hydrogen bonding, cause very little denaturation or adsorption of sensitive biochemical substances because of their hydrophilic nature and the nearly complete absence of charged groups. One serious limitation on the use of agarose gels should be stated: by their very nature the range of applications of these thermally reversible gels is limited. The gel 'melts' at high temperatures, or upon the application of high concentrations of eluants which are capable of breaking down hydrogen bonds (e.g., urea, guanidine–HCl) and can no longer be used. Furthermore, microbial growth leading to clogging of the column has to be carefully controlled. The introduction of a suitable bacteriostatic agent (0·02% sodium azide) is sufficient to curb the latter effect.

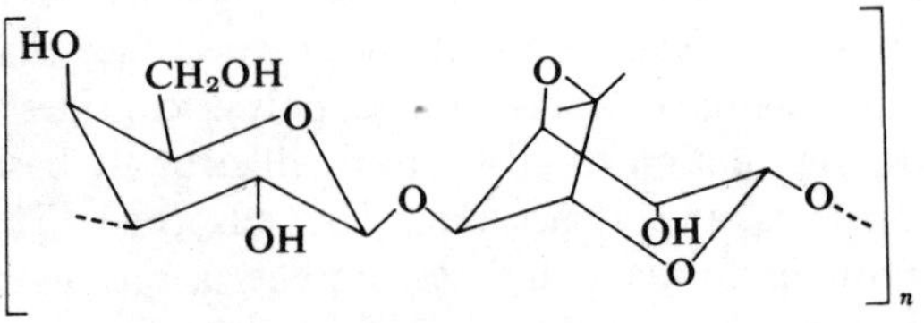

Figure II.3. The structure of the repeating unit of agarose. Reproduced by permission of Bio-Rad Laboratories from *Gel Chromatography*

Consideration of the foregoing should demonstrate that there is no ready rule as to which inert matrix will give the best results for the individual system under study. A largely empirical approach is still, unfortunately, necessary. However, experience has shown that the beaded agarose derivatives are generally more suitable for the purification of enzymes and proteins by affinity chromatography, a fine vindication of the concepts on which the technique is based.

B. LIMITATIONS IMPOSED BY THE MATRIX

The presence of an inert gel matrix to which the affinity ligand is anchored generates two major steric and diffusional problems for successful interaction with the complementary macromolecule. The first relates to the molecular sieving and exclusion effects of the matrix which determine the initial penetration of the protein into the pores of the lattice and the second relates to the influence of the matrix backbone on the micro-environment of the immobilized ligand.

1. Exclusion Effects of the Matrix

The solvent in a column packed with swollen gel particles may be regarded as being in two components:[24] in the spaces between the particles, the *void volume*, V_0, and within the gel particles, the *internal volume*, V_i. Thus if V_t is the total packed volume of the column,

$$V_t = V_0 + V_i + V_g$$

where V_g is the volume occupied by the gel matrix itself.

A solute introduced into the column with equilibrate between the mobile and gel phases, although only a fraction of the volume of the gel phase, represented by the *distribution coefficient*, K_d, is available to the substance. The total volume accessible is thus $K_d \times V_i$, and if the substance is included within the pores then the solute will emerge in the mobile phase after a volume, V_e, the *elution volume*, has been displaced.[25]

$$V_e = V_0 + K_d \times V_i$$

The rate of permeation of the solute through the gel bed is characterized by the distribution coefficient:

$$K_d = \frac{V_e - V_0}{V_i}$$

This parameter is characteristic for chromatography of a given solute on a given gel under specified conditions, but is independent of bed geometry. For solutes completely excluded from the space within the gel, $K_d = 0$ and hence $V_e = V_0$, i.e., the solute emerges in the void volume. For very small molecules with almost free access to the internal volume of the gel, K_d approaches 1. In practice, however, not all the internal volume, V_i, is available even to small molecules, as part is bound to the dextran backbone as solvent of solvation. The actual value is uncertain although in aqueous solvents the true stationary phase volume is about $0{\cdot}8\ V_i$.

The volume of the solvent within the gel particles (V_i) and the volume of the gel matrix (V_g) are difficult to evaluate realistically. Consequently, an alternative means of expressing solute behaviour in terms of an *available distribution coefficient*, K_{av}, is evoked,[26] where the term V_i is replaced by the total volume of the gel ($V_i + V_g$).

$$K_{av} = \left(\frac{V_e - V_0}{V_i + V_g}\right) = \left(\frac{V_e - V_0}{V_t - V_0}\right)$$

In this expression, K_{av} is easily evaluated in terms of the bed parameters, V_t and V_0, and the elution volume, V_e.

There is a close correlation between molecular size and elution behaviour.[27] Solutes included in the gel matrix are ideally eluted in order of decreasing effective molecular size. Fig. II.4 demonstrates the relationship between K_{av} and molecular weight for a series of globular proteins. Below a certain molecular weight, all solutes are eluted together at a volume approaching the total column volume and the curve is almost horizontal.

The central part of the curve is linearly inclined downwards and, in this range, differences in molecular weight generate differences in elution volume. This region of the curve represents the working range of the gel where resolution of protein mixtures according to size is possible. At higher molecular weights, the curve is again almost horizontal and all solutes move with the void volume. The point where the curve reaches $K_{av} = 0$ or $V_e = V_0$ is termed the *exclusion limit* of the gel.

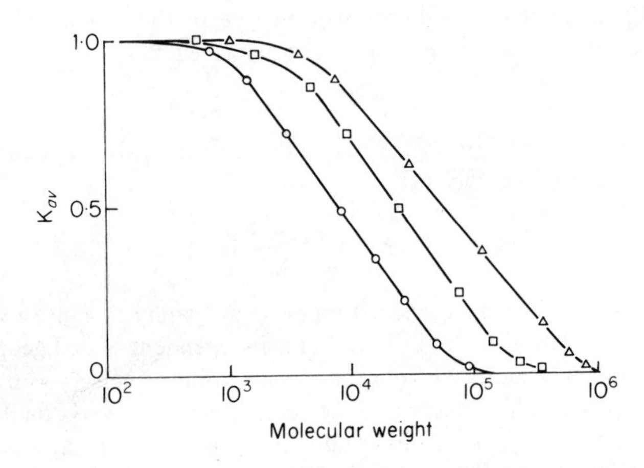

Figure II.4. The relationship between elution behaviour and molecular properties: K_{av} versus molecular weight, low globular proteins on different types of Sephadex. △ G-200, □ G-100, ○ G-75

The first attempt to give an explanation of this phenomenon was that of Flodin,[28] who considered the partition of a solute between the gel and the liquid to be entirely governed by steric factors. The gel matrix forms a network of varying density; large molecules can only penetrate into regions where the meshes in the net are large. Small molecules can, on the other hand, penetrate more closely knit regions of the network nearer to the cross-links. In this way, larger molecules can only penetrate a limited part of the gel.

Most experimental evidence supports this assumption. Thus evidence for a predominance of steric effects rests on the changes in elution behaviour that follow steric changes such as denaturation of proteins[29,30] and on the temperature dependence of chromatographic behaviour. Since steric effects are essentially thermodynamic entropy effects, the partition coefficient will have very little temperature dependence. For proteins this has been found to be the general rule.[31] However, in some cases, especially with tightly cross-linked gels, pronounced temperature effects have been observed;[32] invariably, however, these substances behave abnormally in other ways and indicate a marked interaction with the gel matrix.[33] The dextran chains in 'Sephadex' can attract solutes that interact with glucose or polyglucans. Thus, dextranases,[34] certain polyglucanases such as cellulase,[35] α-amylase,[36] α-glucosidase[37] and glucokinase[38] complex with the matrix backbone itself and lead to an alteration in K_{av} values.

Clearly, when a specific ligand is immobilized within a gel lattice the porosity of the beads will affect the accessibility of the ligand towards interactions with its complementary macromolecule. The effect of the degree of porosity of the Sephadex matrix on the binding of dehydrogenases to immobilized NAD^+ has been demonstrated by Lowe and Dean.[2] NAD^+ was coupled directly to Sephadex G-25, G-50, G-100 and G-200 by the cyanogen bromide technique of Axen *et al.*[21] The behaviour of several pyridine nucleotide-dependent dehydrogenases of different molecular weights was dependent on the extent of their exclusion from the gel filtration media. Thus, with synthetic mixtures of L-malate dehydrogenase, L-lactate dehydrogenase and bovine serum albumin (BSA), the enzymes appeared in the void volumes of columns comprising NAD^+ covalently attached to the highly cross-linked gel, Sephadex G-25, in which most, if not all, of the immobilized ligand was inaccessible to the enzymes (Fig. II.5). Malate dehydrogenase was included in NAD^+-Sephadex G-100 whilst the higher molecular weight lactate dehydrogenase was not. Both lactate and malate dehydrogenases were bound to NAD^+-Sephadex G-200. Furthermore, the strength of the interaction between the enzyme and the immobilized NAD^+, as measured by the concentration of KCl required to elute the enzyme on a linear KCl gradient, was increased when the pore size of the gel was increased. This behaviour has been related to the molecular weight of the enzymes.[2]

The exclusion of a solute from part of a solvent by the presence of a neutral polymer can also increase the chemical potential of the solute. Studies by Laurent[39] on the reaction catalysed by lactate dehydrogenase have shown that dextran affects the initial reaction rate to a much larger extent than the final equilibrium. This finding is ascribed to the decrease in mobility of reactants, which must occur in a three-dimensional network. The decrease in diffusion coefficients of the solutes results in a decrease in collision frequency between reactants and thus a decrease in reaction rates.

For some kinds of enzyme–ligand interactions the porosity of the bead or matrix may not be a limiting factor. This may be the case where the capacity of the adsorbent is not a limiting factor, where high-affinity systems are being purified or where one of the interactants is an extremely large complex. Thus,

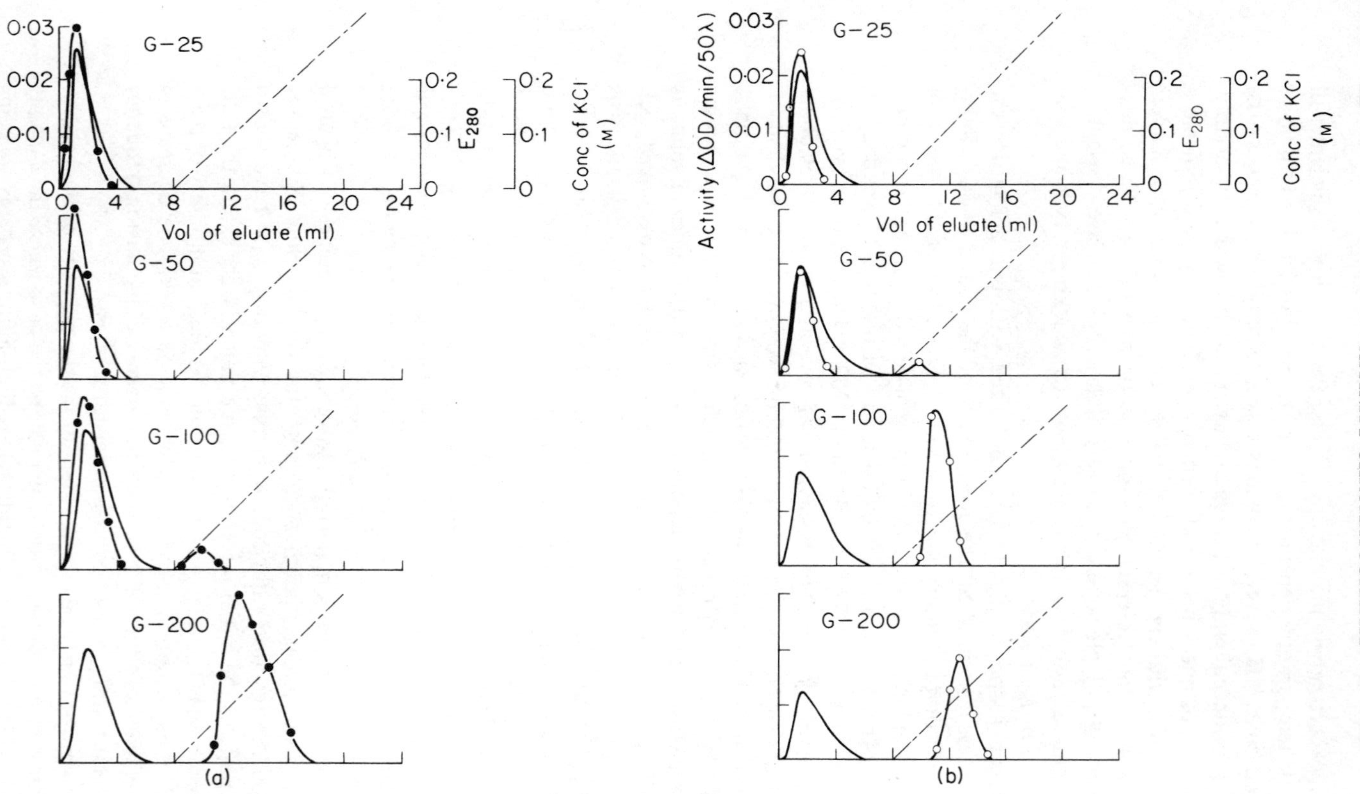

Figure II.5. The effect of bead porosity on the interaction of (a) lactate dehydrogenase and (b) malate dehydrogenase with NAD^+-Sephadex of various pore sizes. A 50 μl sample containing 1·85 units lactate dehydrogenase (or 0·335 units malate dehydrogenase) and 0·8 mg bovine serum albumin (BSA) was applied to a 5 mm × 20 mm column of the appropriate NAD^+-Sephadex equilibrated with 10 mM KH_2PO_4-KOH buffer, pH 7·5. Non-adsorbed protein was washed off with the same buffer and the column eluted with a 0–0·5M KCl gradient in 10 mM KH_2PO_4-KOH buffer, pH 7·5, 20 ml total volume. Lactate dehydrogenase (●), malate dehydrogenase (○) and BSA (——) were assayed in the effluent. Reproduced with permission from C. R. Lowe and P. D. G. Dean, *FEBS Lett.*, 19, 21 (1971).

polysomes, ribosomes, intact cells, organelles or membrane fractions may not reasonably be expected to penetrate the pores of the bead. In general, interactions of high affinity and specificity occur between the ligand immobilized on the *surface* of the bead and the macromolecules or complex in the mobile phase. Consequently, large spherical beads with large spaces *between* them are well suited to this purpose.

Hapten-polyacrylamide derivatives have been successfully employed in selectively adsorbing lymphocytes capable of recognizing the hapten.[40] Similarly, insulin-agarose retains fat-cell ghosts containing specific receptor sites for the hormone[41] and glucagon-agarose will selectively adsorb liver membranes.[42]

2. Spacer Molecules and Steric Considerations

The critical role of the matrix in the determination of the accessibility of the coupled ligand to the macromolecule has already been emphasized. Nevertheless, irrespective of the porosity of the polymeric support if the ligand is coupled directly to the matrix backbone, steric hindrance with the ligand interaction will be experienced. It is thus obvious that, for successful purification by affinity chromatography, the chemical groups critical in the interaction with the macromolecule must be sufficiently distant from the solid matrix support.[12,22] Steric considerations appear to be most important with proteins of high molecular weight and with systems of low affinity.[43] The problem can be approached by placing the ligand at the end of a long chain or 'arm' such that it protrudes into the solvent. Several general procedures are now available for the preparation of gels which contain such extension arms of varying length.[22] Fig. II.6 illustrates diagrammatically the principle of accessibility to the ligand.

The importance of interposing a hydrocarbon chain between the ligand and the matrix backbone is shown by the relative ineffectiveness of insolubilized D-tryptophan methyl ester compared to ε-aminohexanoyl-D-tryptophan methyl ester in the purification of α-chymotrypsin.[1] Furthermore, this necessity for an extension arm is also illustrated in studies on the purification of *E. coli* β-galactosidase[12] with agarose derivatives containing the relatively weak competitive inhibitor, *p*-aminophenyl-β-D-thiogalactopyranoside (Fig. II.7). Direct attachment of the ligand to the matrix (Fig. II.7(a)) resulted in an adsorbent that displayed no affinity for the enzyme. By placing the inhibitor at a moderate distance (~10 Å) from the solid support (Fig. II.7(b)), β-galactosidase activity was slightly retarded and emerged just after the void volume of the column had passed. Insertion of a very long spacer molecule, 3-aminosuccinyl-3′-aminodipropylamine (~21 Å) between the ligand and the matrix (Fig. II.7(c)) strongly retained the enzyme from several sources.

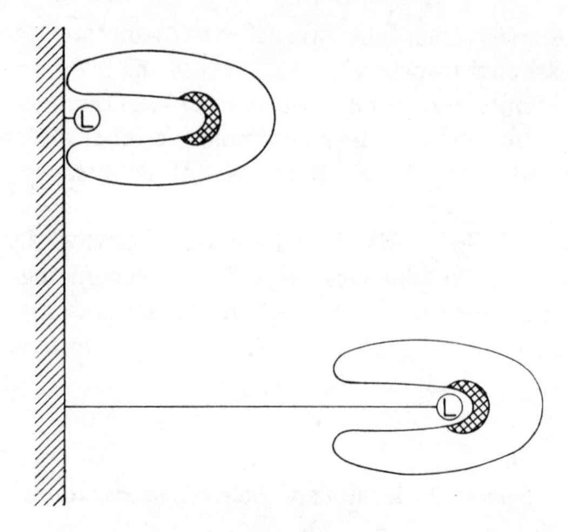

Figure II.6. The principle of accessibility and the use of spacer arms in affinity chromatography. The ligand binding site is cross-hatched

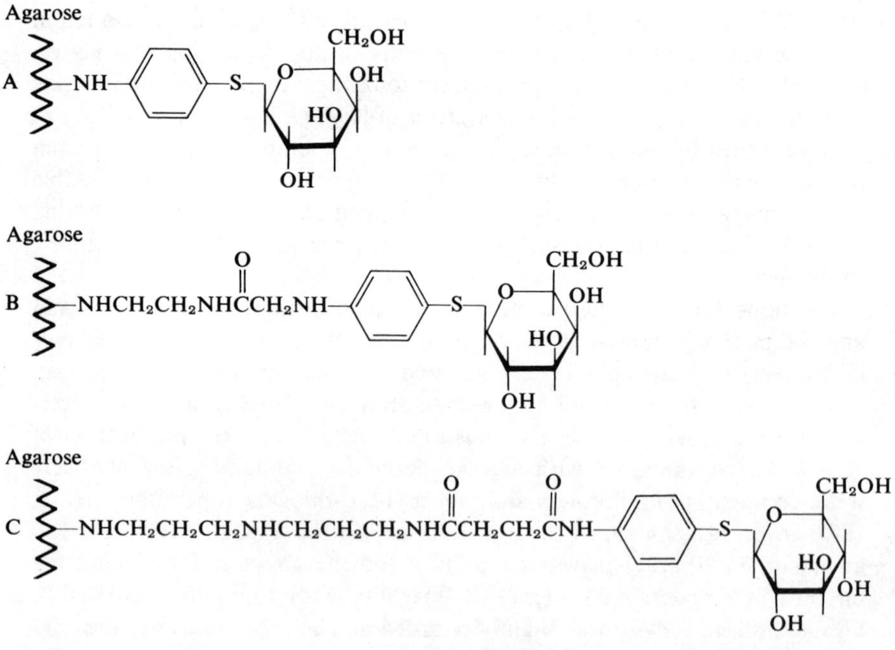

Agarose

A —NH—(phenyl)—S—(galactoside: CH_2OH, OH, HO, OH)

Agarose

B —$NHCH_2CH_2NHCCH_2NH$—(phenyl)—S—(galactoside: CH_2OH, OH, HO, OH) [C=O]

Agarose

C —$NHCH_2CH_2CH_2NHCH_2CH_2CH_2NHCCH_2CH_2CNH$—(phenyl)—S—(galactoside: CH_2OH, HO, OH, OH) [C=O, C=O]

Figure II.7. Agarose adsorbents prepared for the selective purification of β-galactosidase from several sources by affinity chromatography. Reproduced with permission from P. Cuatrecasas, *Adv. Enzymol.*, **36**, 29 (1972)

Many other examples of the use of extension arms to increase the steric availability of the ligand are to be found in the literature. Until recently, however, there has been no systematic study of the effect that increasing the length of the extension arm has on the strength of the interaction between the complementary macromolecule and the immobilized ligand. Lowe *et al.*[43] have made a thorough study of this effect with the interaction of kinases and pyridine nucleotide-dependent dehydrogenases with insoluble derivatives of ATP and NAD^+. Fig. II.8 illustrates the effect of increasing the chain length

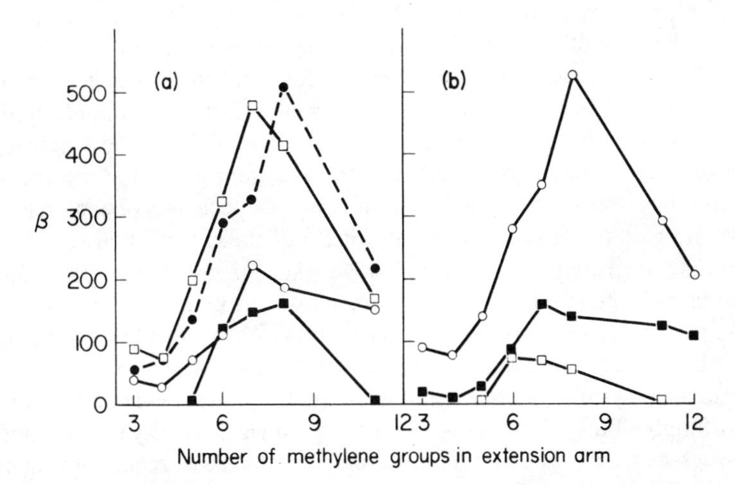

Figure II.8. The effect of extension arm length on the binding of dehydrogenases and kinases to immobilized nucleotides. (a) □, Lactate dehydrogenase from rabbit skeletal muscle, M_4; ●, lactate dehydrogenase from pig heart muscle, H_4; ○, malate dehydrogenase; ■, glucose 6-phosphate dehydrogenase; immobilized NAD^+. (b) □, Hexokinase; ■, 3-phosphoglycerate kinase; ○, glycerokinase; immobilized ATP. β Represents a measure of the strength of the interaction of the enzyme with the immobilized nucleotide and is the concentration of KCl (mM) required to elute the enzyme. Reproduced with permission from C. R. Lowe *et al., Biochem. J.*, **133**, 499–506 (1973)

on the interaction of several kinases and dehydrogenases with agarose derivatives containing ATP and NAD^+ attached to the matrix with polymethylene extension arms of various lengths. The system allows the effects of affinity and molecular size to be investigated using the same ligand under essentially identical conditions. Several important points emerge from these results. The binding of all the kinases and dehydrogenases was weak to zero when the extension arm contained up to four methylene groups and the nucleotide was in close proximity (0–5 Å) to the matrix backbone. When the length of the extension arm was increased from 5 to 10 Å, by increasing the number of interposed methylene groups from four to eight, there was a substantial increase in the strength of binding of the enzymes in that a significantly higher concentration of salt was required to elute them from the columns. However, when extension arms containing more than eight methylene groups (> 10 Å) were interposed between the matrix skeleton and the ligand, there was a concomitant fall in the strength of the interaction. This relationship between the length of the extension arm and the strength of the interaction of the macromolecule with the immobilized ligand was of the same form for all the enzymes studied. Maximum binding was observed when the length of the extension arm was in the region of 8–10 Å. At first sight this is surprising since, in the case of the dehydrogenases, the apparent dissociation constant for NAD^+ in free solution differed by 10^3-fold and the molecular weight by 5-fold over those enzymes investigated. However, further examination of Fig. II.8 reveals that the increase in the strength of binding as the length of the extension arm is increased is related to the apparent molecular weight of the enzyme. Thus, for low molecular weight proteins, such as 3-phosphoglycerate kinase and malate dehydrogenase, the length of the extension arm was not as critical as for proteins with molecular weights greater than 70,000, such as lactate dehydrogenase or glycerokinase. However, contributions from other factors such as the conformation or accessibility of the nucleotide binding site cannot be entirely ruled out in the interpretation of these results.

Interaction systems of weak affinity require a longer extension arm than those of high affinity. Thus, enzymes displaying weak affinity for immobilized ATP, such as hexokinase and 3-phosphoglycerate kinase, require an extension arm containing at least five methylene groups before any significant interaction is observed (Fig. II.8). Furthermore, D-glucose 6-phosphate dehydrogenase has less affinity for NAD^+ than its preferred cofactor, $NADP^+$. No retardation on immobilized NAD^+ was observed until at least five spacer atoms were interposed between the nucleotide and the matrix, whilst with its preferred cofactors only four were required to retard the enzyme.[43]

Cuatrecasas[22] has suggested that the dramatic effects of increasing the extension-arm length may in part be explained by the relief of steric restrictions imposed by the matrix and in part by the increased flexibility and mobility of the ligand as it protrudes further into the solvent. Fig. II.8 demonstrates that at least four methylene groups are required before significant affinity chromatography is achieved. It is suggested that the use of an extension arm at least 5 Å long enables the nucleotide to traverse a barrier imposed by the micro-environment of the hydrophilic polymer.[43] This may be due to an ordered layer of water molecules surrounding the matrix backbone which restricts diffusion in this region, or to the vibrational motion of the lattice. In any case, the region of solvent in close proximity to the matrix backbone represents a real barrier to the interaction of the macromolecule with its complementary ligand, especially if the interaction is of low affinity.

The flexibility and folding of the extension arm could account for the apparent decrease in binding observed when more than eight methylene bridges are interposed between the ligand and the matrix. Thus an extension arm containing eleven methylene groups, for example, would become equivalent to a distance of some 7 Å. Alternatively, on such a long extension arm, the ligand could be placed in close proximity to a region dominated by a more distant part of the lattice. It is perhaps significant that the binding of malate

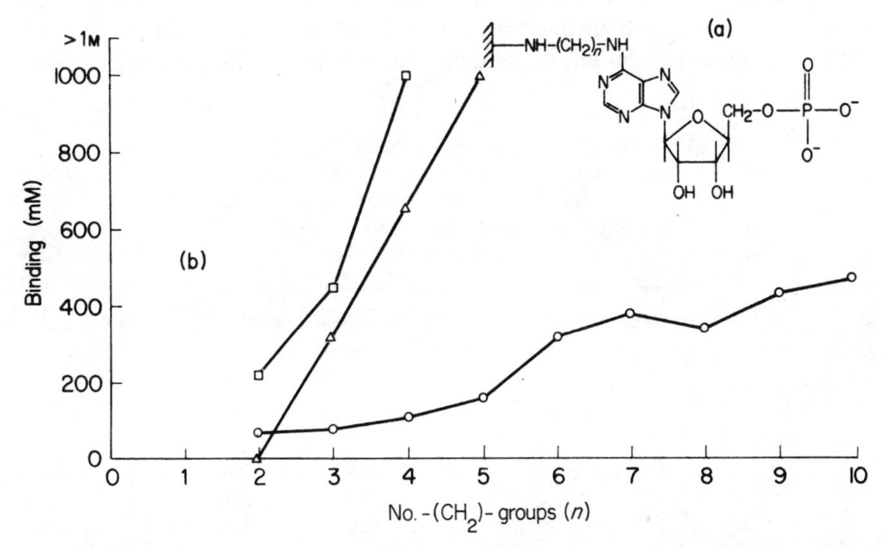

Figure II.9. The effect of extension arm length on the binding of lactate and malate dehydrogenases to N^6-(ω-aminoalkyl) derivatives of adenosine 5′-monophosphate attached to agarose. (a) The structures of the homologous series of derivatives. (b) The binding of malate dehydrogenase (○), lactate dehydrogenase from pig heart muscle, H_4 (△) and lactate dehydrogenase from rabbit skeletal muscle, M_4 (□) to adenosine 5′-monophosphate attached to agarose by a series of polymethylene bridges of various lengths. β (Binding) Represents a measure of the strength of the interaction of the enzyme with the immobilized AMP and is the concentration of KCl (mM) required to elute the enzyme

dehydrogenase and 3-phosphoglycerate kinase was virtually unaffected when the length of the extension arm was increased beyond eight methylene groups, suggesting that the extended ligand was unavailable to those enzymes of higher molecular weight. More recent studies with a homologous series of N^6-alkyl derivatives of adenosine monophosphate[44] (Fig. II.9(a)) have supported the previous conclusions. The binding of lactate and malate dehydrogenases to N^6-(ω-aminoalkyl)-AMP-Sepharoses in which the number of methylene groups in the extension arm was varied from two to ten is shown in

Fig. II.9(b). The strength of the binding of pig heart muscle (H_4) and rabbit skeletal muscle (M_4) lactate dehydrogenases increases sharply when the number of methylene groups increases from two to four, and with extension arms longer than five methylene groups requires nucleotide to effect elution. In contrast, the binding of malate dehydrogenase is almost independent of chain length beyond about six to seven methylene groups.

The effect of interposing polyglycine extension arms between NAD^+ and the agarose matrix on the binding of the two lactate dehydrogenase isoenzymes is shown in Table II.2. The strength of the interaction of the enzyme with the immobilized ligand was virtually independent of the number of glycine units used. This effect could be related either to the more hydrophilic nature of the polyglycine system compared with the polymethylene series or to the increased rigidity of the peptide linkage.

Table II.2. The effect of interposing glycine extension arms between NAD^+ and the matrix backbone on the binding of rabbit muscle (M_4) and pig heart muscle (H_4) lactate dehydrogenases

Extension	Binding (mM)[a]	
	H_4	M_4
Glycine	175	220
Diglycine	205	290
Triglycine	190	220
Tetraglycine	220	190

[a] Binding refers to a measure of the enzyme-immobilized nucleotide interaction and is the concentration of KCl (mM) at the centre of the enzyme peak when the enzyme is eluted with a linear gradient of KCl.

The increase in accessibility of the extended ligand towards interaction with macromolecules has one other major consequence in that it results in a concomitant increase in the capacity of the adsorbent. Thus, the capacity of affinity adsorbents for Staphylococcal nuclease was increased as the inhibitor was coupled at greater distances from the matrix backbone.[1]

Although the general necessity for spacer arms has been well established their widespread use prompts two further considerations. O'Carra *et al.*[45] point out that the length of an extension arm may not be the only consideration involved in their use. It is recognized that spacer arms may interfere with the affinity of a ligand for a macromolecule in a negative fashion, i.e., by generating a local steric interference. Recent studies have indicated, however, that spacer arms can cause positive interference, such as non-specific adsorption, even when they do not contain ionic groups.

Thus, O'Carra *et al.*[45] prepared gels analogous to those used by Steers *et al.*[12] for the affinity chromatography of β-galactosidase but containing no attached ligand. This non-specific adsorbent strongly retained β-galactosidase and behaved in a manner remarkably similar to that described by Steers *et al.*[12] Similar observations have been made by Er-El *et al.*[46] Glycogen was immobilized by activating Sepharose with CNBr, followed by reaction with octamethylenediamine to give ω-aminooctanyl-Sepharose and then coupling to CNBr-activated glycogen. Phosphorylase *b* was adsorbed to glycogen immobilized in this fashion and could readily be eluted by deforming buffers such as imidazolecitrate[47] which are known to disrupt the tertiary structure of proteins. Sepharose 4B did not adsorb phosphorylase *b*, although a control polymer of ω-aminooctanyl-Sepharose did bind the enzyme. When octamethylene diamine was replaced by tetramethylene diamine, no adsorption of phosphorylase occurred. These observations suggested the possibility that the length of the hydrocarbon extension arm might endow the modified Sepharose with the capacity to adsorb glycogen phosphorylase *b*.

A series of alkyl-Sepharoses (Seph-$NH(CH_2)_nH$) which differed in the length of their hydrocarbon side-chains was synthesized. It was found that D-glyceraldehyde 3-phosphate dehydrogenase was not adsorbed or retarded by any of these columns. However, in contrast Seph-C_2 retarded phosphorylase *b* and the higher alkyl-Sepharoses (Seph-C_4 to Seph-C_6) adsorbed it (Fig. II.10). Elution could not be effected by increasing the ionic strength but by using a deforming buffer the enzyme could be eluted from Seph-C_4, though not from Seph-C_6. Adsorption to Seph-C_6 was so strong that recovery of enzyme was possible only in the denatured form. Furthermore, out of several proteins tested, lysozyme, serum albumin, bovine γ-globulin and D-glyceraldehyde 3-phosphate dehydrogenase, only phosphorylase *b* was retained by Seph-C_4.

In view of the fact that all the alkyl-Sepharoses used in this study were similar in structure, except for the length of the side-chain, and since they are not substrate analogues or effectors of the enzyme, it seems reasonable to assume that retention of phosphorylase *b* occurs through hydrophobic interactions. This may arise by interaction with hydrophobic pockets in the enzyme, possibly those between protomers, and explain why deforming agents, which generate conformational changes and dissociation into protomers, elute the enzyme from the column.

These observations raise the possibilities that hydrophobic interactions may contribute significantly to the tightness of binding of an enzyme to an immobilized ligand, by a mechanism that does not involve the biospecific

Abbreviation	Structure
Seph-C_1	$-NH-CH_3$
Seph-C_2	$-NH-CH_2-CH_3$
Seph-C_3	$-NH-CH_2-CH_2-CH_3$
Seph-C_4	$-NH-CH_2-CH_2-CH_2-CH_3$
Seph-C_5	$-NH-CH_2-CH_2-CH_2-CH_2-CH_3$
Seph-C_6	$-NH-CH_2-CH_2-CH_2-CH_2-CH_2-CH_3$
Seph-C_8	$-NH-CH_2-CH_2-CH_2-CH_2-CH_2-CH_2-CH_2-CH_3$

(a)

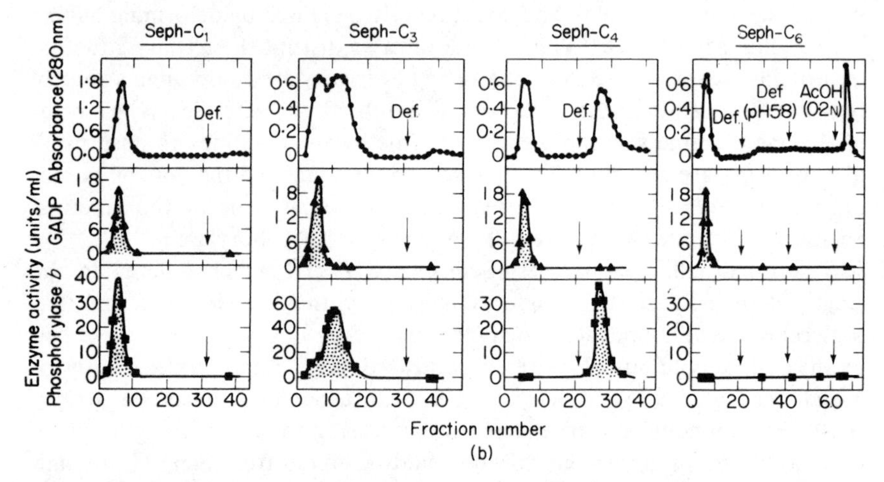

(b)

Figure II.10. Affinity chromatography on hydrocarbon-Sepharoses. (a) The structures of the alkyl-Sepharoses of various lengths. (b) Preferential adsorption of phosphorylase *b* on hydrocarbon-coated Sepharoses varying in the length of their alkyl chains. The following samples were applied on the columns (8 × 0·9 cm) Seph-C_1, 0·75 ml containing 5·0 mg GAPD and 5·1 mg phosphorylase *b*; Seph-C_3, 0·9 ml containing 4·0 mg GAPD and 4·2 mg phosphorylase *b*; Seph-C_4, 1·0 ml containing 4·9 mg GAPD and 4·1 mg phosphorylase *b*; Seph-C_6, 1·0 ml containing 4·9 mg GAPD and 4·1 mg phosphorylase *b*. The columns were equilibrated at 22° with a buffer composed of sodium β-glycerophosphate (50 mM), 2-mercaptoethanol (50 mM) and EDTA (1 mM), pH 7·0. Fractions of 1·6 ml were collected and their absorption at 280 nm (—●—) as well as their GAPD activity (—▲—) and phosphorylase *b* activity (—■—) were monitored. Non-adsorbed protein was washed off with the buffer mentioned above and then elution with a deforming buffer (0·4M imidazole, 0·05M 2-mercaptoethanol adjusted to pH 7·0 with citric acid) was initiated (arrow). In the case of Seph-C_6, attempts were made to elute phosphorylase using also a more acidic deformer (pH 5·8) and subsequently 0·2N CH_3COOH. GAPD: D-glyceraldehyde 3-phosphate dehydrogenase. Reproduced with permission from Z. Er-El, Y. Zaidenzaig and S. Shaltiel, *Biochem. Biophys. Res. Commun.*, **49**, 383 (1972)

recognition of the ligand *per se*. Furthermore, any extension arms that may not be connected to ligand may bind not only the desired protein but other proteins with hydrophobic pockets, thus reducing the efficiency of the specific adsorbent.

Similar effects have also been described by Yon.[48] Hydrophobic affinity chromatography has been used as a step in the purification of aspartate aminotransferase from wheat germ. This enzyme has a high affinity for the C_{10} chain of *N*-(3-carboxypropionyl) aminodecyl-agarose.[49] A crude preparation from wheat germ was applied to a column of this adsorbent whence 80% of the total protein was eluted in the void volume. A second inert protein peak was eluted with 0·2M NaCl, whilst 40% of the enzyme activity and 5% of the protein was eluted with 0·2% (w/v) sodium deoxycholate. An 8-fold increase in specific activity was achieved.

The second consideration with respect to spacer arms relates to their length. The possibility of introducing non-specific hydrophobic interactions with long spacer arms may annul the general trend towards long arms. Furthermore, the length of a suitable extension arm does not necessarily bear a simple relationship to the parameters of the system under study, and seems to vary from one ligand to another and even for one enzyme. For example, studies with lactate dehydrogenase have shown that a spacer arm about 8 Å long is optimal for interaction with a pyruvate analogue,[50] whereas 11 Å is necessary for a ligand representing the AMP portion of NAD^+ when the linkage is through the N^6 of adenine.[45] A spacer arm 15 Å long is necessary when attachment is through the C8 position of the adenine residue.[45] Further studies on the interaction of immobilized coenzymes with lactate dehydrogenase may resolve these difficulties.

O'Carra *et al.*[45] suggest that the effective length of a spacer arm depends on the nature of both ligand and extension arm. Hydrophobic ligands attached via hydrophobic extension arms give very disappointing results, a phenomenon attributed to 'folding back' of the hydrophobic ligand onto the hydrophobic arm. Similar observations have been made by Lowe *et al.*[43] where extension arms greater than 10 Å long result in a decreased affinity of dehydrogenases and kinases for immobilized NAD^+ and ATP respectively. Molecular models suggest that this assumption may be justified,[51] and in cases where the ligand is hydrophobic, hydrophilic extension arms may be preferable. Such considerations may explain why there is little or no increase in affinity for NAD^+ when it is attached to a Sepharose backbone by sesqui-peptides of glycine (Table II.2). Clearly, the nature of the extension arm can play a crucial role in determining the affinity of the macromolecule for the immobilized ligand.

The generalization that long spacer arms are essential to the retention of enzymes that display weak affinity for the immobilized ligand may also not be a universal one. Thus Harris *et al.*[52] demonstrated that whilst saccharo-1,4-lactone had a high affinity for β-glucuronidase ($K_i \sim 5{\cdot}4 \times 10^{-7}$M) it was necessary to place the inhibitor at the end of a 3,3′-diaminodipropylamine extension arm of length 12 Å to achieve adequate adsorption. Thus these observations demonstrate the need for a more rational approach in the design and use of extension arms for affinity adsorbents.

C. CONSIDERATIONS IN THE SELECTION OF THE LIGAND

The ligand to be covalently attached to the inert matrix must be one that displays special and unique affinity for the macromolecule to be purified. It can be a substrate, inhibitor, allosteric effector, cofactor, hormone or an analogue of any of these. Table II.3 illustrates the diversity of potential ligands and prompts consideration of the principles behind the selection of a competent ligand for immobilization as an affinity adsorbent.

Table II.3. Some ligands that have been immobilized and used for affinity chromatography

Ligands	References
Substrates and substrate analogues	1, 22
Inhibitors	53, 54
Allosteric effectors	55
Cofactors	56–58, 76
Nucleic acids	59
Nucleotides	60
*t*RNA	61
Plant hormones	62
Steroids	63
Antibiotics	64
Enzymes	65, 66
Protease inhibitors	67
Hydrophobic ligands	46, 48
Chromophores	68, 69

1. The Nature and Mechanism of the Ligand–Macromolecule Interaction

In many cases, a careful consideration of the nature and mechanism of the macromolecule–ligand interaction could be advantageous in the evaluation of prospective ligands for immobilization. Thus, for monosubstrate enzymic reactions of the type represented in Fig. II.11(a), ligand A, product P or an analogue of either will satisfy the requirements for affinity chromatography. Bisubstrate enzymic reactions can be of two types: (i) ordered mechanisms in which ligand A compulsorily binds before ligand B can interact with the resultant binary complex and (ii) random mechanisms where ligands A and B are bound independently of each other in any order. For the latter case the choice of ligand will be determined to a large extent by the affinity for the enzyme and its relative ease of immobilization. For ordered mechanisms (Fig. II.11(b)) immobilization of ligand A will be subject to the usual limitations of affinity chromatography. In contrast, immobilization of ligand B will

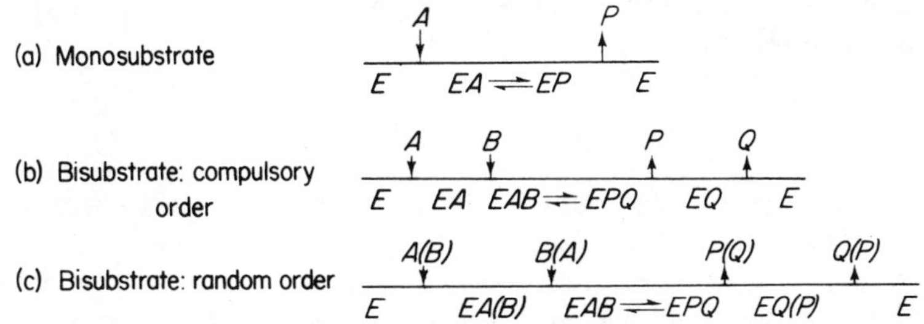

(a) Monosubstrate

$A \downarrow$ $P \uparrow$
$E \quad EA \rightleftharpoons EP \quad E$

(b) Bisubstrate: compulsory order

$A \downarrow$ $B \downarrow$ $P \uparrow$ $Q \uparrow$
$E \quad EA \quad EAB \rightleftharpoons EPQ \quad EQ \quad E$

(c) Bisubstrate: random order

$A(B) \downarrow$ $B(A) \downarrow$ $P(Q) \uparrow$ $Q(P) \uparrow$
$E \quad EA(B) \quad EAB \rightleftharpoons EPQ \quad EQ(P) \quad E$

Figure II.11. Some common enzyme mechanisms: E represents enzyme, while A, B, P and Q are reactants and products respectively

generate an incompetent adsorbent for the complementary macromolecule unless ligand A is included in the irrigating buffer. The presence of ligand A is essential for the binding of the enzyme to the adsorbent since it is the binary ligand A-enzyme complex that binds ligand B. This process introduces a preliminary selection into the chromatography and subsequent negative elution by removal of ligand A will achieve a secondary effect. Examples of this technique will be discussed in detail in Section F of this chapter.

2. The Affinity of the Ligand for the Macromolecule

The affinity of the ligand for the macromolecule is an important consideration in the design and preparation of affinity adsorbents. The interaction of an enzyme E with an immobilized ligand L is non-productive in that no product is formed and hence

$$E + L \overset{K_L}{\rightleftharpoons} EL$$

The dissociation constant (K_L) of the enzyme–ligand complex (EL) is given by

$$K_L = \frac{[E][L]}{[EL]} = \frac{[E_0 - EL][L_0 - EL]}{[EL]}$$

where E_0 and L_0 are the initial concentrations of enzyme and immobilized ligand respectively. When $L_0 \gg E_0$, then

$$K_L = \left(\frac{E_0 - EL}{EL}\right)L_0$$

The chromatographic distribution coefficient, K_d, is defined by

$$K_d = \frac{\text{Bound enzyme}}{\text{Free enzyme}} = \left(\frac{EL}{E_0 - EL}\right) = \frac{L_0}{K_L}$$

and
$$V_e = V_0 + K_d \times V_0$$

where V_e and V_0 are the elution volume of the protein and the void volume of the gel bed respectively. Thus, expressing the retardation of the specific enzyme in terms of column volume units,

$$\frac{V_e}{V_0} = 1 + \frac{L_0}{K_L} \tag{II.1}$$

and is determined by the ratio of the initial ligand concentration to the dissociation constant of the enzyme–immobilized ligand complex. A similar relationship has been reported by O'Carra *et al.*[45] Thus, for significant retardation of the complementary enzyme, $L_0 \gg K_L$. It is thus possible, using equation (II.1), to estimate the maximum K_L between an immobilized ligand and an enzyme to effect an operationally useful retardation of the enzyme from the raw material. For example, a retardation equivalent to 10 void volumes would be adequate to resolve the specific enzyme from the non-adsorbed inert proteins appearing in the void volume of the column. Thus, if the concentration of the immobilized ligand in the matrix is 10^{-3}M, then providing that the enzyme concentration in the volume containing the gel matrix is low compared with the ligand concentration, calculation shows that approximately 10^{-4}M is obtained as the upper limit of the K_L for an effective ligand. It is possible in theory, however, to prepare adequate adsorbents for systems with dissociation constants greater than 10^{-4}M if a sufficiently large amount of ligand can be coupled to the solid support.

3. The Mode of Attachment

The reciprocal of the dissociation constant (K_L) is a measure of the affinity of the enzyme for the immobilized ligand. The closer the structure of the immobilized ligand to that of the substrate, the greater the possibility that K_L will approach the dissociation constant for the enzyme–substrate interaction in free solution, K_S. The small molecule to be insolubilized must therefore possess chemical groups that can be modified for linkage to the solid support without abolishing or impairing the interaction with the complementary protein. It is the latter restriction that makes the choice of ligand somewhat empirical in practice. Reference to Fig. II.12 will show that of the five possible binding points to the small ligand, only that denoted by *e* is not involved in the interaction with the complementary macromolecule. Hence, attachment of the ligand to the matrix by this point should yield an effective affinity adsorbent. Attachment by any other point could lead to an adsorbent that is either only partially effective or totally ineffective. These restrictions imposed on the preparation of biospecific adsorbents also apply to the synthesis of active site-directed irreversible inhibitors[70] and in both cases it is important to know what part of the ligand can be modified without being detrimental to the interaction with the macromolecule. This principle is illustrated by the purification of the tyrosine-sensitive isoenzyme of 3-deoxy-D-arabinoheptulosonate 7-phosphate (DAHP) synthetase from *Saccharomyces cerevisiae* on an adsorbent of its appropriate allosteric effector, L-tyrosine-agarose.[55] The enzyme was retarded but not strongly adsorbed. The apparent low affinity observed between the DAHP synthetase and the tyrosine-Sepharose was not due to an intrinsically weak binding of tyrosine to the

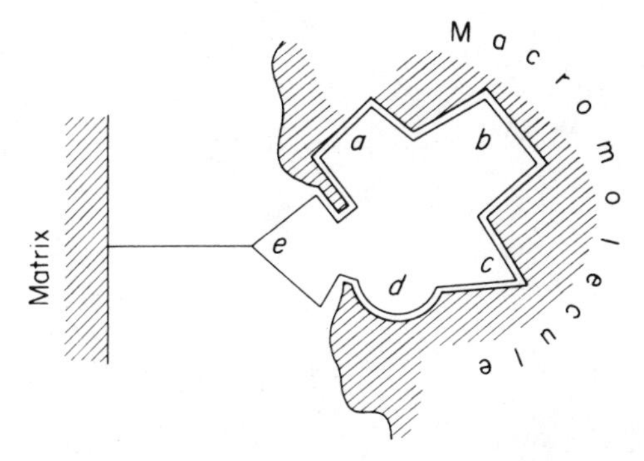

Figure II.12. The importance of the mode of attachment of the ligand to the matrix

enzyme.[71] It is suggested that steric hindrance caused by the proximity of the matrix backbone or the loss of the ionic character of the amino group involved in linkage of the ligand to the matrix may be responsible for the low affinity. This is particularly likely if the charged amino group participates in the binding process.

A small ligand such as an amino acid will have a limited number of functional groups, all of which are probably involved in the interaction with the protein. Immobilization of the ligand by attachment to any of these functional groups would almost certainly destroy the specific interaction and result in a totally ineffective adsorbent. In this case, increasing the distance of the ligand from the matrix backbone would have little effect. Fig. II.13 shows the chromatography of lactate dehydrogenase on affinity adsorbents prepared by coupling 4-amino-salicylate directly to cyanogen-bromide-activated Sepharose and to ϵ-aminocaproyl-Sepharose via a carbodiimide-promoted reaction.[51] It is clear that chromatography on the extended ligand does not lead to the dramatic increase in the strength of binding observed with other systems.[1,22]

Studies by Harvey *et al.*[80] on the effectiveness of immobilized nucleotides for the affinity chromatography of kinases and pyridine nucleotide-dependent dehydrogenases have shown that the point of attachment of the nucleotide to the matrix is of fundamental importance. Thus Fig. II.14 shows the structures of two affinity adsorbents prepared by immobilizing adenosine 5′-monophosphate (AMP) by two different procedures. Adsorbent I, N^6-(6-aminohexyl)-AMP-Sepharose, comprises AMP attached to Sepharose by the N^6 of

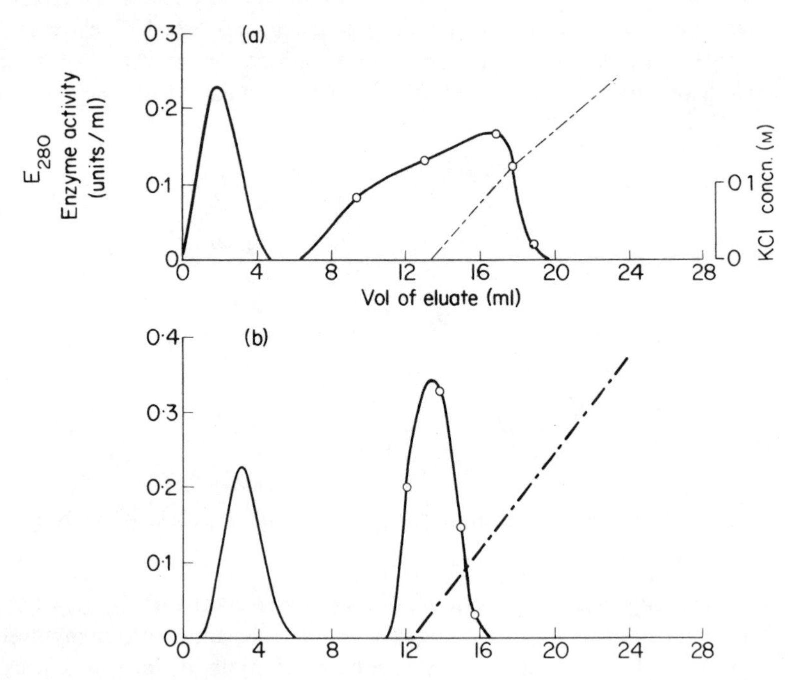

Figure II.13. Chromatography of lactate dehydrogenase on 4-amino salicylate (a) linked directly to cyanogen-bromide-activated Sepharose, and (b) attached to ϵ-aminohexanoyl-Sepharose with a carbodiimide-promoted reaction

the adenine moiety, whilst adsorbent II, P^1-(6-aminohexyl)-P^2-(5′-adenosine)-pyrophosphate-Sepharose, is linked through the 5′-phosphate. The binding of several kinases and dehydrogenases to these adsorbents has been examined and the results are presented in Table II.4. Alcohol dehydrogenase and glycerokinase bound to N^6-(6-aminohexyl)-AMP-Sepharose but not to P^1-(6-aminohexyl)-P^2-(5′-adenosine)-pyrophosphate-Sepharose, whilst glyceraldehyde 3-phosphate dehydrogenase showed no affinity for the former adsorbent but bound so strongly to the latter that a pulse of NADH was required to effect elution. Hexokinase was not bound to either adsorbent. These results probably reflect the nature of the enzyme–nucleotide interaction and would suggest that while a free 5′-phosphate group was essential to the binding of yeast alcohol dehydrogenase and glycerokinase, it has a quite different role in the interaction with glyceraldehyde 3-phosphate dehydrogenase. The adenosine moiety may be the vital portion of the molecule in this

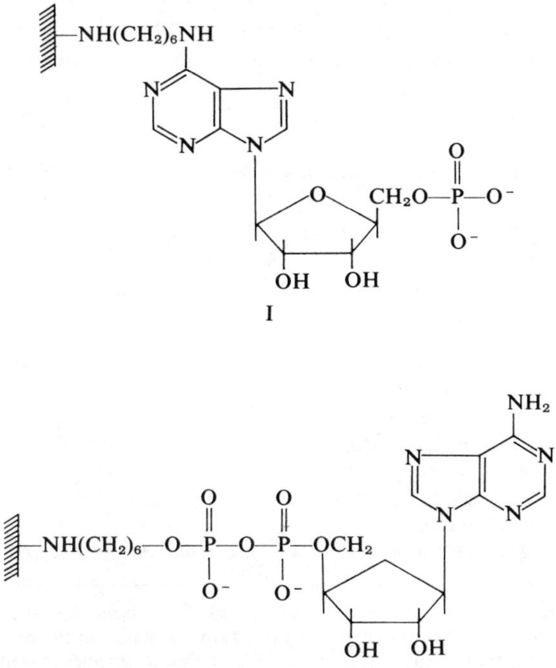

Figure II.14. The structures of two immobilized-AMP adsorbents: (I) N^6-(6-aminohexyl)-AMP-Sepharose and (II) P^1-(6-aminohexyl)-P^2-(5′-adenosine)-pyrophosphate-Sepharose

case. The lack of binding of hexokinase to both immobilized-AMP derivatives probably reflects the more stringent binding requirements for this enzyme.

As a general rule, the larger the ligand, the more points of attachment between it and the complementary macromolecule, and hence the greater the degree of latitude in preparing the affinity adsorbent. The utilization of protein–protein interactions for affinity chromatography has been widely employed for the purification of antigens and antibodies,[72–74] for the isolation of proteolytic enzymes[75] and for the purification of human milk lactose synthetase[60] and serum retinol binding protein.[77] In each case, effective adsorbents were obtained by coupling the protein directly to Sepharose by the cyanogen bromide technique. An important consideration in this respect is that the biologically active protein should retain its native tertiary structure. For this reason, the macromolecule should be covalently attached to the matrix backbone by the fewest possible bonds. Proteins react with cyanogen-bromide-activated Sepharose through the unprotonated form of their free amino groups. Since most proteins are richly endowed with lysyl residues on their surfaces, when the coupling reaction is performed at pH values of 9·5 or greater such molecules will be attached to the matrix by more than one point.

Table II.4. A comparison of the binding of several enzymes to N^6-(6-aminohexyl)-AMP-Sepharose (A) and P^1-(6-aminohexyl)-P^2-(5′-adenosine)-pyrophosphate-Sepharose (B)

	Enzyme	Source	Binding (mM KCl)[a] A	Binding (mM KCl)[a] B
EC.1.1.1.1	Alcohol dehydrogenase	Yeast	400	0
EC.1.2.1.12	D-Glyceraldehyde 3-phosphate dehydrogenase	Rabbit muscle	0	> 1M[b]
EC.2.7.1.1	Hexokinase	Yeast	0	0
EC.2.7.1.30	Glycerokinase	*Candida mycoderma*	122	0

[a] Binding refers to a measure of the strength of the enzyme–immobilized nucleotide interaction and is the concentration of KCl (mM) at the centre of the enzyme peak when the enzyme is eluted with a linear gradient of KCl. Enzyme samples (5 units) were applied to columns (5 mm × 50 mm) containing 1 g of each adsorbent.

[b] Pulse of 5×10^{-3}M NADH required for elution.

This problem may be circumvented by coupling at less favourable pH values. Thus, for example, it has been demonstrated that the capacity of an immunoadsorbent for antigen can be considerably enhanced by the coupling of antibodies at pH 6·0–6·5.[78] Furthermore, the low capacity of collagen-Sepharose for the adsorption of human skin, rheumatoid synovial and tadpole collagenases[79] may conceivably be improved by immobilizing the protein under less favourable conditions.

Two good examples of the rationale behind the choice of a ligand for immobilization will be given to illustrate these principles.

a. Chorismate Mutase

Chorismate mutase from *Claviceps paspali* is an allosteric enzyme that binds its substrate, chorismic acid, the inhibitors, L-phenylalanine and L-tyrosine, and its allosteric activator, L-tryptophan.[81] The suitability of

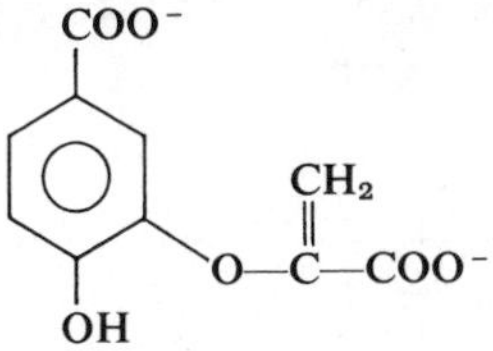

Chorismic acid

chorismic acid for affinity chromatography is restricted by its lability. The allosteric activator, L-tryptophan, was chosen for coupling to Sepharose because of its strong affinity for the enzyme. L-Tryptophan activation has been observed at concentrations above 10^{-7}M. Furthermore, analogues and homologues of the effector are also capable of interacting with the inhibitor and activator sites. In this respect it is especially fortuitous that *N*-amino-substituted tryptophan compounds are also allosteric effectors. Thus L-tryptophan covalently attached to Sepharose by its amino group should yield a competent adsorbent, especially if a high ligand concentration is also employed. Fig. II.15 demonstrates that chorismate mutase is indeed retarded by L-tryptophanyl-Sepharose. Elution of the enzyme can be effected by the addition of 10^{-3}M L-tryptophan to the eluant buffer.

b. Penicillinase

Studies on the secretion of penicillinase by *Bacillus licheniformis* necessitated a rapid and mild procedure for isolating microgram amounts of the enzyme. Crane *et al.*[64] have shown that affinity chromatography using the antibiotic and substrate analogue, cephalosporin *C*, covalently attached to agarose was effective in purifying penicillinase from crude extracts of the Bacillus.

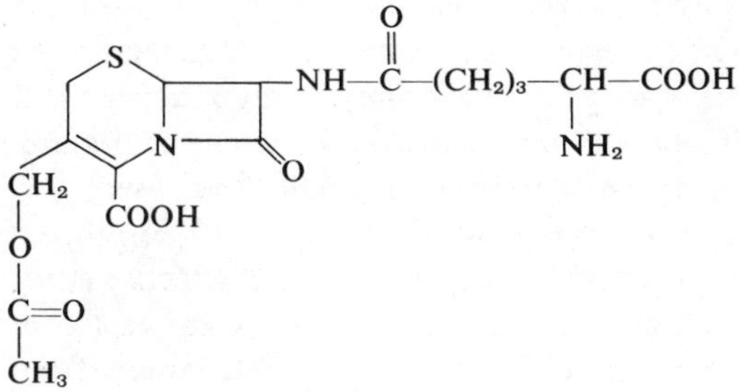

Cephalosporin *C*

Cephalosporin *C* has several properties which make it an excellent ligand for the affinity adsorbent. It has a high affinity for the penicillinase ($K_m \simeq 10^{-5}$M) but is hydrolysed at a very low rate. Furthermore, the amino-adipoyl side-chain presents a non-essential and distant amino group for coupling to the matrix and, finally, the ligand is stable to non-enzymic hydrolysis. Thus, adsorbents prepared with this ligand can be used repeatedly over long periods of time with no detectable loss of potency.

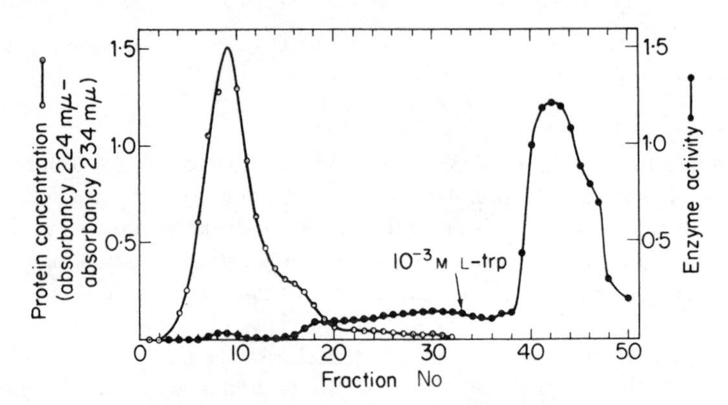

Figure II.15. Chromotography of a partially purified preparation of chorismate mutase from *Claviceps paspali* on L-tryptophan-substituted Sepharose. The column (1·5 × 9 cm) was equilibrated with 0·01M potassium phosphate buffer, pH 6·9. The sample consisted of 3 ml *Claviceps* extract (9 mg protein) also in the same buffer. 3 ml fractions were collected. After 100 ml buffer had passed through the column, 10^{-3}M L-Trp-solution was added to elute chorismate mutase. Reproduced with permission from B. Sprossler and F. Lingens, *FEBS Lett.*, **6**, 232 (1970)

4. The Ligand Concentration

Despite optimization of the mode of attachment of the ligand to the matrix, the interaction with the complementary protein is almost invariably impaired. Immobilization of a specific ligand can decrease its affinity for the protein by up to three orders of magnitude. The important parameter is thus the effective experimental affinity, i.e., that displayed between the protein and the immobilized ligand under the experimental conditions chosen.

For interacting systems of low affinity ($K_L \geqslant 10^{-4}$M) the ligand concentration is a critical parameter in the preparation of effective adsorbents. Thus, to achieve a retardation equivalent to 10 void volumes for a protein whose affinity for the immobilized ligand, $K_L = 10^{-3}$M, requires from equation (II.1) a ligand concentration of about 10^{-2}M, assuming that no impairment of binding is observed on immobilization. In practice, however, it has proved difficult to prepare adequate adsorbents for enzyme–ligand systems whose dissociation constants under optimal conditions in free solution are greater than 10^{-3}M. This is partly because sufficiently high ligand concentrations are as yet unattainable, and partly because of the decreased affinity of the immobilized ligand for the protein. However, the latter restriction is not necessarily a universal phenomenon. Thus, Gawronski and Wold[82] have found that, within the limits of the method used, there was no change in the dissociation constant of the ribonuclease *S*-protein and *S*-peptide interaction when one of the components was immobilized on Sepharose. Furthermore, Lowe *et al.*[83] have found that the affinity of lactate dehydrogenase and glycerokinase for adenosine 5′-monophosphate is actually increased by about three orders of magnitude on immobilization. However, quantitative data on the binding of proteins to immobilized ligands are as yet unavailable.

Lowe *et al.*[43] have demonstrated that the concentration of the immobilized ligand is a fundamental parameter that characterizes the strength of the interaction between the complementary macromolecule and the ligand. The strength of binding of lactate dehydrogenase increased as the concentration of immobilized NAD^+ increased. Furthermore, the strength of binding decreased when Sepharose-bound nucleotide was diluted over a 21-fold range with unmodified Sepharose or with ε-aminohexanoyl-Sepharose. Similar results have been observed with other affinity adsorbents[63,85] and with the binding of glycerokinase and lactate dehydrogenase to N^6-(6-aminohexyl)-AMP-Sepharose diluted up to 200-fold with underivatized Sepharose.[80]

Dilution of the enzyme sample by up to 21-fold had no effect on the binding of glycerokinase to immobilized ATP or of lactate dehydrogenase to immobilized NAD^+, or on the enzyme recovery or elution volume. Likewise, no effect of diluting the same enzymes up to 200-fold was observed in their binding to N^6-(6-aminohexyl)-AMP-Sepharose. Thus, dilution of the enzyme sample does not have the same effect on binding as dilution of the bound ligand. The equilibrium that exists between the bound ligand and the free enzyme is presumably being altered in favour of the free enzyme when the ligand concentration is decreased.

When the concentration of the immobilized ligand is fixed and a sample containing the complementary macromolecule is allowed to penetrate the column, enzyme molecules entering the gel will collide with the immobilized ligand and be slightly retarded. Enzyme molecules subsequently arriving in the neighbourhood of this prior engagement will increase the concentration of the specific enzyme in this region. Thus, as the effective concentration of the enzyme increases, the concentration of the binary complex (ES) is also increased. Hence, as the enzyme concentration increases, the strength of the interaction with the immobilized ligand also increases and the downward migration of the enzyme molecules is decreased and ultimately reduced to zero if the interaction is strong enough. The specific enzyme thus forms in a narrow zone at the top of the column. This precept was readily confirmed by the demonstration that fluorescently labelled lactate dehydrogenase was adsorbed to the top of a column of immobilized NAD^+ as a sharp band.[51]

One consequence of the above hypothesis is that for low-affinity systems or where it is technically difficult to generate adsorbents with high ligand

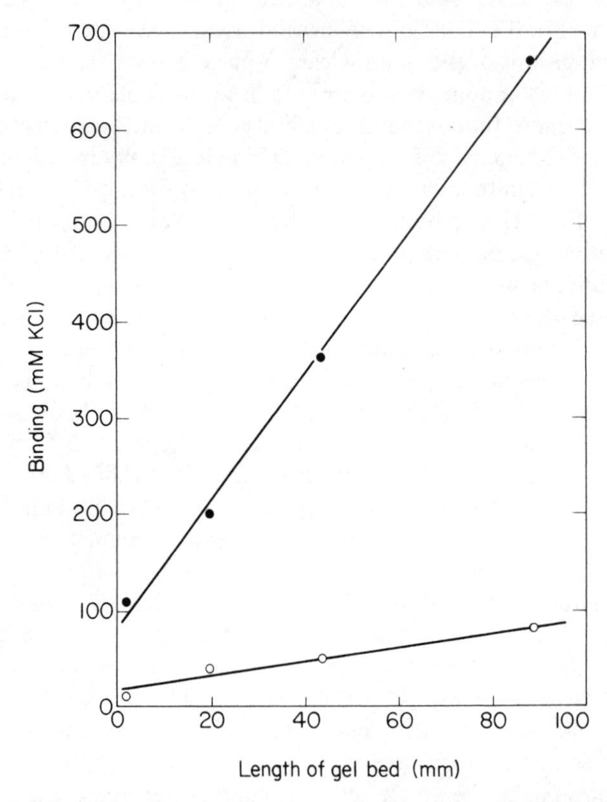

Figure II.16. The effect of column length on the binding of lactate dehydrogenase and glycerokinase to N^6-(6-aminohexyl)-AMP-Sepharose. The binding is a measure of the strength of the enzyme–immobilized ligand interaction and is the concentration of KCl (mM) where the peak of enzyme activity is eluted on a linear gradient

concentrations, the length of the column through which the enzyme percolates is an important factor that can determine the strength of the interaction. Fig. II.16 demonstrates the effect of increasing the column length of an adsorbent comprising N^6-(6-aminohexyl)-AMP-Sepharose containing 125 mμmoles AMP/ml Sepharose on the binding of lactate dehydrogenase and glycerokinase; the longer the column the greater the strength of the interaction.[80]

For high-affinity systems or with adsorbents containing a high ligand concentration, the length of the column assumes less importance. Thus, when the binding of lactate dehydrogenase to N^6-(6-aminohexyl)-AMP-Sepharose was investigated, using columns of different lengths containing the same total amount of ligand (300 mμmoles AMP) but of different concentrations (75–1500 mμmoles AMP/ml Sepharose), the binding of the enzyme was related to the ligand concentration and not the column length.

These considerations relating to the ligand concentration have several important practical consequences. Firstly, columns containing high ligand concentrations will concentrate dilute enzyme samples. Secondly, for high-affinity systems, where elution of the adsorbed macromolecule is difficult without denaturation, dilution of the polymer with unsubstituted gel or reduction of the ligand concentration will effect a more facile elution under milder conditions. Finally, for interacting systems of low affinity, column geometry may be an important parameter in that long columns may be preferable to short ones for the resolution of specifically adsorbed proteins from non-adsorbed proteins.

It has tacitly been assumed that as high a ligand concentration as possible is desirable in all cases. However, Schmidt and Raftery[86] have pointed out that unnecessarily high ligand concentrations can lead to chromatographic aberrations when the immobilized ligand is charged. Adsorbents containing covalently linked cholinergic ligands bind acetylcholinesterase and, to a lesser extent, acetylcholine receptor proteins. Kalderon *et al.*[87] noticed a loss of specificity towards acetylcholinesterase when the ligand concentration was increased from $1{\cdot}5 \times 10^{-4}$M to $1{\cdot}6 \times 10^{-3}$M. A critical ligand concentration of approximately 10^{-3}M was found in this case and confirmed by the parallel studies of Schmidt and Raftery.[86] If it is assumed that the ligands are uniformly distributed throughout the matrix and positioned at the corners of a cubic lattice, the nearest-neighbour distance is about 100 Å. When the ligand concentration is below 10^{-3}M, the ligands may be spaced sufficiently far apart to prevent non-specific proteins from interacting with more than one charged group at a time. Hence, under these conditions, only those macromolecules capable of interacting specifically with the immobilized ligand will be retarded by the resin. Thus in this case the high selectivity of an affinity adsorbent can only be achieved by lowering the concentration of bound inhibitor. The concomitant loss of capacity can be recovered by scaling up the chromatographic procedures.

D. OTHER CONSIDERATIONS IN THE PREPARATION OF AFFINITY ADSORBENTS

The conditions under which the ligand is coupled to the solid support must be sufficiently mild to be tolerated well by both the ligand and the matrix. This could be particularly significant where highly labile ligands, such as cofactors or proteins, are attached to the matrix. Furthermore, the derivatized gel must be washed exhaustively to ensure complete removal of the material not covalently bound and any products formed during the coupling process. In some cases, especially with aromatic or heterocyclic compounds, complete removal of adsorbed material is difficult and may require extensive washing for many days. To this end organic solvents may be required; ethanol, methanol, butanol, ethylene, glycol, aqueous pyridine (80% v/v) and aqueous dimethylformamide (50% v/v) are particularly useful in accelerating the washing of agarose derivatives. Dimethyl sulphoxide, on the other hand, is known to disrupt the structure of agarose.

It is generally instructive to devise an accurate method for determining the amount of ligand covalently attached to the solid support. This can be done conveniently by determining the amount of ligand (by radioactivity, absorbance, fluorescence, etc.) released from the substituted matrix either by acid or alkaline hydrolysis or, in some cases, by exhaustive enzymic digestion. Alternatively, direct absorbance or fluorescence measurements on the column matrix can give a rough estimate, especially in conjunction with a determination of the amount of ligand not recovered in the final washings. This has limitations when a very large excess of ligand is added during the coupling procedure or when appreciable unspecific adsorption to the matrix occurs and necessitates the use of large volumes of solvent for thorough washing. Cuatrecasas has suggested that the degree of ligand substitution on the solid matrix be expressed in terms of concentration, such as micromoles of ligand per millilitre of packed gel.[22] This circumvents the more accurate but operationally less useful expression on the basis of dry weight of gel. These aspects are discussed in more detail in Chapter V.

E. CONSIDERATIONS AFFECTING ADSORPTION OF PROTEINS

Assuming that the correct choices of matrix and ligand have been made and an affinity adsorbent prepared, the next aspect to be considered relates to the adsorption of the complementary macromolecule.

Optimal conditions for adsorption are dictated not only by the nature of the protein to be purified but also by its interaction with the insolubilized ligand. The buffer used to equilibrate the affinity column, and its ionic strength and pH, should reflect the conditions optimal for the enzyme–ligand interaction. It is important to realize that the optimum pH for binding of the substrate or inhibitor may not necessarily be optimal for catalytic activity.

The sample should be dialysed against the same buffer used to equilibrate the column, prior to its application to the column. The passage of several void volumes of starting buffer through the column will ensure that non-adsorbed proteins are eluted. Subsequent alteration in the irrigant buffer will elute the specifically bound protein.

However, in order to achieve satisfactory adsorption to the affinity column several basic concepts should be considered.

1. The Nature of the Adsorption Isotherm

The amount of a particular substance taken up, or adsorbed, by the stationary phase depends on the concentration in the mobile phase. The curve obtained by plotting the amount adsorbed against the concentration, at constant temperature, is termed an *adsorption isotherm.*

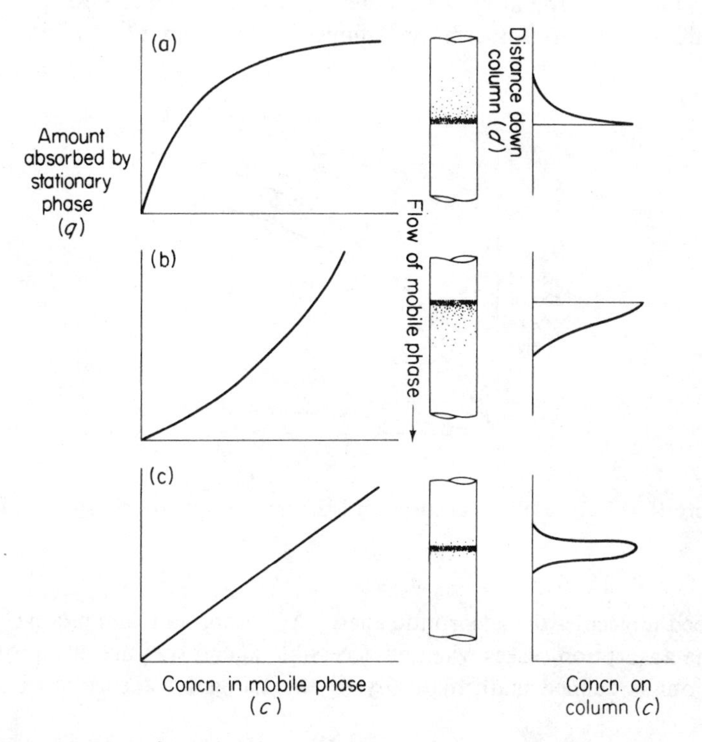

Figure II.17. Adsorption isotherms and chromatographic behaviour

Fig. II.17(a) shows a typical adsorption isotherm together with the characteristics of the chromatographic zone associated with it when a column is developed by elution analysis. The bands are depicted at an arbitrary stage in their migration down the column. Bands with sharp fronts and long diffuse tails are said to exhibit 'tailing', an undesirable effect in adsorption chromatography. The steeper the isotherm near the origin, the more diffuse the tail.

Fig. II.17(b) shows an isotherm that is associated with 'fronting' and found more frequently in partition chromatography.

With linear isotherms, such as that in Fig. II.17(c), the chromatographic zones are compact and symmetrical; this type of isotherm is experienced primarily in partition chromatography.

The free energy, $\Delta G'$, of the protein–ligand interaction $E + L \rightarrow [EL]$ is given by the expression $\Delta G' = -RT \ln K_L$, where K_L is the apparent dissociation constant for the enzyme–ligand system, R is the universal gas constant and T is the absolute temperature. Substitution for K_L of a mean value of 10^{-5}M in the expression for $\Delta G'$ yields a value of 6–7 kcal/mole, a relatively high value for chromatographic conditions. Fig. II.18 represents a hypothetical adsorption isotherm for affinity chromatography, where for all

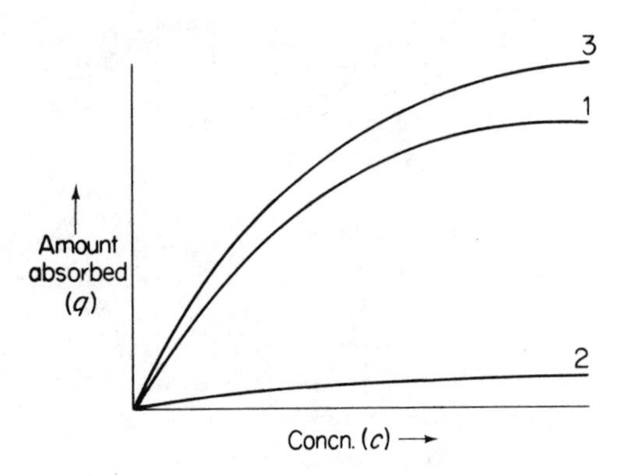

Figure II.18. Generalized adsorption isotherm for affinity chromatography

adsorbed molecules the adsorption energy $\Delta G'$ is constant and relatively large and the adsorption ceases when all accessible ligand sites are occupied. This behaviour is defined mathematically by the Langmuir adsorption isotherm

$$q = \frac{k_1 c}{1 + k_2 c}$$

where q represents the amount of substance adsorbed (x) per unit mass of adsorbent (m), c is the equilibrium concentration and k_1 and k_2 are constants. At very high concentrations, q has a finite upper limit, k_1/k_2, which characterizes the amount of solute adsorbed per unit mass of adsorbent. At sufficiently low concentrations, where $k_2 c \ll 1$, the equation assumes the form of a linear isotherm,

$$q = k_1 c$$

and, furthermore, since

$$\frac{dq}{dc} = \frac{k_1}{(1 + k_2c)^2}$$

the slope is a decreasing function of c, with a finite value k_1, the *adsorption coefficient*, where $c \to 0$.

It should be noted that where the adsorption is weak the isotherm will tend to have less curvature (Fig. II.18: Curve 2) than when it is strong. This is because the curvature is a function of the ratio of spaces available to the concentration of molecules available to fill them. At low concentrations, there is more space for adsorption but there are also fewer molecules to be adsorbed.

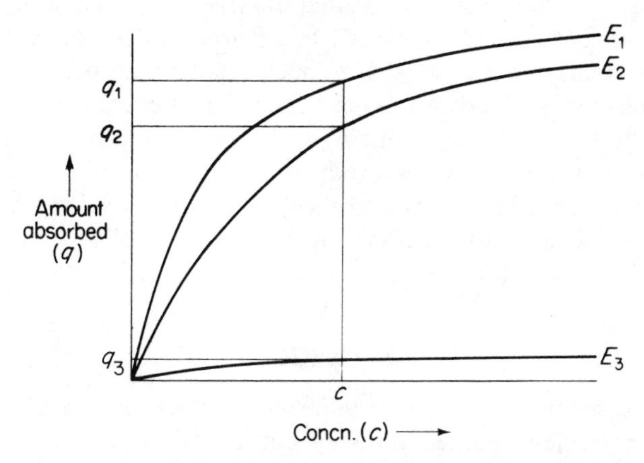

Figure II.19. Adsorption isotherms for three enzymes (E_1, E_2 and E_3) interacting with a single immobilized ligand

Thus all isotherms of the Langmuir type tend to linearity at low concentrations. When the affinity of the molecules for the adsorbent is weak, these considerations extend into higher concentration ranges.

Weak adsorptions often arise from non-specific interactions and consequently the adsorption energy $\Delta G'$ is composed of two terms, one representing the specific interaction (ΔG_{sp}) and a second representing the unspecific adsorption (ΔG_{unsp}), i.e.,

$$\Delta G' = \Delta G_{sp} + \Delta G_{unsp}$$

The adsorption energy of the unspecific adsorption, ΔG_{unsp}, which results from hydrophobic, hydrophilic and/or ionic interactions, is comparable to the adsorption energy in classical chromatography and is critically dependent on the nature of the matrix and the protein. Ideally the value of ΔG_{unsp}

should be as small as possible. It is worth noting that the unspecific adsorption of an already specifically adsorbed macromolecule entails an apparent increase in the affinity of the macromolecule for the ligand. Thus the gross adsorption isotherm for affinity chromatography (Fig. II.18: Curve 3) is defined as the sum of the specific (Curve 1) and unspecific (Curve 2) adsorption isotherms.

The specific shape of the isotherm used in affinity chromatography has other important consequences. It has generally been assumed that only one enzyme in the raw material shows marked affinity for the immobilized ligand. However, there are cases (see, for example, Chapter III) where two or more enzymes have small K_L values for the ligand used. For example, if the second enzyme has an affinity $K_{L_2} \geqslant 10^{-3}$M then only small amounts of this enzyme will be retained along with the desired enzyme. However, if $K_{L_2} \leqslant 10^{-4}$M then a mixture of the two enzymes will be adsorbed, even though the K_L value of the desired enzyme (K_{L_1}) might be much smaller than that of the second. Fig. II.19 illustrates the adsorption isotherms for three enzymes, two with K_L values about 10^{-5}M and one with $K_L > 10^{-3}$M. If, for simplicity, the concentrations of all three enzymes are the same, c, then inspection of the three curves shows that the amounts adsorbed are q_1, q_2 and q_3 respectively. This phenomenon has a profound effect on the behaviour of groups of enzymes on insolubilized cofactors.

2. Flow Rate

In some cases the adsorption equilibrium between the specific ligand and the macromolecule is reached at a very slow rate since, on the one hand, the collision numbers of large molecules are relatively small and, on the other hand, close alignment of the active site of the protein to the affinity ligand is essential for specific adsorption. Thus, approximation to equilibrium conditions can be an important prerequisite for complete retention of the desired enzyme by a given amount of the affinity matrix.

Anchoring of the ligand and the restriction of static films on the surface and within the pores of the support means that diffusion may become a factor in the overall kinetics of the reaction. The diffusional limitations imposed by the nature and mechanism of the chromatographic process can be broadly classified into three types.

(i) Longitudinal diffusion: this is classical Fickian molecular diffusion due to a concentration gradient which can occur in both the radial and axial directions. It is probably of minor importance under the conditions of ordinary affinity chromatography, although it could assume importance at low flow rates for weakly interacting systems.[88]

(ii) Eddy diffusion: this could be the most important factor at high flow rates and is caused by the irregularities in the flow generated by the particles of the gel in the bed.[88] A solute molecule in a fast stream path will be less likely to interact with an immobilized ligand than one in a slow stream path if the rate of attainment of equilibrium is a slow one.

(iii) Restricted diffusion: this is the restriction of molecular diffusion within the pores of the gel matrix which could seriously hinder the correct approach of a macromolecule to the bound ligand.[89] The contribution of this phenomenon is difficult to assess in practice but can be minimized by the use of a very porous support for affinity chromatography.[84]

In view of the requirement for equilibrium conditions the lowest flow rate acceptable from a practical point of view is desirable. On studies with Staphylococcal nuclease, Cuatrecasas *et al.*[1] have shown that small amounts of nuclease appeared in the void volume together with the protein impurities if very fast flow rates were used, especially when the total protein concentration in the sample was also relatively high. However, even with such high flow rates, nuclease could be completely retained if more dilute samples were applied. Similar results were obtained by Lowe *et al.*[83] with the adsorption of heart muscle lactate dehydrogenase to N^6-(6-aminohexyl)-AMP-Sepharose when samples containing high concentrations of enzyme were applied at high flow rates. However, in a parallel series of experiments, but using lower protein concentrations, increasing the flow rate of an ϵ-aminohexanoyl-NAD^+-Sepharose column from 6·7 to 41 ml/hour (6·5 to 40 void volumes/hour) had no effect on the binding of lactate dehydrogenase or an inert protein, bovine serum albumin. Furthermore, no effect of equilibration time on the binding of either lactate dehydrogenase to immobilized NAD^+ or of glycerokinase to immobilized ATP could be found when left for 1, 5 and 20 hours in contact with their respective nucleotide columns prior to elution. This observation could prove particularly useful in the storage of enzymes which depend on the presence of substrate or cofactor for stability.

3. Protein Concentration

The dynamic nature of the processes involved in affinity chromatography would imply that concentration of the applied sample occurs until a sufficiently strong interaction with the ligand can take place to arrest the downward migration of the enzyme through the column. This process is critically dependent on the ligand concentration and is almost independent of the initial concentration of free macromolecule. The latter statement is subject to the limitation that sub-saturating amounts of complementary macromolecule are applied to the column and that the flow rate is suitably adjusted.

The observation that protein appears in the void volume eluate when high flow rates are used with highly loaded columns can be rationalized in terms of two related phenomena. The first concerns a process termed secondary exclusion.[90] The diffusion rate of a molecule in a gel depends not only on its size, as it does in the bulk phase, but also on the ratio of its molecular to pore size. When a concentrated solution of large molecules is applied to the gel, some molecules will diffuse into the available pores. Subsequent molecules arriving will find many pores occupied and their probability of diffusing into an occupied pore will be reduced, depending on the reduction of available pore size. This can result in effective exclusion from a pore, and hence an immobilized ligand, which would otherwise have been accessible.

This brings us to the second consideration; the steric hindrance to succeeding molecules by the protein molecules that have already been adsorbed. An estimate of the maximum molar ratio between pyranose units and ligands such that no steric interference is produced is, surprisingly, as high as 100:1. A globular, medium-sized protein molecule like haemoglobin covers an area of about 2500 Å^2, whilst the surface area of a pyranose molecule is only about 25 Å^2. A major contribution by mutual steric hindrance is therefore unlikely.

4. The Effect of Temperature

In general, adsorption phenomena decrease with increasing temperature. This means, according to the Le Chatelier principle, that the adsorption process is accompanied by a decrease in enthalpy, i.e., an evolution of heat, the *heat of adsorption*, ΔH^0. Under chromatographic conditions an increase in temperature will move the equilibrium towards a higher relative concentration in the mobile phase. Thus, a temperature rise will usually lead to a faster migration, and the more exothermic the adsorption the greater the migration for a given temperature increment.

Quantitatively, the dependence of the distribution coefficient, K_d, on the absolute temperature, T, is given by:

$$K_d = e^{\Delta S^\circ/R}\, e^{-\Delta H^\circ/RT} \simeq \text{constant} \times e^{-\Delta H^\circ/RT}$$

where ΔS^0 and ΔH^0 are the standard entropy and enthalpy (heat) of adsorption respectively and R is the gas constant.

The effect of temperature on the binding of lactate dehydrogenase to N^6-(6-aminohexyl)-AMP-Sepharose is shown in Fig. II.20(a). The concentration of NADH required to elute the enzyme decreased with increasing temperature. The decreased binding in the temperature range 5–10°C is of considerable practical significance since this range of temperature is that generally experienced in a typical laboratory coldroom.

The behaviour of lactate dehydrogenase was characterized by a linear Arrhenius plot (Fig. II.20(b)) corresponding to an enthalpy of adsorption of 54·6 kJ/mole (13·06 kcal/mole). However, since the affinity of lactate dehydrogenase for NADH in free solution decreases with increasing temperature

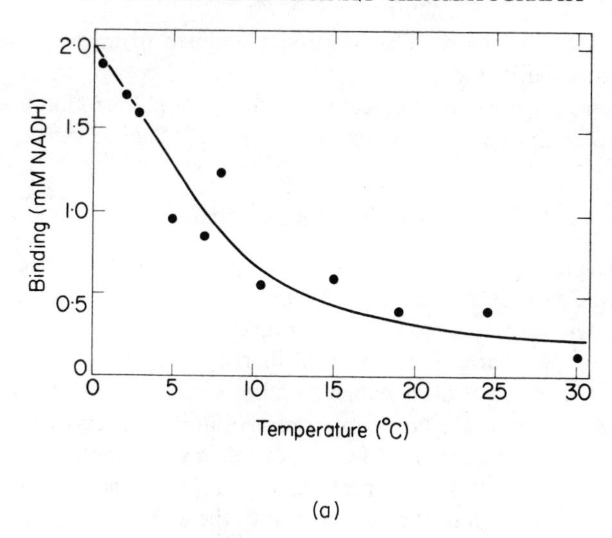

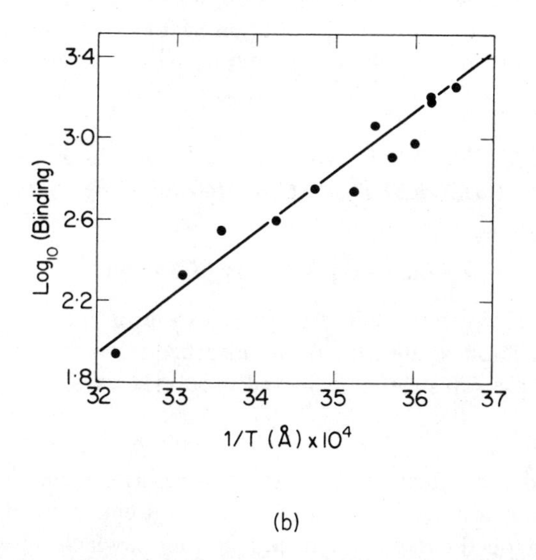

Figure II.20. The effect of temperature on the binding of lactate dehydrogenase to N^6-(6-aminohexyl)-AMP-Sepharose. The enzyme sample (5 units) containing 1·5 mg of bovine serum albumin (100 μl) was applied to a column (5 mm × 50 mm) containing 0·5 g N^6-(6-aminohexyl)-AMP-Sepharose (1·5 μM AMP/ml). (a) Binding as a function of temperature; the concentration of NADH required to elute the peak of enzyme activity on a linear gradient of NADH (0–5 mM; 20 ml total volume). (b) Arrhenius plot of the above data

(11 kcal/mole), the enthalpy of adsorption calculated from the slope of Fig. II.20(b) is a minimal value.

Elevated temperatures also decrease the affinity of glycerokinase and yeast alcohol dehydrogenase for N^6-(6-aminohexyl)-AMP-Sepharose.

5. Batchwise Adsorption

Affinity purification need not be restricted to column procedures and, indeed, in some cases it may be desirable to use a batchwise technique. Thus, for example, when small amounts of a protein are to be extracted from a mixture containing a large proportion of inert protein with an adsorbent of high affinity, the purification is achieved more readily by adding a slurry of the specific adsorbent to the crude mixture. Column procedures can often be hampered by a deterioration in flow rate caused by the application of crude samples. In some cases it may be preferable to apply the sample to a packed column, thoroughly wash it, then proceed with the elution by dismantling the column and incubating the matrix in an appropriate buffer. This is a particularly useful method in cases involving very high affinity interactions, such as those involving antigen–antibody systems. In general, this method of elution requires less severe conditions than the column technique and the yields can be significantly higher, since higher dilution of the insoluble ligand, more thorough mixing and more facile control of time and temperature are possible.

F. THE ELUTION OF SPECIFICALLY ADSORBED MACROMOLECULES

1. The General Theory of Elution

The principal motivation behind the application of elution techniques is to drive the adsorption equilibrium of the adsorbed solute from the stationary to the mobile phase. This can be achieved in several ways.

a. Frontal Analysis

In this method, a solution of the mixture to be investigated is added continuously until the column is saturated. If the concentration of the effluent is monitored a stepped curve is obtained (Fig. II.21), which reflects the composition of the original mixture. Sharp steps are generated only if the system exhibits isotherms of the type shown in Fig. II.17(a) since, if the isotherms are of this type, the chromatographic bands will have sharp fronts.

When a binary mixture comprising components A and B is added continuously to a column of adsorbent, component A will emerge ahead of component B if A is less strongly adsorbed than B. The first step will consist of pure A, whilst the second step will contain both A and B. The number of steps is thus equal to the number of components in the original mixture, although only the first step contains a pure component.

One of the few examples of frontal analysis in affinity chromatography was provided by Lerman to prove the retention of mushroom tyrosinase by percolating enzyme mixtures through columns containing different cellulose derivatives.[3] Lowe *et al.*[43] have used the technique to demonstrate the heterogeneity of nucleotide adsorbents.

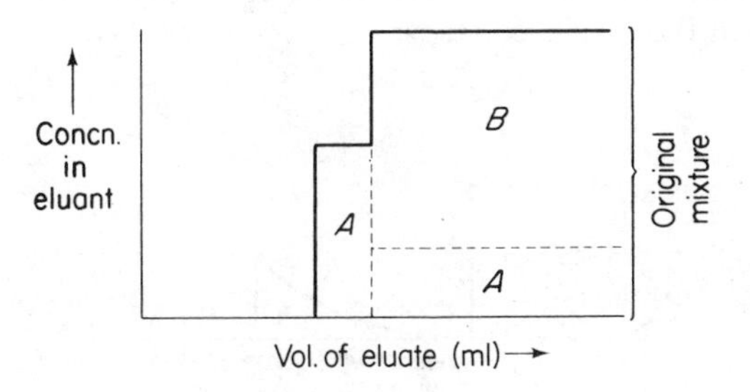

Figure II.21. Frontal analysis

b. Displacement Analysis

Displacement analysis requires that the distribution of the adsorptive on the adsorbent be reversible and that all solutes be adsorbed on the same kind of site on the adsorbent, so that the more strongly adsorbed ones will displace those less strongly adsorbed.

The introduction of a small amount of substance A to a column of adsorbent will produce a zone with a sharp front at the top of the column. In this method a solvent is chosen that is not strongly adsorbed relative to the solute, so that it acts as a carrier and does not interfere appreciably with the formation of the initial zone. If an excess of solution in the same solvent system of a *displacing agent*, D, is now applied to the column, it is so strongly adsorbed that it expels substance A from the adsorbent. The displacer, acting like a 'piston', forces ahead of its front the compact zone of substance A (Fig. II.22). After all substance A has been displaced ahead of the displacer D then, as more displacer is added, the velocity of movement of the lengthening zone of displacer and of the front of substance A will be the same.

For a multi-component mixture, the sample resolves itself into bands of pure components, in an order depending on the strength of adsorption on the matrix. Each pure band acts as a displacer for the component ahead of it, and the last most strongly adsorbed band is pushed along ahead of the displacer. The elution of the components from the adsorbent is shown in Fig. II.23(a) and is similar to that obtained by frontal analysis. However, in this case the steps represent pure components.

As in frontal analysis, isotherms of the type shown in Fig. II.17(a) are necessary to give sharp steps. The concentration of displacer (C_d) and the form of the isotherms for the components of the mixture determine the concentrations of the various components (A and B) in the steps, as shown in Fig. II.23(b). Curve D represents the isotherm for the displacer, and if a concentration C_d is used, the displacer line, obtained by joining the corresponding point on the isotherm to the origin, cuts the isotherms of A and B at points corresponding to the concentrations C_a and C_b. These are the concentrations in the displaced steps.

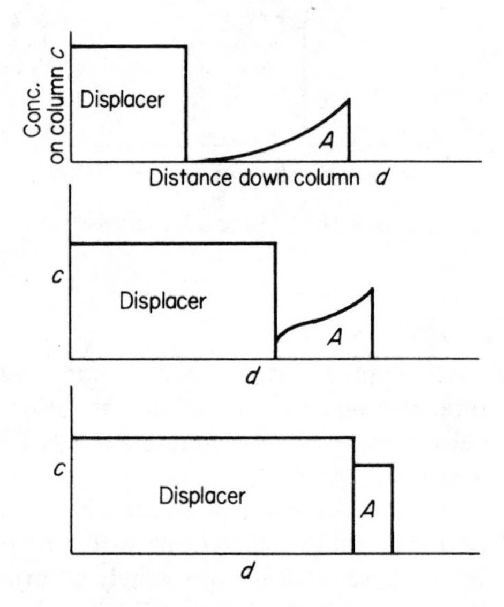

Figure II.22. A displacing agent overtaking a previously adsorbed band

It should be noted, however, that the above conclusions are only valid for 'Langmuir isotherms'. Furthermore, for complete displacement of a component, the isotherm must be depressed so much in the presence of the displacer that the displacer line no longer cuts it.

c. Pulse Elution

In this modification of displacement analysis, the displacing agent is applied to the column in a small volume and washed through with the carrier solvent.[91] A 'pulse' of displacing agent, either in high concentration or of a different kind to the carrier solute, migrates through a chromatographic bed as a compact zone. Conversion to a new buffer or eluant medium can therefore be accomplished within a fraction of the total bed volume.

d. Elution Analysis

When sample components migrate through a chromatographic bed with a velocity approaching that of the eluant front, they become clustered and poorly resolved. On the other hand, components which move slowly require excessive elution times and concomitantly the zones broaden to a point where detection may become difficult. Thus Fig. II.24(a) demonstrates how two sample components, 1 and 2, are eluted with a weak solvent. Component 1 is

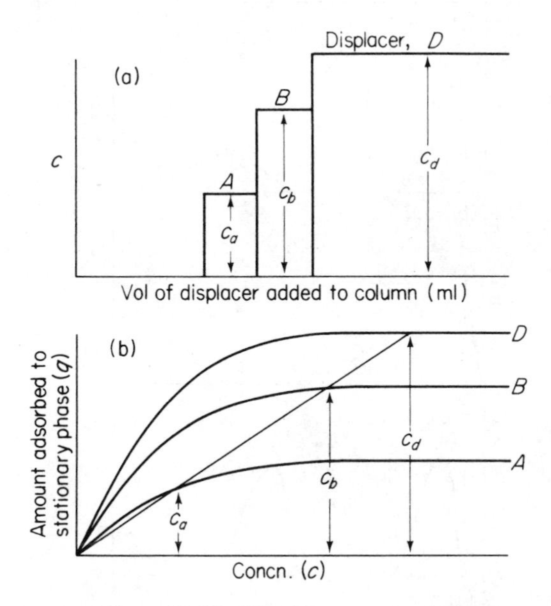

Figure II.23. Displacement analysis

eluted in a convenient time, whilst component 2 requires excessive time and is very broad. The use of a stronger solvent (Fig. II.24(b)) results in incomplete resolution of the two components, whilst an intermediate solvent will improve the resolution. The technique of 'stepwise' elution, where solvents of gradually increasing strength are successively used for elution of the weak component 1 and then the strongly adsorbed component 2, can solve these problems. This method is particularly useful for routine separations under linear isotherm conditions, but where unknown samples or non-linear isotherm systems are used several complications arise. The use of stepwise elution may generate spurious peaks (band splitting) by collecting the tail end of a previous peak at the change of solvent and producing the illusion of an additional peak.

In gradient elution (Fig. II.24(d)) the elution strength of the solvent is increased continuously. This generates a concentration gradient down the column such that the rear portion of the chromatographic zone is in contact with a stronger eluting solution than the front portion, and consequently tailing of the band is reduced. A balance must be achieved between an increased resolution with increased width of eluted peaks caused by a shallow gradient and a decreased resolution, but with sharper peaks produced by a steep gradient (Fig. II.24(e)).

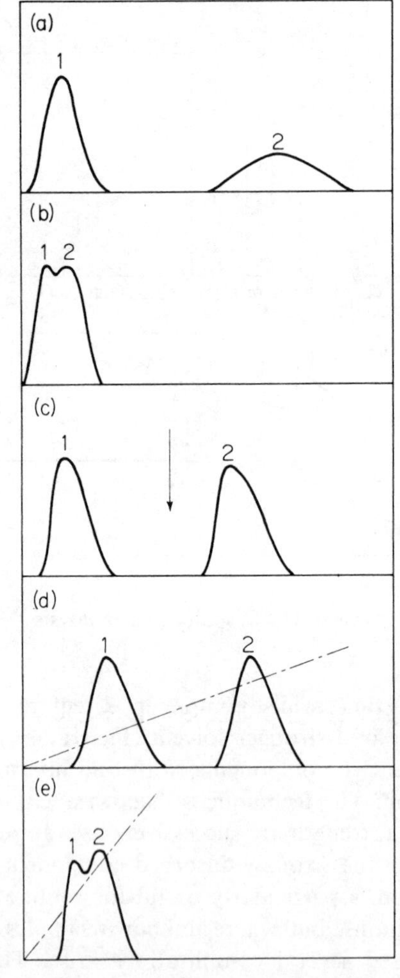

Figure II.24. Techniques of elution analysis

2. The Elution of Specifically Adsorbed Macromolecules

When a sample containing the complementary macromolecule is applied to a column of the selective adsorbent, and the column is washed with the equilibrating buffer, several elution profiles are possible. The nature of the elution pattern depends on the effectiveness of the adsorbent under the experimental conditions selected. If the matrix is unsubstituted or if the adsorbent is totally ineffective, the complementary protein will emerge with the inert protein in the void volume (Fig. II.25(a)). The desired macromolecule may be retarded relative to the void volume by subsequent partition down the column, and result in one of several elution patterns (Fig. II.25(b), (c), (d)). Ideally, the enzyme will be adsorbed as a concentrated zone at the top of the column, and a change in buffer will be necessary to elute the enzyme as a sharp concentrated peak (Fig. II.25(e)). In the latter case, the conditions required for elution may be more extreme than would be anticipated from the conditions necessary for adsorption. Thus a change in pH or ionic strength leading to a 5-fold decrease in effective affinity for the immobilized ligand may not induce a rapid downward migration in the mobile phase, although such a decrease in affinity may prevent adsorption to the column in the first place. Consequently, one of the best ways of eluting a strongly adsorbed enzyme is to remove the top part of the column and dilute the gel with a large volume of an appropriate buffer.

a. Non-specific Methods of Elution

Protein specifically adsorbed to a solid carrier matrix will eventually emerge from the column without altering the nature of the buffer if the interacting system is of relatively low affinity. This type of elution suffers the disadvantage that the protein is recovered in dilute form and under such conditions its stability could be adversely affected. Chan and Takahashi [55] report that 3-deoxy-D-arabinoheptulosonate 7-phosphate (DAHP) synthetase is eluted from L-tyrosine-Sepharose without altering the properties of the buffer and Dean *et al.*[92] report the retardation of the phosphatases on phospho-cellulose and glyceryl-phospho-cellulose. Some retardation of an enzyme involved in the inactivation of the antibiotic gentamicin was also observed on columns of immobilized gentamicin or kanamicin *A*.[93]

In most cases, however, elution of the protein will require changing either the pH, ionic strength, dielectric constant or temperature of the buffer. Ideal elution of a strongly adsorbed protein should utilize a solvent which sufficiently alters the conformation of the protein appreciably to decrease the affinity of the protein for the immobilized ligand, without being so severe as to cause partial or complete denaturation of the enzyme. In most cases, the alteration of a single physical variable suffices to effect elution, although in some cases such as the elution of glutamic-oxaloacetic transaminase from N'-(ω-aminohexyl)-pyridoxamine phosphate-Sepharose it was found that simultaneous changes in pH and ionic strength were more effective than alterations in either alone.[56]

Dissociation of proteins from very high affinity adsorbents may require

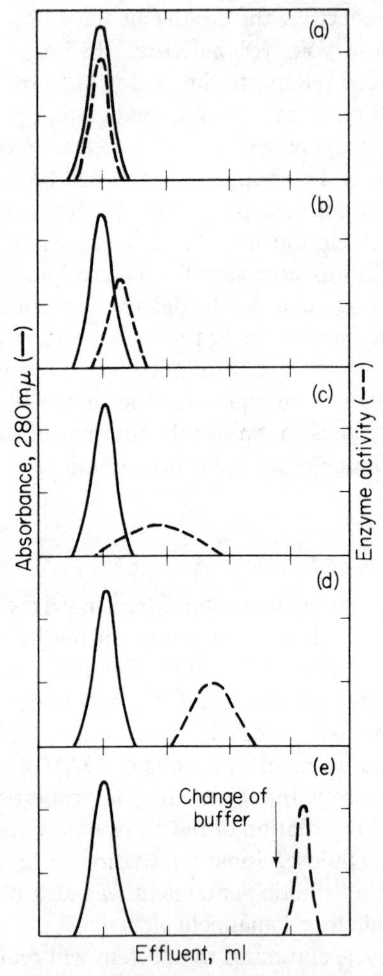

Figure II.25. Theoretical elution profiles for the purification of a specific enzyme (- - - -) from a crude protein mixture by affinity chromatography. The specific enzyme is eluted in the void volume of columns of adsorbent (a) together with the inert non-adsorbed proteins, but is retarded to various extents by adsorbents (b) to (d). Successful application of affinity chromatography is depicted in (e) where a change of buffer is required to elute the specifically adsorbed enzyme. Reproduced with permission from P. Cuatrecasas, *Adv. Enzymol.*, **36**, 29 (1972)

extremes of pH and/or protein denaturants such as guanidine–HCl or urea. In these cases, prompt restoration of the native protein structure may be affected by neutralization, dilution or dialysis. Thus, elution of egg-white avidin from Biocytin-Sepharose requires a combination of 6M guanidine–HCl and pH 1·5.[94]

The elution technique utilized can often be an additional means of enhancing the separation and/or purification of the adsorbate, particularly in cases where non-specific adsorption is a serious problem. Thus, the application of gradients of pH, ionic strength or dielectric constant may achieve a secondary resolution by virtue of the different sensitivities of the adsorbed enzymes, even though their affinities may have been similar under the original chromatographic conditions. This technique could overcome a major limitation of affinity chromatography, i.e., the lack of specificity for enzymes with K_L values below 10^{-4}M. There are, however, few recorded examples of the deliberate application of gradient elution following specific adsorption. The chromatography of L-threonine dehydrogenase on NAD^+-cellulose reported by Lowe and Dean illustrates the considerable advantages of such a system.[57] In such cases, where several enzymes are adsorbed to group specific matrices, the application of gradients of pH or temperature may offer considerable advantages over stepwise elution procedures. Fig. II.26 illustrates the resolution of yeast alcohol dehydrogenase, glycerokinase, hexokinase and lactate dehydrogenase on N^6-(6-aminohexyl)-AMP-Sepharose using a linear temperature gradient.[95] It is relevant to note that glycerokinase and yeast alcohol dehydrogenase were eluted in the order expected from their apparent energies of adsorption with almost quantitative recovery (70–90%) whilst lactate dehydrogenase even at 40°C still required a pulse of 5 mM NADH for elution. However, lactate dehydrogenase can be eluted by exploitation of a characteristic property of the enzyme–nucleotide interaction that is distinct from either its temperature or ionic strength dependence.[83] Fig. II.27 depicts the separation of a complex mixture of nicotinamide nucleotide-dependent dehydrogenases on N^6-(6-aminohexyl)-AMP-Sepharose by application of a pH gradient. The technique could be particularly useful for strongly bound enzymes which could otherwise only be eluted by specific procedures.

b. Special Methods of Elution

The methods described above are generally applicable to the elution of specifically adsorbed macromolecules. There are, however, several additional techniques which are not related to the biological function of the macromolecule, but may be applied in some circumstances.

Since such approaches are peculiar to the enzyme–ligand system under investigation, several examples to illustrate the idea will be given.

Steers *et al.*[12] describe the purification of β-galactosidase from *E. coli* by affinity chromatography on a derivative of Sepharose containing covalently attached *p*-aminophenyl-β-D-thiogalactopyranoside. The enzyme was adsorbed tightly at neutral pH to the substituted matrix but could not be eluted by buffers containing high concentrations (50 mM) of the substrates, lactose, isopropyl-β-D-galactopyranoside or *o*-nitrophenyl-β-D-galactopyranoside. Furthermore, columns equilibrated with tris-HCl buffer pH 7·5 were washed in a stepwise manner with tris-HCl buffers of pH 8·5, 9·0 and 9·5 without elution of the bound enzyme. The total enzymatic activity could, however, be eluted with 0·1 M sodium borate buffer pH 10·0, despite the fact that the pH of

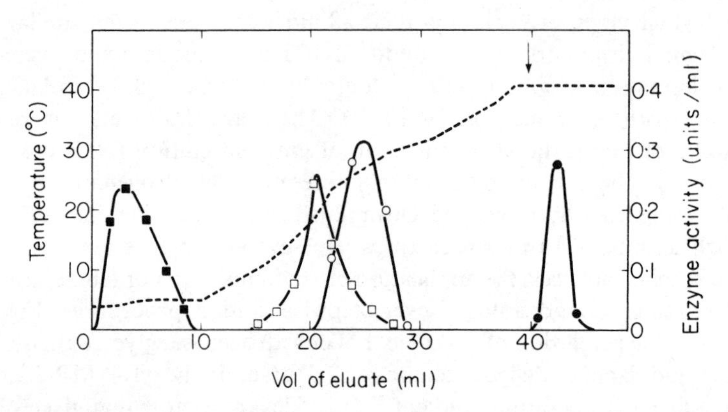

Figure II.26. The resolution of an enzyme mixture on N[6]-(6-aminohexyl)-AMP-Sepharose by a temperature gradient. The enzyme sample (100 μl), containing 5 units of each enzyme and bovine serum albumin (1·5 mg), was applied to a column (5 mm × 50 mm) containing 0·5 g N[6]-(6-aminohexyl)-AMP-Sepharose (1·5 μM AMP/ml) at 4·7°C. The column was equilibrated at each individual temperature for 5 minutes prior to elution with 1·6 ml equilibration buffer, 10 mM tricine-KOH, pH 7·5, containing 10 mM glycerol, 5 mM $MgCl_2$, 1 mM EDTA and 0·02% sodium azide. A 'pulse' (200 μl) of 5 mM NADH in the equilibration buffer was added as indicated by the arrow. Bovine serum albumin was located in the initial column wash (0–4 ml) and hexokinase (■), glycerokinase (□), yeast alcohol dehydrogenase (○) and pig heart lactate dehydrogenase (●) were assayed in the effluent

the leading edge of the eluted peak was measured as 8·5–9·0. It is suggested that the borate interacts specifically with the ligand to prevent re-adsorption of the desired enzyme occasioned by the high pH.

A similar application of the special properties of borate has been reported by Barker *et al.*[60] The galactosyltransferase (UDP-galactose: *N*-acetylglucosamine galactosyltransferase) of lactose synthetase from bovine milk is readily adsorbed on UDP-Sepharose, especially when the buffers contain manganous ions. Under such conditions, the binding of the enzyme was unchanged in the presence of high concentrations of sodium chloride (0·5–1·0M). Borate buffers at pH 8·5 eluted the enzyme in good yield. It is suggested that the borate forms an addition compound with the ribose moiety of uridine 5′-phosphate and thereby weakens the binding of the enzyme.

Mapes and Sweeley[96] report the elution of ceramide trihexosidase from an affinity column prepared by coupling melibiose to succinylated-Sepharose. The mixture of α-galactosidases prepared from Cohn fraction IV–I of human

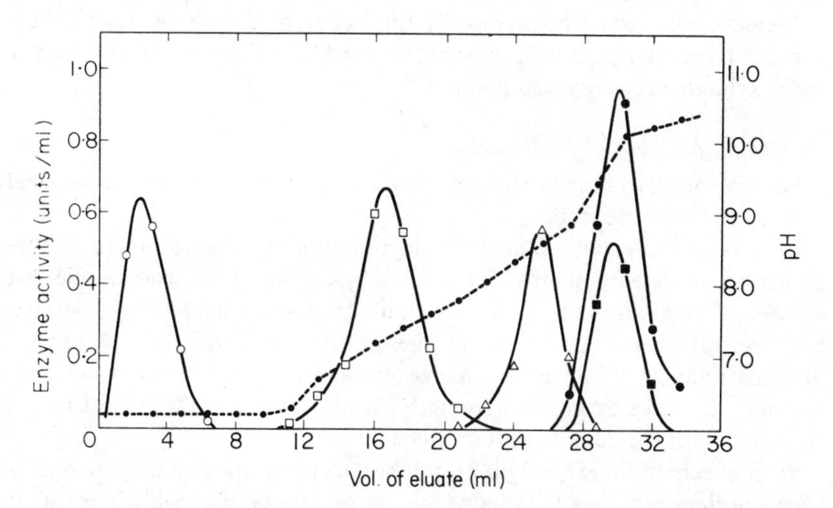

Figure II.27. The resolution of a mixture of dehydrogenases on N^6-(6-aminohexyl)-AMP-Sepharose by a pH gradient. The enzyme mixture (100 μl), containing bovine serum albumin (1·5 mg) and 5 units of each enzyme, was applied to a column (5 mm × 50 mm) containing 0·5 g N^6-(6-aminohexyl)-AMP-Sepharose equilibrated with 10 mM KH_2PO_4-KOH, pH 6·0. The column was washed with equilibration buffer, pH 6·0, prior to development with a pH gradient (pH 6–10; 10 ml equilibration buffer against 10 ml 30 mM K_2HPO_4-KOH, pH 11·0 in a linear gradient apparatus). Bovine serum albumin (○), malate dehydrogenase (●), glucose 6-phosphate dehydrogenase (□), pig heart lactate dehydrogenase (■) and yeast alcohol dehydrogenase (△) were assayed in the effluent

plasma was eluted by a non-ionic detergent (0·1% (w/v) Triton X-100) added to the equilibrating buffer.

An alternative approach to eluting the specifically bound protein is to cleave the matrix–ligand bond selectively, thus removing the intact ligand–protein complex. Excess ligand can then be removed by dialysis or by Sephadex gel filtration. This approach can be applied where ligands are attached to Sepharose by susceptible bonds such as those involved in azo, thiol or alcohol ester linkage. Furthermore, in principle the photo-destruction of the ligand or linkage would be equally effective.

The azo-linked derivatives can be cleaved by reduction with sodium dithionite at pH 8·5.[22] This method is particularly useful for high-affinity systems where the complementary macromolecule is denatured irreversibly by exposure to extremes of pH or by protein denaturants such as urea or guanidine–HCl. Thus, the serum oestradiol-binding protein binds oestradiol very tightly ($K_i \sim 10^{-9}$M) and is particularly susceptible to denaturation. The protein can readily be removed in active form from an oestradiol-agarose gel by reductive cleavage of the azo-linkage with dithionite.[22]

Ligands attached to Sepharose by thiol ester or carboxylic ester linkages can be released by short exposure to pH 11·5 or by treatment with 1N hydroxylamine for about 30 minutes.[97]

c. Specific Methods of Elution

Despite optimization of the conditions of elution, there are cases where specific elution is desirable.

(i) Charged Ligands. Where the ligand itself is charged there is often reduction of ligand affinity for the complementary macromolecule with increasing ionic strength. This behaviour of charged ligands is undesirable as it results in contamination of the purified material with non-specific proteins retained by an ion-exchange mechanism. Such contaminants also respond to increases in ionic strength during chromatography and may be co-eluted with the complementary macromolecule.

Such effects of ionic strength on inhibitor affinity are well documented for acetylcholinesterase and explain the chromatographic behaviour of the enzyme on resins containing covalently bound quaternary ammonium functions.[87]

(ii) Group Specific Adsorbents. Where the immobilized ligand shows affinity for more than one macromolecule, some additional method of increasing selectivity is desirable.

Elution of a specifically adsorbed macromolecule can often be achieved by supplementing the equilibrating buffer with a high concentration of specific inhibitor or substrate. Fig. II.28 illustrates the possible molecular species of a given enzyme (E) within the bed of a matrix to which a specific ligand of the enzyme (S) is covalently attached.[98] The presence of an unfixed ligand (I) in the mobile phase can affect the migration of the enzyme through the column in one of three ways.

(a) Competitive Effect. When the ternary complex (ESI) is less stable than the EI complex, an increase in the concentration of I increases the fraction of enzyme in the mobile phase and so reduces the retardation of the enzyme through the column. A special case of this phenomenon is when the free ligand (I) and the immobilized ligand (S) are one and the same.

(b) Non-competitive Effect. When the stability of the ternary complex (ESI) is the same as the binary complex (EI), the binding of I does not affect the affinity of the enzyme for *S*, and *I* has no effect on the retardation of the enzyme.

(c) Uncompetitive Effect. Where the stability of the ternary complex (*ESI*) is greater than the binary complex (*EI*) the effect of the free ligand *I* is to decrease the fraction of enzyme in the mobile phase, and thus to enhance binding. When the ternary complex is less stable than the binary complex, increasing concentrations of the free ligand invariably lead to elution of the enzyme from the column. Many examples of this specific type of elution are to be found in the literature; a selection is given in Table II.5. Two examples will suffice here to demonstrate the principles of the method. Kaufman and

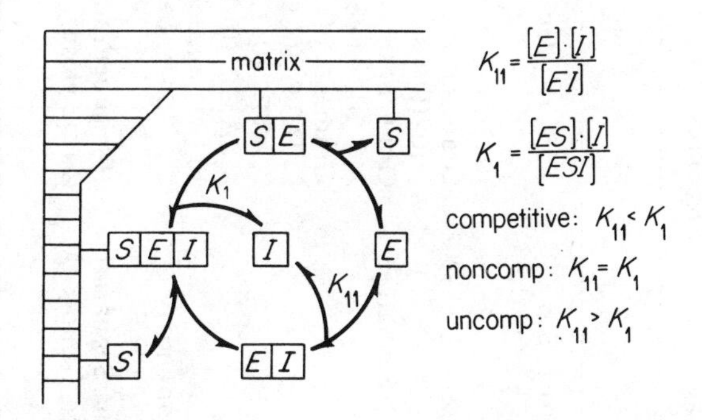

Figure II.28. Possible molecular species of a given enzyme within the affinity matrix. *S*: covalently fixed ligand to matrix; *I*: unfixed ligand; *E*: enzyme. Reproduced with permission from H. Akanuma *et al.*, *Biochem. Biophys. Res. Commun.*, **45**, 27 (1971)

Pierce[99] have applied the idea of specific elution to achieve a 250-fold purification of chicken liver dihydrofolate reductase on columns of Sepharose containing methotrexate (4-amino-10-methylpteroylglutamic acid) coupled via a six carbon chain. Elution of the enzyme readily occurs in the presence of dihydrofolate. Furthermore, Kristiansen *et al.*[100] have purified L-asparaginase from *E. coli* by specific adsorption and desorption. Fig. II.29 shows the elution of L-asparaginase from a specific adsorbent prepared by linking D-asparagine via a spacer molecule to the carbohydrate matrix, when 10^{-3}M D-asparagine is added. Evidence for specific interaction between the enzyme and the adsorbent was provided by the fact that no selective retardation of asparaginase activity was obtained in the presence of 10^{-3}M D-asparagine in the equilibrating buffer, whereas 10^{-3}M L-glutamine barely affected retardation.

It is important to note that for elution of enzymes from affinity matrices

Table II.5. Specific elution procedures

Enzyme or protein	Adsorbent	Eluant conditions	Reference
Soybean agglutinin	Sepharose-*N*-ϵ-aminocaproyl-β-D-galactopyranosylamine	0·5% D-galactose	82
L-Asparaginase	D-Asparagine-Sepharose	10^{-3}M D-asparagine	100
Carbonic anhydrase *B*	Sulphanilamide-Sephadex	Gradient of acetazolamide (0–10^{-4}M)	84
Lactate dehydrogenase	ϵ-Aminohexanoyl-NAD^+-Sepharose	Gradient of NAD^+ (0–5 × 10^{-2}M)	103
Oestradiol-17β-dehydrogenase	Oestrone-Sepharose	2 × 10^{-4}M oestrone hemisuccinate	63
Ribonucleotide reductase	dATP-Sepharose	Gradients of: dATP (0–1 × 10^{-3}M) ATP (0–1 × 10^{-2}M) dAMP (0–2 × 10^{-2}M)	76
Threonine deaminase	Isoleucine-Sepharose	10^{-3}M isoleucine	117
Thrombin	*p*-Chlorobenzylamido-ϵ-amino-hexanoyl-Sepharose	2 × 10^{-1}M benzamidine	110
Trypsin	*p*-Aminophenyl-guanidine-Sepharose	10^{-3}M benzamidine	109
Xylosidase	Sepharose-*p*-aminobenzyl-1-thio-β-D-xylopyranoside	2M D-xylose	54

by competitive effects, the free ligand should either be present in higher concentrations than that bound to the matrix or display an intrinsically higher affinity for the enzyme. In many cases, the elution of a specifically adsorbed protein with a buffer containing a high concentration of a competing inhibitor results in a greater dilution of the protein than when elution is occasioned by changes in pH or ionic strength. This effect is particularly obvious with high-affinity interactions and, in these cases, elution of the bound enzyme may

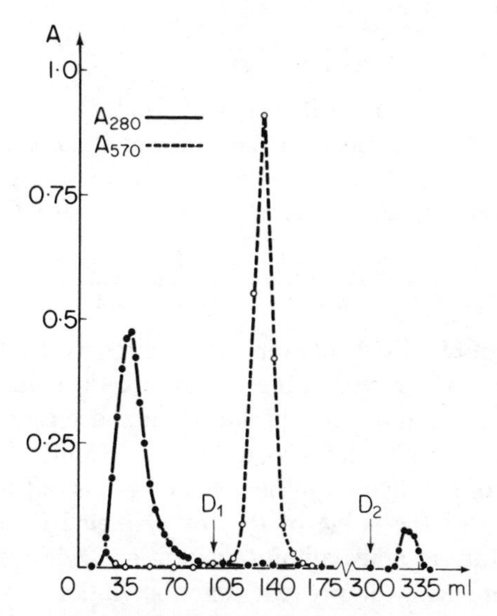

Figure II.29. Elution of L-asparaginase from specific adsorbent. Buffer during adsorption phase: 0·05M borate, containing 0·3M NaCl and 0·006% NaN_3, pH 8·6. Desorption with 10^{-3}M D-asparagine in buffer started at D_1, desorption of residual non-specifically bound material with 2M NaCl in buffer started at D_2. Broken line indicates asparaginase activity. Reproduced with permission from T. Kristiansen *et al.*, *FEBS Lett.*, 7, 294 (1970)

be a time-dependent process, despite utilization of high concentrations of competing ligand in the buffer.

For a generalized enzyme–ligand interaction:

$$E + L \underset{k_{-1}}{\overset{k_1}{\rightleftharpoons}} EL$$

the factors governing the elution of specifically bound enzymes are the rate of dissociation of the complex (k_{-1}) and the concentration of the enzyme immobilized on the stationary phase, $[EL]$:

$$\mathrm{d}[E]/\mathrm{d}t = k_{-1}[EL]$$

Clearly, the rate of dissociation of the enzyme from the stationary phase is a first-order process, dependent only on the concentration of the complex, and is independent of the concentration of free competing ligand. The free substrate or inhibitor will reduce the tendency of the enzyme to reassociate with the immobilized ligand by forming the soluble complex in preference, provided that the free competing ligand is present in an excessive concentration.

If the affinity (k_1/k_{-1} or K_L) of the complex is very high ($K_L \leqslant 10^{-7}$M), the time required for complete dissociation may be quite appreciable. Integration of the equation above gives

$$[E] = [E_0]\,e^{-k_{-1}t}$$

where $[E]$ is the concentration of free enzyme, $[E_0]$ is the initial concentration of bound enzyme, k_{-1} is the dissociation constant and t is the time. The time required for the concentration of the bound enzyme to decrease to half its original value, the *half-life* $t_{\frac{1}{2}}$, is given by the expression:

$$t_{\frac{1}{2}} = \frac{\ln [E_0]/[E]}{k_{-1}} = \frac{\ln 2}{k_{-1}} = \frac{0{\cdot}693}{k_{-1}}$$

This disadvantage of specific elution can be circumvented in several ways.

(i) By percolation of the competing ligand into the column thus stopping the flow through the column for a period of time determined by the nature of k_{-1}. In practice, however, this has to be done empirically since the effect of immobilization of the ligand on the value of k_{-1} is difficult to assess.

(ii) By alteration of the value of k_{-1} by changing either the pH, ionic strength or temperature. Since affinity adsorption is decreased at increased temperatures, dramatically improved inhibitor elutions can be affected at room temperature.*

These principles are well illustrated by reference to Fig. II.30 depicting the efficiency of different eluant systems on ox heart lactate dehydrogenase specifically bound to N^6-(6-aminohexyl)-AMP-Sepharose.[102] The first curve demonstrates the relative effectiveness of 0·5 mM NADH in quantitatively eluting the enzyme within approximately one hour. On the other hand, when the equilibrating buffer was supplemented with 0·5 mM NAD^+ and 0·5 mM L-lactate, as little as 30% of the total activity was recovered after 20 hours. Furthermore, Ohlsson *et al.*[102] subsequently demonstrated the effectiveness of stopping the flow through a column to elute a specifically bound protein.

Uncompetitive effects represent a convenient method of enhancing the binding of an enzyme, and hence of increasing the selectivity of the adsorption process. This is advantageous in the resolution of the specifically adsorbed protein from non-specifically adsorbed proteins. Thus, Akanuma *et al.*[98] have

* For example, Jackson *et al.*[101] have shown that dihydroneoptein triphosphate synthetase, the first enzyme in folate biosynthesis, can be readily eluted from a modified GTP-Sepharose derivative with GTP at room temperature.

shown that bovine carboxypeptidase *B* is not markedly retained on a column of phenylalanine-Sepharose even at low ionic strength in the pH range 7–9. The addition of 20 μM ε-aminocaproic acid (a lysine analogue) to the eluant, however, results in the enzyme being tightly bound to the phenylalanine-coupled matrix. Furthermore, an exponential decrease in the concentration of ε-aminocaproic acid was found to elute the enzyme as shown in Fig. II.31. The concentration of the lysine analogue required to elute the enzyme was in good agreement with the kinetic inhibitor constant of ε-aminocaproic acid towards the carboxypeptidase *B* catalysed hydrolysis.

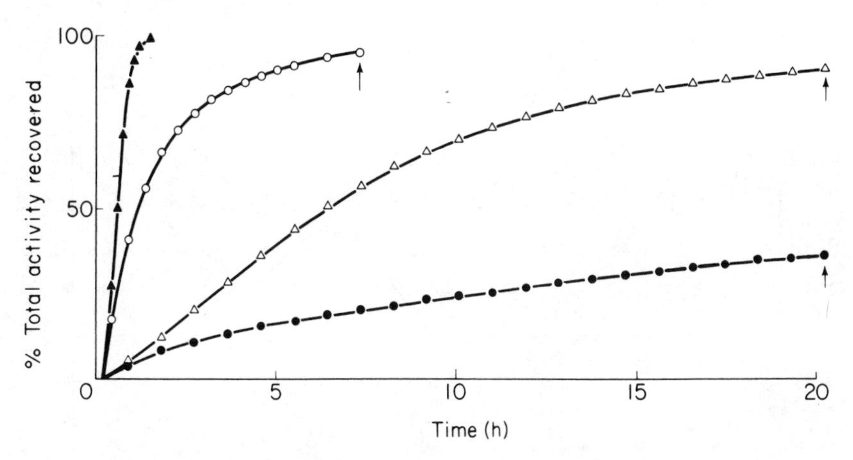

Figure II.30. Efficiency of different eluant systems on ox heart lactate dehydrogenase (LDH) bound to a AMP-Sepharose column (40 mm × 15 mm, containing 1·0 g of wet gel). 0·1 mg of LDH in 0·5 ml 0·1M phosphate buffer, pH 7·5, was applied. The following systems in the same buffer were used: 0·5 mM NAD^+ + 0·5 mM L-lactate (●–●–●), 0·5 mM NAD^+ + 0·5 mM pyruvate (△–△–△), 0·5 mM oxidized NAD-pyruvate adduct (○–○–○) and 0·5 mM NADH (▲–▲–▲). The arrows indicate a pulse of 2·0 ml of 10 mM NADH to permit elution of the remaining bound enzyme. Corrections are made for the inhibition effects in enzyme assays; 5·5 ml fractions were collected at a rate of 6 ml/hour. Reproduced with permission from R. Ohlsson, P. Brodelius and K. Mosbach, *FEBS Lett.*, **25**, 234 (1972)

Lowe and Dean[103] have demonstrated that the binding of pig heart lactate dehydrogenase to ε-aminohexanoyl-NAD^+-Sepharose can be considerably enhanced in the presence of 1 mM sodium sulphite. The ternary NAD^+–sulphite–enzyme complex is more stable than the corresponding NAD^+–enzyme binary complex, and can only be displaced in the presence of NADH. Thus, elution of the enzyme can be effected by inclusion of 5 mM NADH in the buffer.

For bi- and multi-substrate enzymes an extra degree of selectivity can be introduced. For example, many pyridine nucleotide-dependent dehydrogenases have compulsory ordered kinetic mechanisms in which the pyridine nucleotide binds first (Fig. II.11). Such a mechanism has been proposed for lactate dehydrogenase and consequently affinity chromatography on immobilized ligands analogous to lactate or pyruvate is unlikely to succeed. However, this superficial disadvantage may be exploited by addition of a low concentration of the pyridine nucleotide to the irrigating buffer. Thus, O'Carra and Barry[50] have demonstrated that in the presence of 100 μM NADH lactate dehydrogenase is strongly retarded by immobilized oxamate,

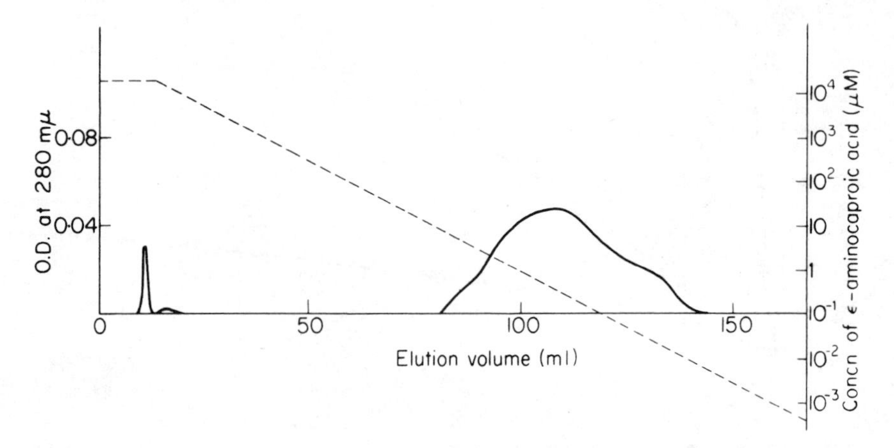

Figure II.31. Elution of carboxypeptidase *B* from ε-aminocaproyl-D-phenylalanine-coupled CM-Sephadex column with the effect of decreasing concentration of ε-aminocaproic acid. Amount of enzyme applied: 5·0 mg, column size: 1·0 cm × 6·8 cm, solid line: optical density at 280 mμ, eluant: 10 mM sodium phosphate buffer containing 100 mM NaCl, an exponential gradient from 20 to +0 mM ε-aminocaproic acid was applied as indicated by the dotted line, pH 7·05 throughout. Reproduced with permission from H. Akanuma *et al.*, *Biochem. Biophys. Res. Commun.*, **45**, 27 (1971)

an analogue of pyruvate, and that subsequent removal of the nucleotide elicits prompt elution of the enzyme. It is clear that more specificity is inherent in the 'negative' elution achieved by discontinuation of the complementary ligand than could be affected by 'positive' elution with the counter ligand. Thus, not only is the adsorption of lactate dehydrogenase dependent on the specific complementary ligand, resulting in a preliminary separation from proteins with no affinity for oxamate, but also the subsequent elution should leave behind any other adsorbed enzymes. The specificity implicit in this procedure is thus equivalent to two independent affinity chromatography

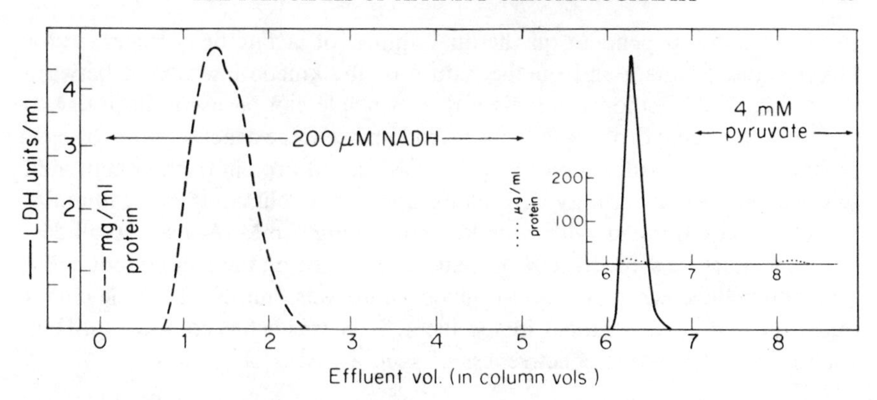

Figure II.32. Chromatographic isolation of lactate dehydrogenase (LDH) from a crude placental extract on a column of insolubilized oxamate. Human placenta, mixed with an equal volume of 0·5M NaCl in 0·02M phosphate buffer, pH 6·8, was homogenized (Sorval Omni mixer, setting 8, 1 minute) and centrifuged at 35,000 g for 30 minutes. The supernatant was adjusted to pH 6·8, 0·5M NaCl and 200 μM NADH before application. Addition of NADH to irrigant was commenced shortly before application of sample. Volume of applied sample was 1·6 ml (0·4 of a column volume) but larger samples may be applied. Recovery of LDH is almost quantitative (>98%). Reproduced with permission from P. O'Carra and S. Barry, *FEBS Lett.*, **21**, 281 (1972)

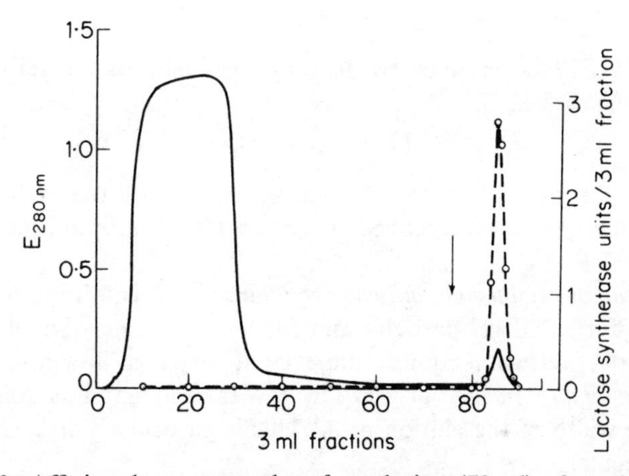

Figure II.33. Affinity chromatography of a solution (72 ml) of partially purified lactose synthetase *A* protein on a column (1·1 cm × 23 cm) of Sepharose-α-lactalbumin equilibrated with 0·01M tris-HCl, pH 7·5, containing 0·04M KCl and 3 mM NAG. Elution was continued with the same buffer after application of the sample until 76 fractions had been collected (arrow), then elution was with the tris buffer containing 0·04M KCl only. ——, $E_{280\,nm}$; ○ - - - - ○, lactose synthetase activity. Reproduced with permission from P. Andrews, *FEBS Lett.*, **9**, 297 (1970)

steps, since it is dependent on the dual affinity of lactate dehydrogenase for NADH and oxamate and on the nature of the kinetic interaction between them. Fig. II.32 demonstrates how this principle can be used effectively to isolate lactate dehydrogenase from a crude placental extract.

Similarly, the purification of lactose synthetase *A* protein from human milk has been effected by affinity chromatography on a column of α-lactalbumin-Sepharose equilibrated with a buffer containing 3 mM *N*-acetyl-D-glucosamine.[104] Lactose synthetase *A* protein was retained on the column but could be readily eluted when *N*-acetyl-D-glucosamine was omitted from the eluant buffer. A 40-fold enrichment in specific activity was achieved and Fig. II.33 shows the considerable effectiveness of such a means of elution.

3. The Effectivity of Elution Procedures

The broadening of peaks, which is brought about by incomplete establishment of diffusion equilibrium, is frequently a limiting factor for effectivity during affinity chromatography. The more the equilibrium is removed from ideality the larger the broadening. Calculation of the height equivalent to the theoretical plate (*HETP*) of a column is an exact measure of this phenomenon.[105] The number of plates (N) for a molecularly uniform substance chromatographed on the gel column is calculated from the width of the peak (w) and the elution volume (V_e) (Fig. II.34)

$$N = [(4V_e)/w]^2$$

whence the *HETP* is obtained by dividing the length of the gel bed by the number of plates, i.e.,

$$HETP = (L/16)(w/V_e)^2$$

The smaller the base of the elution curve, the smaller the *HETP* and the greater the effectivity of the gel bed. There are three important factors which affect the *HETP*.

(a) *The rate of establishing diffusion equilibrium:* the equilibrium is reached more easily for small gel particles and for low flow rates. Small molecules establish perfect diffusion equilibrium even at very high flow rates.

(b) *Longitudinal diffusion:* at very low flow rates longitudinal diffusion can increase the width of the elution peak. This is particularly true for proteins of low molecular weight.

(c) *Bed geometry: HETP* increases in columns not packed uniformly and is affected by errors in experimental design such as excessive dead volumes.

G. THE CAPACITY OF AN AFFINITY ADSORBENT

The efficiency of a biospecific adsorbent depends partly on the extent to which the matrix can be loaded with complementary macromolecules and partly on its ability to achieve adequate and worthwhile purifications on subsequent elution.

The capacity of a selective adsorbent is thus determined principally by two interdependent parameters: (i) the correct choice of matrix, extension arm and ligand in the initial design and preparation of the adsorbent to optimize the enzyme–ligand interaction; and (ii) the way in which the capacity is determined by such dynamic factors as flow rate, equilibration time and method of adsorption.

The capacity of a specific adsorbent is determined by the concentration of ligand that is freely available for interaction with the complementary enzyme.

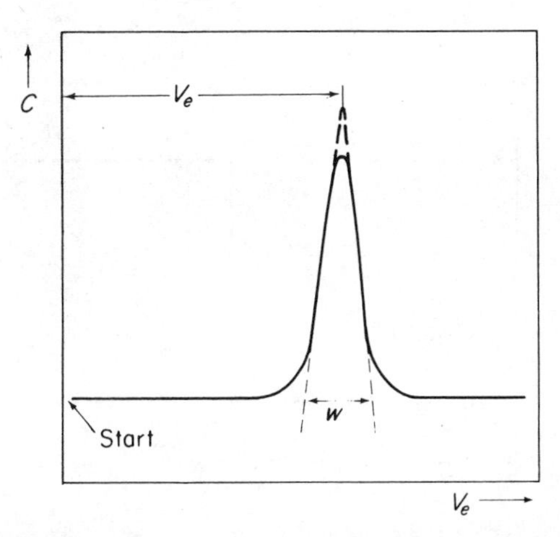

Figure II.34. Approach to the calculation of the height equivalent to one theoretical plate (HETP) from an elution diagram of a uniform substance: V_e = elution volume, w = width of the elution peak. The number of theoretical plates (N) may then be calculated from $N = ((4 \times V_e)/w)^2$. Reproduced with permission from H. Determann, *Gel Chromatography*, Springer-Verlag, New York, 1968, p. 70

The exclusion of macromolecules from extensively cross-linked Sephadexes has already been mentioned (Section B) and Robinson *et al.*[85] have demonstrated the effects of exclusion from Sepharose polymers on the capacity of a specific adsorbent for β-galactosidase. Even with such loose porous matrices the capacity is determined by the freedom of the ligand from steric constraints imposed by the matrix backbone. Furthermore, the maximum enzyme binding capacity for several affinity columns containing different amounts of ligand was proportional to the gel inhibitor concentration over the range tested.[85]

However, even when these considerations are optimized in the design of the adsorbent, the definition of 'capacity' is meaningless unless the conditions of adsorption are specified. Using the emergence of enzyme in the void volume as an index of the operational capacity of the column is misleading since it may be possible to add many times more enzyme whilst losing significant quantities of enzyme in the breakthrough peak.

A balance should be achieved between the rate of addition of the enzyme sample, the rate of elution and the volume of the sample at any particular concentration. Since affinity adsorption exhibits very fast kinetics, it is practicable to use a contact time that allows most of the nominal or equilibrium capacity to be used under column chromatographic conditions. If time is allowed for diffusion into the more inaccessible regions of the adsorbent, the rate of subsequent elution should not exceed that permitted by the rate of desorption from the gel. Excessive rates of elution compared to contact

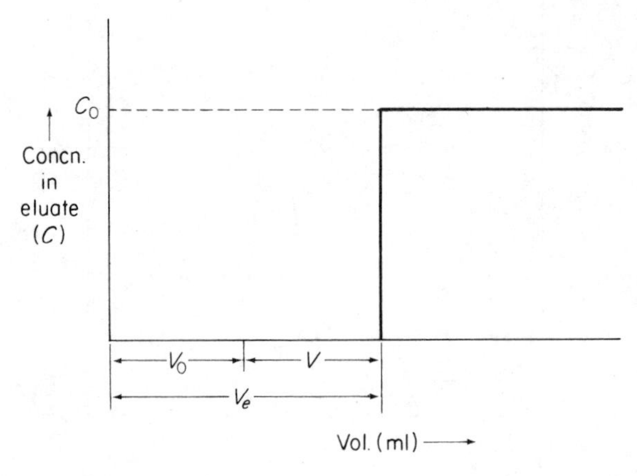

Figure II.35. Determination of the capacity of an affinity adsorbent by frontal analysis

time during column loading can result in zone distortion and loss of resolution. Alternatively, the available capacity may be increased by increasing the volume of the sample, rather than by adding the sample more slowly. Thus, with affinity adsorbents it may not be desirable to apply the sample in a small volume and high concentration; although this may be better if a high concentration is required to establish a more favourable equilibrium.

When the species required is selectively adsorbed on a continuous basis until the bed is 'saturated' the capacity determined will be a total 'equilibrium' capacity. As the bed becomes saturated with the adsorptive, the solution breaks through at the same concentration as it had on entering the column (Fig. II.35).

The volume that appears up to the 'step', where the concentration of the effluent increases rapidly over a small volume, is called the *retention volume*

(V_e). It is made up partly of the volume present in the interstices (V_0) and also the volume of solution from which the adsorptive was removed by the adsorbent (V), i.e.,

$$V_e = V_0 + V$$

It is often convenient to measure V_e in terms of the void volume V_0 by dividing V_e by V_0. This gives the number of void volumes required to break through and is dimensionless and hence independent of bed geometry.

If m is the total weight of adsorbent in grams, the specific amount of adsorptive adsorbed per gram q, is $q = (V/m)C_0$, and the total amount adsorbed by the bed is qm, i.e., VC_0.

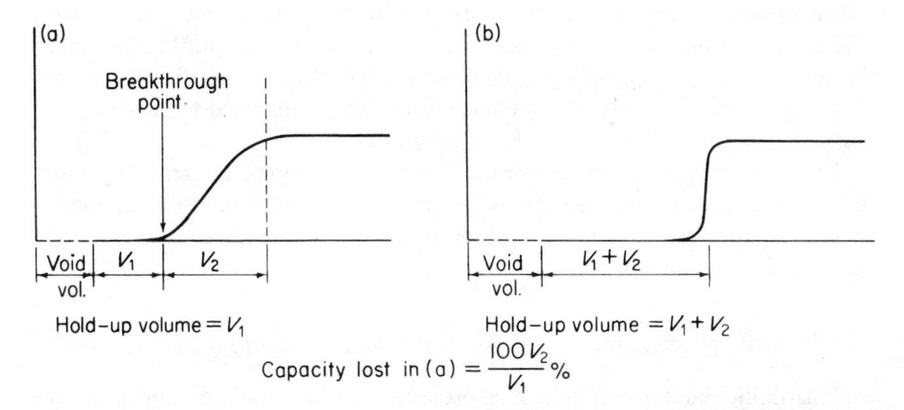

Figure II.36. The effect of flow rate on the determination of the capacity of an affinity adsorbent by frontal analysis; (a) fast and (b) slow flow rate

The ease with which q is measured by the frontal analysis technique for any given value of C_0 makes this one of the best ways of determining an isotherm; q is measured for several predetermined values of C_0 and the isotherm plotted.

This method is usually quicker and simpler than the method of shaking a known weight of adsorbent (m) with a known volume of solution of concentration C_0 and subsequently, after equilibrium has been reached, measuring the new lower concentration c; q is then calculated from $q = (C_0 - C)/m$. In the frontal analysis technique equilibrium has been attained when the effluent has reached the same concentration as the applied sample, i.e., C_0, and qm is calculated as the amount of the adsorptive that has disappeared from the effluent.

The saturation point, consistent with the breakthrough point of the required species, is dependent on the rate of application of the original sample. Furthermore, the rate of application also determines the shape of the breakthrough 'front' of the desired species. Fig. II.36 demonstrates that, at high flow rates, equilibrium is not attained and the capacity of the adsorbent is apparently reduced.

H. THE ACHIEVEMENTS OF AFFINITY CHROMATOGRAPHY

The technique of affinity chromatography has recently achieved some spectacular single-step purifications of enzymes and other proteins. Examination of Table II.6 shows that almost quantitative recoveries can be obtained with careful choice of the adsorbent and eluant conditions. Furthermore, purifications of the order of several thousand-fold are conceivable with systems of high affinity, i.e., with dissociation constants $<10^{-6}$M, especially when coupled with specific methods of elution. Interacting systems with dissociation constants in the range of 10^{-4}–10^{-6}M are best purified by superimposition of specific elution techniques on the already specific adsorption. The high purity of the resulting enzymes has been confirmed by their specific activity, immunodiffusion and disc electrophoresis.

The systems which display relatively low purifications are generally those in which unspecific interactions with the immobilized ligand or the matrix play a significant part. Some reasons for these effects are discussed in the next section of this chapter.

I. THE NATURE OF UNSPECIFIC EFFECTS

Unspecific adsorption of inert proteins can seriously compromise the effectiveness of affinity chromatography as a technique for protein purification. Unspecific effects are generally of a relatively weak nature and hence assume importance only with interacting systems of low affinity. By far the most significant and widespread of these effects is that of ion-exchange on charged groups arising by modification of the matrix backbone.

Intrinsically, commercial Sepharose and cellulose exhibit few or no ion-exchange properties. Derivatization and modification of these polymers in ways fundamental to the technique of affinity chromatography can, however, introduce undesirable ion-exchange capacities.

One of the most common causes of this phenomenon is when the ligand showing unique affinity for the macromolecule is itself charged. Immobilization of the ligand will therefore generate an ion-exchange polymer analogous to the ion-exchange celluloses or Sephadexes commercially available. Inert protein will thus be retained on the column. In principle, this problem can be circumvented by using a high ionic strength buffer or a pH unfavourable to the ionization of the charged ligand. Unfortunately, these approaches generally decrease the specific interaction between the charged ligand and

Table II.6. The Achievements of affinity chromatography

Enzyme or protein	Source	Dissociation constant (K_L) (M)	Recovery (%)	Purification[a] (fold)		Reference
				Specific	Unspecific	
Tyrosine hydroxylase	Brain	10^{-4}	89		5	106
Neuraminidase	*Clostridium perfringens*		105		45	107
	Vibrio cholerae		97		420	107
	Influenza virus		91		2	107
Retinol binding protein	Human serum	10^{-7}	90		2	77
Acetylcholinesterase	Electric eel	6×10^{-6}			17	87
3-Deoxy-D-arabinoheptulosonate 7-phosphate synthetase	Yeast	10^{-7}	100		100	55
Dihydrofolate reductase	Phage T_4	10^{-7}	80	6000		108
Chorismate mutase	*Claviceps paspali*	10^{-7}	80	High		81
Dihydrofolate reductase	Chicken liver		100	250		99
	Mammalian skin	10^{-8}	33	3700		58
Avidin	Egg white	10^{-15}	90		4000	94
Vitamin B_{12} binding protein	Granulocytes		92		9860	111

[a]Refers to the method of elution.

the desired complementary protein, since in many cases the charge on the ligand is critical to the interaction with the macromolecule. This is certainly the case with the purification of acetylcholinesterase from the electric eel by chromatography on *N*-(ϵ-aminocaproyl)-*p*-aminophenyl trimethylammonium bromide-Sepharose (Fig. II.37). Kalderon *et al.*[87] found that although this compound was a good inhibitor of acetylcholinesterase with a K_i of 6×10^{-6}M, elution with 1M NaCl achieved only a 17-fold enrichment of specific activity. When resins were employed containing higher concentrations of bound ligand, the specific activity of the enzyme eluted by 1M NaCl steadily decreased, despite the fact that the resin bound larger amounts of enzyme. This phenomenon was attributed to the abundance of bound quaternary ammonium groups yielding a resin of high positive charge-density capable of unspecifically binding negatively charged proteins. Attempts at decreasing non-specific electrostatic interactions by raising the ionic strength were precluded in view of the decreased affinity of acetylcholinesterase inhibitors

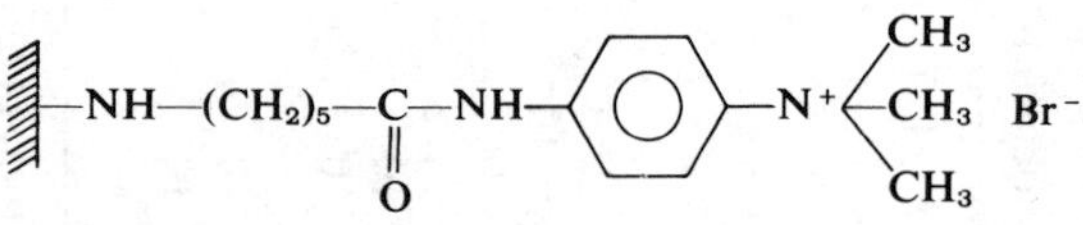

$$-NH-(CH_2)_5-\underset{\underset{O}{\parallel}}{C}-NH-C_6H_4-N^+(CH_3)_3 \quad Br^-$$

Figure II.37. The structure of *N*-(ϵ-aminocaproyl) *p*-aminophenyl trimethylammonium bromide-Sepharose

for the enzyme at high ionic strength. Studies by Uren with proteolytic enzymes also serve to emphasize the dangers in interpretation of retardation by modified matrices in terms of affinity chromatography.[112] For example, Uren has found that trypsin, chymotrypsins *A* and *B*, and carboxypeptidases *A* and *B* are bound to a high-capacity adsorbent, glycyl-D-phenylalanyl-*N*-ethyl-cellulose, when the buffer is of low ionic strength. In buffers of high ionic strength only carboxypeptidase *A* was retained. Consider the case of trypsin and chymotrypsin in more detail. Trypsin is bound to the glycyl-D-phenylalanine adsorbent and released by 0·25M NaCl. Trypsin treated with the active site inhibitor diisopropyl-fluorophosphate (DFP) behaves similarly, a strong indication that the interaction between the enzyme and the adsorbent occurs at a site other than the active one. The ionic nature of this binding was verified by the preparation of an anionic derivative of trypsin, acetyl-trypsin. Whilst this modified enzyme retained most of its proteolytic activity towards casein and benzoyl-L-arginine ethyl ester, in contrast to unmodified trypsin, this anionic protein was not retarded by the adsorbent. Chymotrypsin *B* binding could readily be abolished on inactivation by treatment with 10^{-3}M DFP for one hour. This was taken as evidence for affinity chromatography.

Furthermore, Uren has found that electrostatic interactions can be excluded with this polymer by considering only anionic proteins, and under these circumstances the order in which a group of proteases are eluted is identical to that predicted from their binding characteristics. Several criteria for distinguishing between specific and non-specific interactions are included in Section J of this chapter.

The preparation of derivatives of inert matrices prior to the attachment of the desired ligand can lead to the introduction of ion-exchange phenomena. A detailed investigation into the plethora of chemical methods employed for immobilization of ligands will not be given here. Suffice it to illustrate the problem with reference to a common method for linking ligands to extension arms of various lengths, namely carbodiimide-promoted reactions. The preparation of ω-aminoalkyl and ω-carboxylalkyl derivatives by direct coupling of the corresponding amino compounds to the matrix can lead, by virtue of the incomplete reaction of the carbodiimide reaction, to an anion and a cation exchanger respectively. This undesirable effect could account for the observation of Kaufman and Pierce[99] that methotrexate-Sepharose prepared by linking methotrexate to ω-aminohexyl-Sepharose by a carbodiimide condensation adsorbs large amounts of unknown proteins when attempting to purify dihydrofolate reductase from crude extracts of yeast and bacteria. Similar non-specific interaction with inert proteins were noted by Erickson and Mathews[108] employing a similar adsorbent. The use of the water-soluble carbodiimide, 1-cyclohexyl-3(2-morpholinoethyl) carbodiimide metho-*p*-toluene sulphonate, can in itself lead to the formation of ion-exchange groups.[51] A chemical intermediate formed between the carbodiimide and the extension arm can generate a cation exchanger, especially if the subsequent reaction with the ligand is slow or incomplete. In general, addition of vast excesses of the ligand to be coupled will circumvent this difficulty.

A ligand that has been rendered insoluble by chemical attachment to a polymeric support is surrounded by a microenvironment that is defined, in part, by the nature of the polymeric support. If either the immobilized ligand or the matrix is charged there may be a significant difference between the concentration of a charged macromolecule in the microenvironment of the bound ligand and its concentration in the bulk of the suspending solution. It has been suggested that this unequal distribution of ions, which has been ascribed to the electrostatic interaction, between the charge field of the support and the ionic molecules in the system, can cause a difference in pH and apparent K_m for an enzyme interacting with an immobilized substrate.

From polyelectrolyte theory it is well known that the concentration of hydrogen ions in the domain of a charged polyelectrolyte gel is different from that of the solution with which it is in equilibrium. A ligand embedded in a gel phase will thus be exposed (Fig. II.38) to a pH different from that measured by the standard physicochemical methods for the bulk of the solution. The hydrogen ion concentration within a negatively charged polyelectrolyte gel is higher than in the bulk of the system, while in a positively charged gel it is lower. The effect of the electrostatic potential (ψ) of the microenvironment on the distribution of hydrogen ions can readily be calculated.[113]

The equilibrium condition between the gel Phase I and the external solution II requires that the electrochemical potential of the hydrogen ions in Phase I, $\tilde{\mu}^{\mathrm{I}}_{\mathrm{H}^+}$, equals that of the hydrogen ions in Phase II, $\tilde{\mu}^{\mathrm{II}}_{\mathrm{H}^+}$,

$$\tilde{\mu}^{\mathrm{I}}_{\mathrm{H}^+} = \tilde{\mu}^{\mathrm{II}}_{\mathrm{H}^+} \qquad \text{(II.2)}$$

The electrochemical potentials of the hydrogen ions of the two phases are

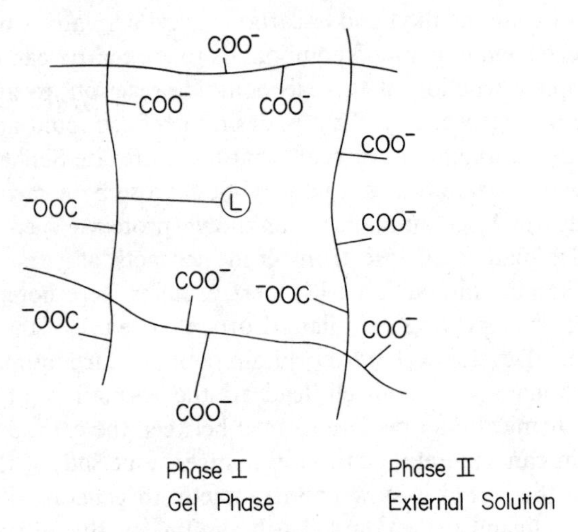

Figure II.38. Schematic representation of an immobilized ligand embedded in a polyelectrolyte carrier

related to the corresponding chemical potentials $\mu^{\mathrm{I}}_{\mathrm{H}^+}$ and $\mu^{\mathrm{II}}_{\mathrm{H}^+}$ by equations (II.3) and (II.4) where ϵ designates the positive electronic charge:

$$\tilde{\mu}^{\mathrm{I}}_{\mathrm{H}^+} = \mu^{\mathrm{II}}_{\mathrm{H}^+} + \epsilon\psi \qquad \text{(II.3)}$$

$$\tilde{\mu}^{\mathrm{II}}_{\mathrm{H}^+} = \mu^{\mathrm{II}}_{\mathrm{H}^+} \qquad \text{(II.4)}$$

Correlating the chemical potentials with the corresponding hydrogen ion activities in Phase I, $a^{\mathrm{I}}_{\mathrm{H}^+}$, and Phase II, $a^{\mathrm{II}}_{\mathrm{H}^+}$, one obtains equations (II.5) and (II.6),

$$\mu^{\mathrm{I}}_{\mathrm{H}^+} = \mu^{0}_{\mathrm{H}^+} + kT \ln a^{\mathrm{I}}_{\mathrm{H}^+} \qquad \text{(II.5)}$$

$$\mu^{\mathrm{II}}_{\mathrm{H}^+} = \mu^{0}_{\mathrm{H}^+} + kT \ln a^{\mathrm{II}}_{\mathrm{H}^+} \qquad \text{(II.6)}$$

in which $\mu^0_{H^+}$ is the standard chemical potential of the hydrogen ion, k is the Boltzmann constant and T is the absolute temperature. Equations (II.2) and (II.3–II.6) reduce to:

$$\ln a^{II}_{H^+} - \ln a^{I}_{H^+} = \epsilon\psi/kT \tag{II.7}$$

or

$$\Delta pH = pH^{I} - pH^{II} = 0{\cdot}43\epsilon\psi/kT \tag{II.8}$$

since $-\log a_{H^+} = pH$ and ΔpH is the difference between the pH of the gel phase (I) and the external solution (II). Goldstein *et al.*[113] have shown that with water-insoluble derivatives of trypsin, ΔpH can be as high as 2·0–2·4 units, corresponding to an electrostatic potential of 125–145 millivolts. The magnitude of ψ has been shown to decrease with increasing ionic strength, in accord with the polyelectrolyte theory.

A similar calculation shows that the dissociation constant for the immobilized ligand, K'_L, is related to the true dissociation constant in free solution, K_L, by an analogous expression:

$$K'_L = K_L\, e^{-z\epsilon\psi/kT} \tag{II.9}$$

or

$$\Delta pK_L = pK'_L - pK_L = 0{\cdot}43z\epsilon\psi/kT \tag{II.10}$$

Equation (II.10) shows that when z ($z\epsilon$ is the charge on the macromolecule) and ψ are of the same sign, i.e., when the macromolecule is positively charged and the polyelectrolyte carrier is negatively charged, or vice versa, $K'_L < K_L$, and the binding of the macromolecule to the affinity matrix is apparently enhanced. On the other hand, when z and ψ are of opposite signs, $K'_L > K_L$, and binding of the macromolecule is decreased. If either z or ψ are zero, i.e., if either the matrix or macromolecule carries no charge, then $K'_L = K_L$ and the binding to the affinity matrix represents the true affinity.

These data demonstrate that a polyelectrolyte carrier might markedly affect the 'local pH' and 'local protein concentration' of the microenvironment in which an immobilized ligand is embedded. It is clear, however, that other parameters of the microenvironment, such as chemical composition, dielectric constant and affinity for the ligand, will influence the nature of the specific interaction. The effect of the difference in pH within the matrix and the effect on the apparent K_L of the system can profoundly influence the behaviour of the protein–ligand interactions. For functional purifications from crude systems these considerations are of little relevance or consequence, but for some of the more analytical uses to which affinity chromatography is now being put, the effects could be highly significant.

J. CRITERIA FOR AFFINITY CHROMATOGRAPHY

The complications and restrictions imposed by the nature of unspecific binding require that some criterion for the designation of the term 'affinity chromatography' be applied to the system under study. The distinction between specific and non-specific adsorption is particularly relevant where quantitative conclusions are being drawn about the strength or nature of the interaction. Unfortunately, the multiplicity and diversity of biological interactions make generalizations difficult and the list of criteria below are intended as a guide only.

(1) The enzyme or protein should not be bound to control matrices to which no ligand has been attached or to matrices to which an inactive substrate analogue has been attached using similar procedures.

(2) The binding characteristics of the affinity matrix should correlate well with the known free solution properties of the ligand–macromolecule system.

(3) Retention of the macromolecule in high ionic strength buffers is a useful criterion for affinity chromatography. However, loss of binding capacity at high ionic strengths is not necessarily indicative of unspecific electrostatic interactions since many protein–ligand interactions are themselves electrostatic in nature.

(4) If complete inhibition of enzymic activity results in a loss in the ability of the enzyme to be bound to the affinity adsorbent, this is strong evidence in favour of a specific interaction. This can be achieved by denaturation or by reaction with active site-directed irreversible inhibitors.

(5) The enzyme should not be enzymically active when immobilized on the column. This criterion is often complicated by the fact that most enzymes have more than one subunit and hence have multiple active sites.

(6) The specific elution of an enzyme bound to an immobilized ligand with substrate, cofactor, activator or allosteric effector is evidence, though not conclusive proof, in favour of affinity chromatography. It is well known that specific substrate elution of enzymes adsorbed to carriers by physicochemical forces can be effected. Thus, dilute pyrophosphate solutions have been used to elute yeast pyrophosphatase from C_{γ} alumina gel.[114]

(7) The increased binding of the enzyme to the immobilized ligand by uncompetitive effects is perhaps the best indication of affinity chromatography, particularly if highly specific for the system under investigation. Likewise, the binding of a complementary macromolecule to an immobilized ligand in the presence of a second obligatory ligand, and subsequent elution of the protein on its removal, is a good foundation for affinity chromatography.

It is thus essential to apply several independent criteria in order to establish the mechanism as affinity chromatography since each of the above considerations is open to ambiguity of interpretation. This is particularly true where the ligand–protein interaction is of low affinity.

K. EXAMPLES OF THE TECHNIQUE

The following examples of the technique of affinity chromatography are included to illustrate some of the concepts outlined above. The first two examples show how effective adsorbents can be prepared by relatively simple chemical procedures to give homogeneous preparations in one step. The third example shows how a macromolecular ligand can be coupled to Sepharose and effectively utilized to purify enzymes to homogeneity from partially purified extracts. The method is, however, less effective when crude enzyme extracts are applied. Furthermore, adsorbents can be prepared in non-aqueous solvent systems and the bound enzyme eluted specifically with the same ligand that is attached to the column.

The final example highlights the contention that specific elution is not necessarily a criterion for affinity chromatography. In this case the potential application of specific effector–enzyme interactions is illustrated by the purification of liver pyruvate kinase on CM-cellulose columns utilizing the positive allosteric effector, fructose-1,6-diphosphate (FDP) as eluant.

1. The Purification of Neuraminidase from *Vibrio cholerae*

N-Substituted oxamic acids have been shown to be potent reversible inhibitors of influenza virus neuraminidase. Cuatrecasas and Illiano[107] have exploited this principle by synthesizing *N*-(*p*-aminophenyl)-oxamic acid and

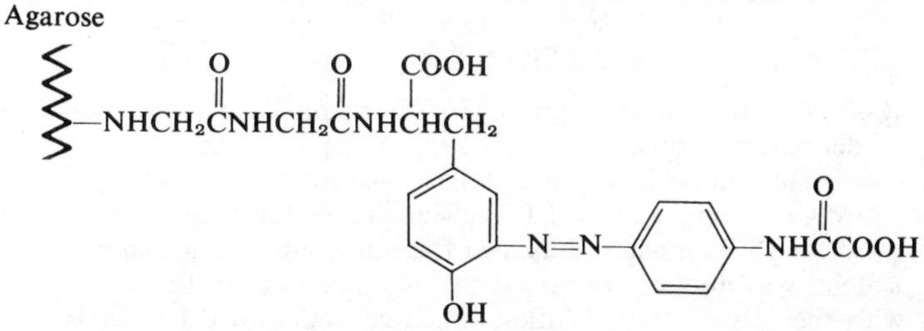

Agarose

—NHCH₂C(=O)NHCH₂C(=O)NHCH(COOH)CH₂—(C₆H₃OH)—N=N—(C₆H₄)—NHC(=O)COOH

Figure II.39. Adsorbent used in the selective purification of neuraminidases by affinity chromatography. The diazonium derivative of *N*-(*p*-aminophenyl) oxamic acid was coupled to agarose-gly-gly-tyr. Reproduced with permission from P. Cuatrecasas and G. Illiano, *Biochem. Biophys. Res. Commun.*, **44**, 178 (1971)

attaching it covalently through an azo linkage to Sepharose-glycylglycyl-tyrosine (Fig. II.39).

Virtually all the neuraminidase activity from a crude filtrate of *Vibrio cholerae* containing 92 mg protein was strongly adsorbed onto a column (0·4 cm × 4 cm) of the specific adsorbent equilibrated with 0·05M sodium acetate buffer, pH 5·5, containing 2 mM $CaCl_2$ and 0·2 mM EDTA. The column could be washed with large volumes of this buffer without loss of enzyme in the effluent. Quantitative elution of the enzyme was effected with 0·1M $NaHCO_3$ buffer, pH 9·1. The resultant preparation contained 0·2 mg protein at a specific activity of 30, representing a 420-fold enrichment over the crude extract (Fig. II.40).

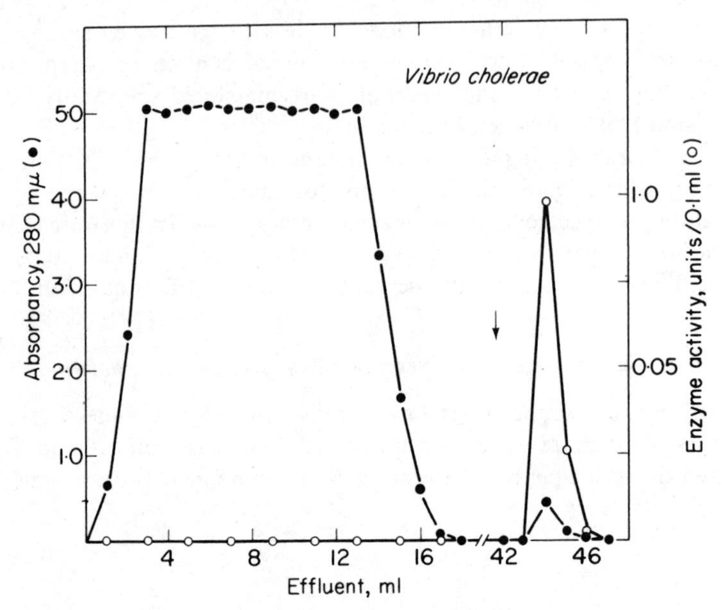

Figure II.40. Affinity chromatography of neuraminidase from *Vibrio cholerae* using the derivative shown in Fig. II.39. 140 mg of a crude filtrate of *Vibrio cholerae* were suspended in 12 ml of 0·05M sodium acetate buffer, pH 5·5, containing 2 mM $CaCl_2$ and 0·2 mM EDTA, centrifuged for 20 minutes at 2000 r.p.m. and dialysed for 18 hours at 4°C against the same buffer. The sample, containing 92 mg protein, was applied to a 0·4 cm × 4 cm column of the adsorbent equilibrated with the same buffer. Elution was achieved with 0·1M $NaHCO_3$ buffer, pH 9·1 (arrow). Reproduced with permission from P. Cuatrecasas and G. Illiano, *Biochem. Biophys. Res. Commun.*, **44**, 178 (1971)

2. The Purification of Staphylococcal Nuclease by Affinity Chromatography on an Immobilized Inhibitor

3′-(4-Aminophenyl-phosphoryl) deoxythymidine 5′-phosphate is a strong competitive inhibitor ($K_i \simeq 10^{-6}$M) of Staphylococcal nuclease (Fig. II.41) that can be readily coupled to agarose by an amino group that is relatively distant from the structural unit recognized by the enzyme binding site.

Columns prepared from this affinity matrix can specifically and strongly adsorb samples of pure and crude nuclease (Fig. II.42). Such columns withstand exhaustive washing with buffer without loss of enzymic activity. Elution can be effected with solutions of low pH (acetic acid, pH 3) or high pH (NH_4OH, pH 11) to give a sharp peak that can be directly lyophilized. Furthermore, these affinity adsorbents will resolve mixtures of active and chemically modified inactive nuclease preparations.

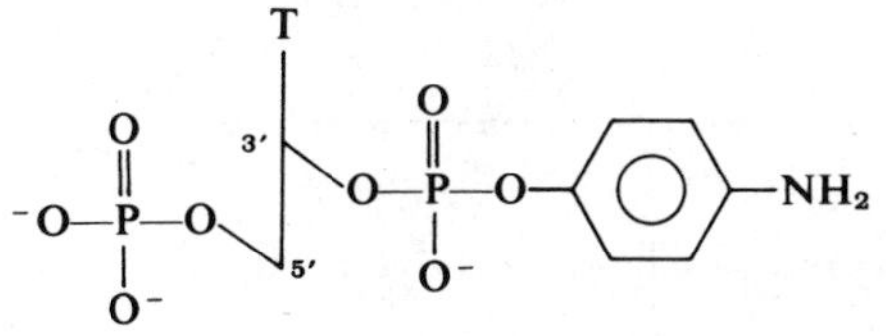

Figure II.41. Structure of the nuclease inhibitor used for attachment to Sepharose. Reproduced with permission from P. Cuatrecasas *et al.*, *Proc. Nat. Acad. Sci., U.S.A.*, **61**, 636 (1968)

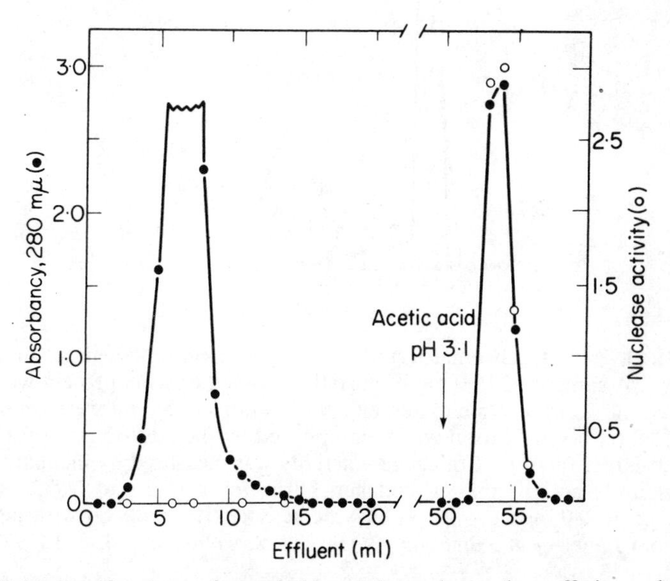

Figure II.42. Purification of staphylococcal nuclease by affinity adsorption chromatography on a nuclease-specific agarose column (0·8 cm × 5 cm). The column was equilibrated with 50 mM borate buffer, pH 8·0, containing 10 mM $CaCl_2$. Approximately 40 mg of partially purified material containing about 8 mg of nuclease was applied in 3·2 ml of the same buffer. After 50 ml of buffer had passed through the column, 0·1M acetic acid was added to elute the enzyme. Nuclease, 8·2 mg, and all the original activity was recovered. The flow rate was about 70 ml per hour. Reproduced with permission from P. Cuatrecasas *et al.*, *Proc. Nat. Acad. Sci., U.S.A.*, **61**, 636 (1968)

3. The Preparation of Three Vertebrate Collagenases in Pure Form

The high affinity of collagenase for its substrate has been exploited for the purification of the collagenase from *Clostridium histolyticum*.[115] The enzyme was bound to native collagen fibrils and subsequently recovered by digestion of the substrate. Attempts to purify the analogous vertebrate enzymes by this method have met with only limited success, although human skin, rheumatoid synovial and tadpole collagenases have been purified to homogeneity by affinity chromatography on collagen-Sepharose.[79]

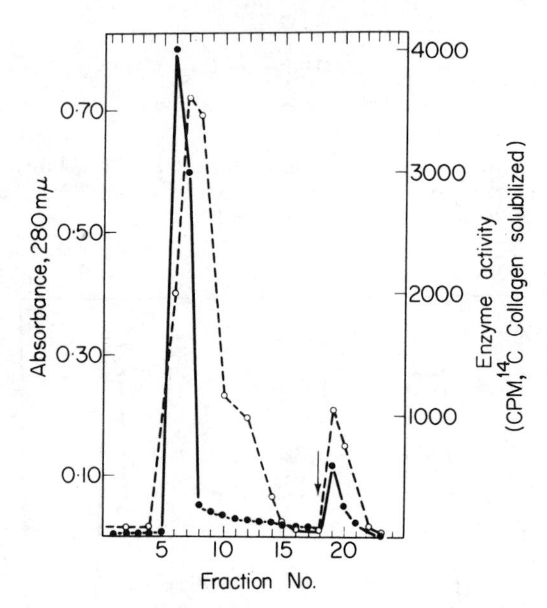

Figure II.43. Affinity chromatography of human skin collagenase on collagen-Sepharose. A sample of 10·0 mg of partially purified enzyme protein was applied to a column (1·2 cm × 1·5 cm) and effluent fractions of 5·0 ml were collected at a rate of 15·0 ml/hour. Elution was accomplished by the addition of 1·0M NaCl to the eluant buffer (arrow). Collagenase activity was measured by incubating 100 μl of the eluant fractions on ^{14}C-collagen fibrils for 6 hours at 37°C. ●——●, absorbance at 280 mμ; ○ - - - ○, enzyme activity. Reproduced with permission from Bauer *et al.*, *Biochem. Biophys. Res. Commun.*, **44**, 813 (1971)

The elution profile of a partially purified human skin collagenase from collagen-Sepharose is shown in Fig. II.43. Similar chromatograms were obtained with rheumatoid synovial and tadpole collagenases. Although incomplete adsorption was observed, the retained activity could be eluted with a buffered solution of 1M NaCl. The eluted protein was electrophoretically homogeneous. With crude enzyme preparations, good yields were obtained from collagen-Sepharose, although a number of contaminating proteins also bound to the adsorbent. Thus, some preliminary purification is desirable before affinity chromatography can be used to advantage.

4. The Purification of the Soluble Estradiol-17β Dehydrogenase from Human Placenta

This enzyme binds its substrate (17β-estradiol or estrone) in the absence of cofactor and will also bind 17β-estradiol with a bulky substitution at carbon-3

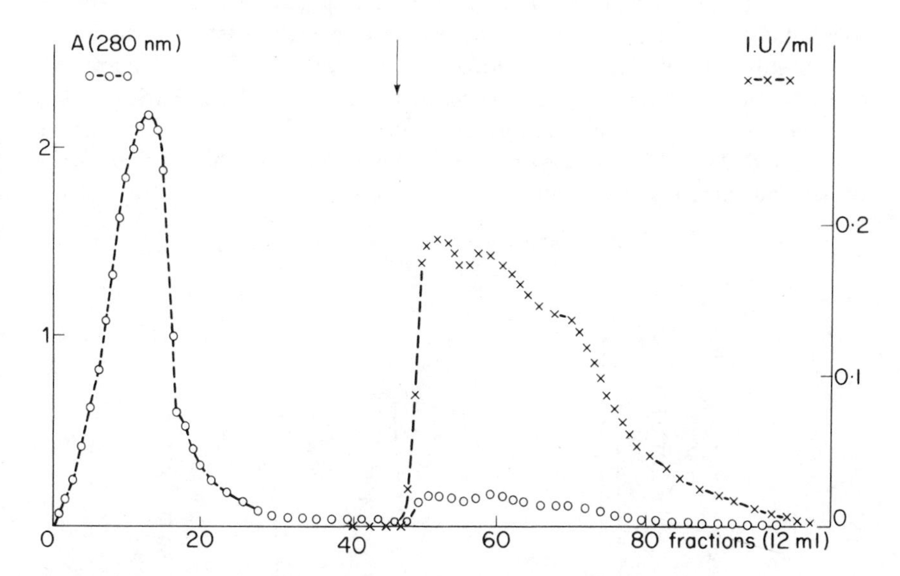

Figure II.44. Affinity chromatography of oestradiol-17β-dehydrogenase on a Sepharose column containing high concentrations of bonded oestrone. Crude enzyme solution (protein 1·5 g, specific activity 0·02 I.U.; in 150 ml 0·3M phosphate buffer, pH 7·2, 0·001M EDTA, 20% glycerol) was applied to a 1 cm × 5 cm column. After a 100 column volume washing with the same buffer, elution was obtained by adding 0·2 mM oestrone hemisuccinate to the buffer (arrow). Fractions 48–75 contained 30 mg of oestradiol-17β-dehydrogenase (specific activity 1 I.U./mg). Reproduced with permission from J. C. Nicolas *et al.*, *FEBS Lett.*, **23**, 175 (1972)

of the steroid nucleus. Consequently, affinity chromatography with estrone linked at carbon-3 to Sepharose seemed the obvious choice of adsorbent. It was prepared by coupling estrone hemisuccinate to aminoalkyl-Sepharose with *N,N'*-dicyclohexyl carbodiimide in dimethylformamide.[63]

A partially purified enzyme preparation from human placenta was applied to a column (1 cm × 5 cm) of estrone-Sepharose in 0·3M phosphate buffer, pH 7·2, containing 1 mM EDTA and 20% glycerol. Non-adsorbed protein was washed through with 100 column volumes of the same buffer and the estradiol-17β dehydrogenase eluted with 10^{-4}M estrone hemisuccinate (Fig. II.44). The homogeneity of the enzyme was confirmed by polyacrylamide electrophoresis.

5. The Specific Elution of Pyruvate Kinase from CM-Cellulose Columns with its Allosteric Effector

A partially purified extract of rat liver 'type L' pyruvate kinase was applied to a column of CM-cellulose and non-adsorbed protein washed through with the equilibrating buffer.[116] The adsorbed pyruvate kinase could be readily eluted with the same buffer supplemented with 0·5 mM FDP with a 60% recovery of activity and an overall purification of 30-fold (Fig. II.45). Further elution with 250 mM tris-maleate buffer, pH 6·0, removed additional adsorbed protein but no enzymic activity.

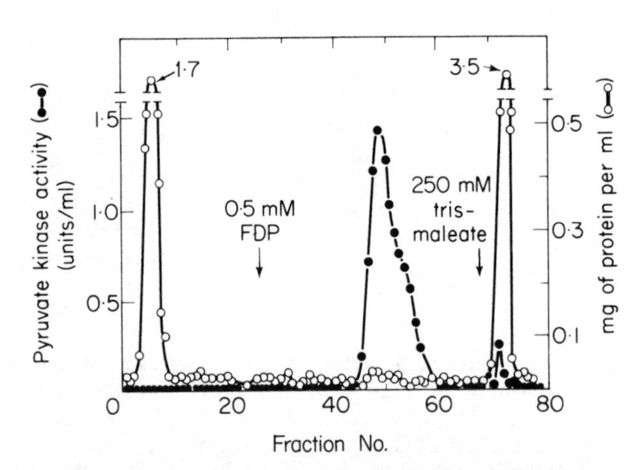

Figure II.45. Elution profile of pyruvate kinase from a CM-cellulose column. A sample containing 15 mg of protein with a specific activity of 1·1 units/mg was applied to the column and elution performed as indicated. Reproduced with permission from H. Carminatti *et al.*, *FEBS Lett.*, **4**, 307 (1969)

REFERENCES

1. Cuatrecasas, P., Wilchek, M., and Anfinsen, C. B. (1968): *Proc. Nat. Acad. Sci. U.S.A.*, **61**, 636.
2. Lowe, C. R., and Dean, P. D. G. (1971): *FEBS Lett.*, **18**, 31.
3. Lerman, L. S. (1953): *Proc. Nat. Acad. Sci. U.S.A.*, **39**, 232.
4. Arsenis, C., and McCormick, D. B. (1964): *J. Biol. Chem.*, **239**, 3093.

5. Arsenis, C., and McCormick, D. B. (1966): *J. Biol. Chem.*, **241**, 330.
6. Sander, E. G., McCormick, D. B., and Wright, L. D. (1966): *J. Chromatog.*, **21**, 419.
7. Bautz, E. K. F., and Holt, B. D. (1962): *Proc. Nat. Acad. Sci. U.S.A.*, **48**, 400.
8. Erhan, S. L., Northrup, L. G., and Leach, F. R. (1965): *Proc. Nat. Acad. Sci. U.S.A.*, **53**, 646.
9. McCormick, D. B. (1965): *Anal. Biochem.*, **13**, 194.
10. Knight, C. S. (1967): *Adv. in Chromatog.*, **4**, 61.
11. Inman, J. K., and Dintzis, H. M. (1969): *Biochemistry*, **8**, 4074.
12. Steers, E., Cuatrecasas, P., and Pollard, H. (1971): *J. Biol. Chem.*, **246**, 196.
13. Carrara, M., and Bernardi, G. (1968): *Biochim. Biophys. Acta*, **155**, 1.
14. Sun, K., and Sehon, A. H. (1965): *Can. J. Chem.*, **43**, 969.
15. Weetall, H. H., and Hersh, L. S. (1969): *Biochim. Biophys. Acta*, **185**, 464.
16. Wigzell, H., and Anderson, B. (1969): *J. Exp. Med.*, **129**, 23.
17. Wigzell, H., and Makela, O. (1970): *J. Exp. Med.*, **131**, 110.
18. Weibel, M. K., Weetall, H. H., and Bright, H. J. (1971): *Biochem. Biophys. Res. Commun.*, **44**, 347.
19. Hjerten, S. (1962): *Arch. Biochem. Biophys.*, **99**, 446.
20. Porath, J., Axen, R., and Ernbäck, S. (1967): *Nature* (*London*), **215**, 1491.
21. Axen, R., Porath, J., and Ernbäck, S. (1967): *Nature* (*London*), **214**, 1302.
22. Cuatrecasas, P. (1970): *J. Biol. Chem.*, **245**, 3059.
23. Araki, C. (1956): *Bull. Chem. Soc. Japan*, **29**, 543.
24. Flodin, P. (1961): *J. Chromatog.*, **5**, 103.
25. Wheaton, R. M., and Baumann, W. C. (1953): *Ann. N.Y. Acad. Sci.*, **57**, 159.
26. Laurent, T. C., and Killander, J. (1964): *J. Chromatog.*, **14**, 317.
27. Andrews, P. (1965): *Biochem. J.*, **96**, 595.
28. Flodin, P. (1962): in *Dextran Gels and their Applications in Gel Filtrations*, Dissertation, Uppsala.
29. Přistoupil, T. I. (1965): *J. Chromatog.*, **19**, 64.
30. Přistoupil, T. I., and Ulrych, S. (1967): *J. Chromatog.*, **28**, 49.
31. Kirret, O., Arro, I., and Heinlo, H. (1966): *Trans. Acad. Sci. Estonian S.S.R.*, **15**, 414.
32. Marsden, N. V. B. (1965): *Ann. N.Y. Acad. Sci.*, **125**, 428.
33. Glazer, A. N., and Wellner, D. (1962): *Nature* (*London*), **194**, 862.
34. Porath, J. (1968): *Nature* (*London*), **218**, 834.
35. Pettersson, G. (1968): *Arch. Biochem. Biophys.*, **123**, 307.
36. Whitaker, J. R. (1963): *Anal. Chem.*, **35**, 1950.
37. Auricchio, F., Bruni, C. B., and Sica, V. (1968): *Biochem. J.*, **108**, 161.
38. Lowe, C. R., and Walker, D. G.: unpublished observations.
39. Laurent, T. C. (1971): *Eur. J. Biochem.*, **21**, 498.
40. Truffa-Bachi, P., and Wofsy, L. (1970): *Proc. Nat. Acad. Sci. U.S.A.*, **66**, 685.
41. Cuatrecasas, P. (1971): *J. Biol. Chem.*, **246**, 7265.
42. Krug, F., Desbuquois, B., and Cuatrecasas, P. (1971): *Nature New Biology*, **234**, 268.
43. Lowe, C. R., Harvey, M. J., Craven, D. B., and Dean, P. D. G. (1973): *Biochem. J.*, **133**, 499.
44. Hipwell, M. C. H., and Dean, P. D. G.: unpublished observations.
45. O'Carra, P., Barry, S., and Griffin, T. (1973): *Biochem. Soc. Trans.*, **1**, 289.
46. Er-El, Z., Zaidenzaig, Y., and Shaltiel, S. (1972): *Biochem. Biophys. Res. Commun.*, **49**, 383.
47. Shaltiel, S., Hedrick, J. L., and Fischer, E. H. (1966): *Biochemistry*, **5**, 2108.

48. Yon, R. J. (1972): *Biochem. J.*, **126**, 765.
49. Yon, R. J. (1971): *Biochem. J.*, **124**, 10P.
50. O'Carra, P., and Barry, S. (1972): *FEBS Lett.*, **21**, 281.
51. Lowe, C. R.: unpublished observations.
52. Harris, R. G., Rowe, J. J. M., Stewart, P. S., and Williams, D. C. (1973): *FEBS Lett.*, **29**, 189.
53. Wilchek, M., and Gorecki, M. (1969): *Eur. J. Biochem.*, **11**, 491.
54. Claeyssens, M., Kersters-Hilderson, H., Van Wauwe, J. P., and De Bruyne, C. K. (1970): *FEBS Lett.*, **11**, 336.
55. Chan, W. W. C., and Takahashi, M. (1969): *Biochem. Biophys. Res. Commun.*, **37**, 272.
56. Collier, R., and Kohlhaw, G. (1971): *Anal. Biochem.*, **42**, 48.
57. Lowe, C. R., and Dean, P. D. G. (1971): *FEBS Lett.*, **14**, 313.
58. Newbold, P. C. H., and Harding, N. G. L. (1971): *Biochem. J.*, **124**, 1.
59. Schaller, H., Nusslein, C., Bonhoeffer, F. J., Kurz, C., and Nietzschmann, I. (1972): *Eur. J. Biochem.*, **26**, 474.
60. Barker, R., Olsen, K. W., Shaper, J. H., and Hill, R. L. (1972): *J. Biol. Chem.*, **247**, 7135.
61. Nelidova, O. D., and Kiselev, L. L. (1968): *Molekul. Biol.*, **2**, 60.
62. Venis, M. A. (1971): *Proc. Nat. Acad. Sci. U.S.A.*, **68**, 1824.
63. Nicolas, J. C., Pons, M., Descomps, B., and Crastes de Paulet, A. (1972): *FEBS Lett.*, **23**, 175.
64. Crane, L. J., Bettinger, G. E., and Lampen, J. O. (1973): *Biochem. Biophys. Res. Commun.*, **50**, 220.
65. Gribnau, A. A. M., Schoenmakers, J. G. C., Van Kraaikamp, M., and Bloemendal, H. (1970): *Biochem. Biophys. Res. Commun.*, **38**, 1064.
66. Wilchek, M. (1970): *FEBS Lett.*, **7**, 161.
67. Stewart, K. K., and Doherty, R. F. (1971): *FEBS Lett.*, **16**, 226.
68. Röschlau, P., and Hess, B. (1972): *Hoppe-Seyler's Z. Physiol. Chem.*, **353**, 441.
69. Böhme, H. J., Kopperschläger, G., and Schulz, J. (1972): *J. Chromatog.*, **69**, 209.
70. Baker, B. R. (1967): in *Design of Active-Site-Directed Irreversible Enzyme Inhibitors*, John Wiley, New York.
71. Lingens, F., Goebel, W., and Uesseler, H. (1967): *Eur. J. Biochem.*, **1**, 363.
72. Cuatrecasas, P. (1969): *Biochem. Biophys. Res. Commun.*, **35**, 531.
73. Wofsy, L., and Burr, B. (1969): *J. Immunol.*, **103**, 380.
74. Omenn, G., Ontjes, D., and Anfinsen, C. B. (1970): *Nature* (*London*), **225**, 189.
75. Feinstein, G. (1970): *FEBS Lett.*, **7**, 353.
76. Barker, R., Trayer, I. F., and Hill, R. L.: *Methods Enzymol.* (Ed. Jakoby and Wilchek), in press.
77. Vahlquist, A., Nilsson, S. F., and Peterson, P. A. (1971): *Eur. J. Biochem.*, **20**, 160.
78. Cuatrecasas, P. (1969): *Proc. Nat. Acad. Sci. U.S.A.*, **63**, 450.
79. Bauer, E. A., Jeffrey, J. J., and Eisen, A. Z. (1971): *Biochem. Biophys. Res. Commun.*, **44**, 813.
80. Harvey, M. J., Lowe, C. R., Craven, D. B., and Dean, P. D. G. (1973): *Eur. J. Biochem.*, submitted for publication.
81. Sprossler, B., and Lingens, F. (1970): *FEBS Lett.*, **6**, 232.
82. Gawronski, T. H., and Wold, F. (1972): *Biochemistry*, **11**, 422.
83. Lowe, C. R., Harvey, M. J., and Dean, P. D. G. (1973): *Eur. J. Biochem.*, submitted for publication.

84. Falkbring, S. O., Göthe, P. O., and Nyman, P. O. (1972): *FEBS Lett.*, **24**, 229.
85. Robinson, P. J. ,Dunnill, P., and Lilly, M. D. (1972): *Biochim. Biophys. Acta*, **285**, 28.
86. Schmidt, J., and Raftery, M. A. (1973): *Biochemistry*, **12**, 852.
87. Kalderon, N., Silman, I., Blumberg, S., and Dudai, Y. (1970): *Biochim. Biophys. Acta*, **207**, 560.
88. Little, J. N., Waters, J. L., Bombaugh, K. J. ,and Pauplis, W. J. (1971): in *Gel Permeation Chromatography* (Ed. Altgelt, K. H., and Segal, L.), Dekker, New York, p. 205
89. Ackers, G. K. (1964): *Biochemistry*, **3**, 723.
90. Altgelt, K. H.: in (see Ref. 88), p. 194.
91. Porath, J., and Fornstedt, N. (1970): J. *Chromatog.*, **51**, 479.
92. Dean, P. D. G., Willetts, S. R., and Blanch, J. E. (1971): *Anal. Biochem.*, **41**, 344.
93. Le Goffic, F., and Moreau, N. (1973): *FEBS Lett.*, **29**, 289.
94. Cuatrecasas, P., and Wilchek, M. (1968): *Biochem. Biophys. Res. Commun.*, **33**, 235.
95. Harvey, M. J., Lowe, C. R., and Dean, P. D. G. (1973): *Eur. J. Biochem.*, submitted for publication.
96. Mapes, C. A., and Sweeley, C. C. (1972): *FEBS Lett.*, **25**, 279.
97. Cuatrecasas, P., and Anfinsen, C. B. (1972): *Methods Enzymol.*, **22**, 345.
98. Akanuma, H., Kasuga, A., Akanuma, T., and Yamasaki, M. (1971): *Biochem. Biophys. Res. Commun.*, **45**, 27.
99. Kaufman, B. T., and Pierce, J. V. (1971): *Biochem. Biophys. Res. Commun.*, **44**, 608.
100. Kristiansen, T., Einarsson, M., Sundberg, L., and Porath, J. (1970): *FEBS Lett.*, **7**, 294.
101. Jackson, R. J., Wolcott, R. M., and Shiota, T. (1973): *Biochem. Biophys. Res. Commun.*, **51**, 428.
102. Ohlsson, R., Brodelius, P., and Mosbach, K. (1972): *FEBS Lett.*, **25**, 234.
103. Lowe, C. R., and Dean, P. D. G. (1973): *Biochem. J.*, **133**, 515.
104. Andrews, P. (1970): *FEBS Lett.*, **9**, 297.
105. Glueckauf, E. (1965): in 'Ion exchange and its applications', *Soc. Chem. Ind.*, **34**, London.
106. Poillon, W. N. (1971): *Biochem. Biophys. Res. Commun.*, **44**, 64.
107. Cuatrecasas, P., and Illiano, G. (1971): *Biochem. Biophys. Res. Commun.*, **44**, 178.
108. Erickson, J. S., and Mathews, C. K. (1971): *Biochem. Biophys. Res. Commun.*, **43**, 1164.
109. Hixson, H. F., and Nishikawa, A. H. (1973): *Arch. Biochem. Biophys.*, **154**, 501.
110. Schmer, G. (1972): *Hoppe-Seyler's Z. Physiol. Chem.*, **353**, 810.
111. Allen, R. H., and Majerus, P. W. (1972): *J. Biol. Chem.*, **247**, 7702.
112. Uren, J. R. (1971): *Biochim. Biophys. Acta*, **236**, 67.
113. Goldstein, L., Levin, Y., and Katchalski, E. (1964): *Biochemistry*, **3**, 1913.
114. Pogell, B. M. (1962): *Biochem. Biophys. Res. Commun.*, **7**, 225.
115. Gallop, P. M., Seifter, S., and Meilman, E. (1957): *J. Biol. Chem.*, **227**, 891.
116. Carminatti, H., Rozengurt, E., and Jimenez de Asua, L. (1969): *FEBS Lett.*, **4**, 307.
117. Rahimi-Laridjani, I., Grimminger, H., and Lingens, F. (1973): *FEBS Lett.*, **30**, 185.

Chapter III
Group Specific Adsorbents

The inherent advantages of affinity chromatography are often lost in the narrow specificity of most adsorbents. The selective and largely unexplored nature of the ligand interaction with its complementary macromolecule often precludes the preparation of affinity adsorbents on a rational basis. The choice of the ideal ligand, in each individual case, is still largely empirical and often involves extensive experimentation to achieve satisfactory separations. Furthermore, even where an ideal ligand is available, practical considerations may prevent its effective use in an affinity system. One attempt to circumvent these difficulties would be to prepare group specific adsorbents capable of interacting with a wide range of complementary proteins. This would reduce the tedious process of screening all the ligands which interact with the macromolecule in order to find that which is best suited to the particular separation in question. Furthermore, although selectivity is lost in the adsorption phase of affinity chromatography, careful choice of eluant conditions could effectively regenerate the specificity. Thus, a single affinity matrix could become competent for adsorption of a wide range of macromolecules. An obvious example of the use of this approach would be the resolution of enzymes that share a common feature in their catalytic action, such as those enzymes that interact with coenzymes or those that contain serine or sulphydryl groups at their active sites. The oxidoreductases, transferases and hydrolases are thus particularly vulnerable to this approach.

It has already been pointed out (Chapter II, Section G.1) that the specific shape of the adsorption isotherm has important consequences where the immobilized ligand displays affinity for more than one complementary macromolecule. Consider the adsorption isotherms for four enzymes each showing different affinities for the immobilized ligand, as shown in Fig. III.1. Enzyme 1 exhibits very high affinity for this adsorbent with a dissociation constant of 10^{-7}–10^{-8}M; enzymes 2 and 3 have dissociation constants $\simeq 10^{-5}$M, and enzyme 4 displays very weak affinity with a dissociation constant $> 10^{-3}$M.

For a generalized Langmuir adsorption isotherm,

$$q_i = \frac{k_1 k_2 C_i}{1 + k_1 C_i} \tag{III.1}$$

where q_i is the specific amount of substance i adsorbed, C_i is the concentration and k_1 and k_2 are constants. For low concentrations of C_i, equation (III.1) reduces to $q_i = k_1k_2C_i$, and to $q_i = k_2$ for high concentrations of C_i. Thus, in general,

$$q_i = f(C_i)^n$$

where $n = 0$–1. Hence, it follows that if the ligand concentration is sufficiently high such that the capacity of the adsorbent is not a limiting factor, then the specific amount of substance i adsorbed (q_i) is dependent on its concentration in the mobile phase (C_i) and not on its affinity for the ligand. Thus, a sample containing equimolar amounts of the four enzymes will result in the adsorption of the amounts q_1, q_2, q_3 and q_4 respectively of each of the enzymes.

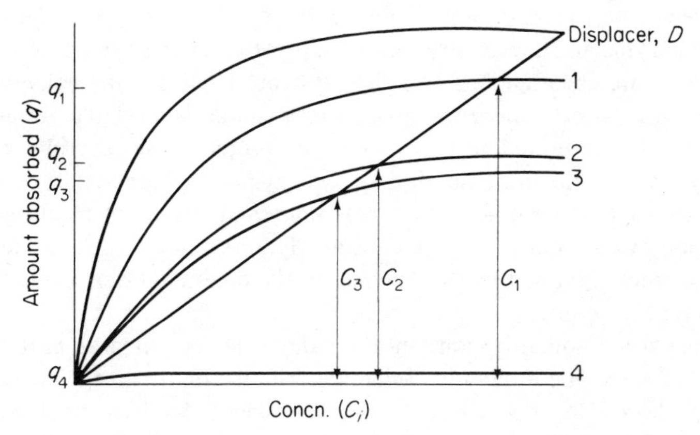

Figure III.1. Adsorption isotherms for four enzymes interacting with a single immobilized ligand

Furthermore, displacement analysis with a concentration of displacer D will elute enzymes 1, 2 and 3 at concentrations of C_1, C_2 and C_3. Enzyme 4 will emerge ahead of the displacing solution, since its adsorption isotherm is not cut by the displacer line. The high affinity enzyme will not displace one less tightly bound even if, subsequent to the initial adsorption, further amounts of the high-affinity enzyme are added. If the capacity of the adsorbent is exceeded then both high- and low-affinity enzymes will appear in the void volume eluate and not only the weakly adsorbed one. This important consideration has one major consequence with respect to the resolution of enzymes which display affinity for general ligands. Additional means of resolution are required to resolve mixtures of enzymes applied to such adsorbents. Thus, a preliminary fractionation step prior to adsorption on the specific matrix may be necessary to remove some of the contaminating proteins. Alternatively, changing the conditions of adsorption, such as pH, ionic strength, temperature, flow rate or dielectric constant could specifically exclude some enzymes. Furthermore, inhibitors or other ligands could be added to prevent adsorption of particular enzymes, or a matrix of low porosity could be used to exclude proteins of large molecular weight. Increased selectivity could also be achieved by using specific methods of elution to resolve those enzymes adsorbed. Such techniques have been discussed elsewhere in this book.

A. AFFINITY CHROMATOGRAPHY ON IMMOBILIZED COENZYMES

Many enzymes catalyse reactions of their substrates only in the presence of a specific non-protein organic molecule termed a *coenzyme*. The presence of both enzyme and coenzyme is required before catalysis can proceed. Where coenzymes are essential, the complete system or *holoenzyme* consists of the protein moiety or *apoenzyme* plus a heat-stable, dialysable, non-protein *coenzyme* that is bound to the apoenzyme protein. The reactions that frequently require the participation of coenzymes include oxido-reductions, group transfer and isomerization reactions, and reactions resulting in the formation of covalent bonds. In contrast, lytic reactions, such as the hydrolytic reactions catalysed by the enzymes of the digestive tract, are not known to require coenzymes.

Coenzymes frequently contain B vitamins as an integral part of their structure, since nicotinamide, thiamine, riboflavin, pantothenic acid and lipoic acid are important constituents of coenzymes for biological oxidations and reductions. Folic acid and cobamide coenzymes function in one-carbon metabolism and many enzymes concerned with the metabolism of amino acids require enzymes containing vitamin B_6.

In many ways, it is helpful to regard the coenzyme as a second substrate, i.e., a cosubstrate, since the chemical changes in the coenzyme exactly counterbalance those taking place in the substrate. Indeed, the reactions of the coenzyme may actually be of greater significance than those of the substrate.

Coenzymes thus function as cosubstrates for a wide range of enzyme-catalysed reactions. Such enzymes will therefore have at least two specific binding sites, one for the coenzyme, which will be common to them all, and one or more for the substrate. The latter will depend on the nature of the substrate and the reaction catalysed.

Thus, in terms of affinity chromatography, immobilized coenzymes should selectively adsorb those groups of enzymes which utilize them in bi- or multi-substrate reactions. However, two other considerations must be applied to multi-substrate reactions, in addition to those already outlined in Chapter II. Firstly, some knowledge of the order of binding of the reactants to the enzyme is desirable. Thus, for example, an ordered reaction involving binding of the coenzyme before binding of the second substrate can proceed is advantageous to the successful application of this principle. Many enzymic reactions proceed via random-order mechanisms in which the prior binding of one or the other substrate is immaterial to the binding of the second substrate. Secondly, the effect of the second substrate could be turned to advantage if this respectively increases or decreases the binding of the coenzyme. The former would be a useful adjunct to the selectivity of the adsorption phase of the chromatography, whilst its subsequent removal from the buffer system could facilitate elution of the enzyme from the matrix. On the other hand, if the effect of the second substrate is to reduce the binding of the coenzyme, then this can be employed as a means of selective elution of the enzyme.

Thus, although group separations on immobilized cofactors decrease the selectivity of the adsorption phase, considerable specificity of elution can be achieved by careful control of the eluant conditions such as to employ the interactions of the coligands to advantage. Thus, ideally one affinity matrix can be utilized to adsorb several complementary proteins which can be sequentially resolved by altering the eluant conditions.

In the following discussion of the use of insolubilized cofactors for affinity chromatography, the full potential of the technique has not been exploited in depth. The spasmodic reports of their use in the literature perhaps reflects the lack of information on their binding to their respective apoenzymes and the difficulties experienced in their chemical derivatization for the preparation of affinity adsorbents. The complexity of the chemistry involved is often a barrier to the preparation of insolubilized cofactor analogues that retain the specificity of their biological interactions. Furthermore, the relative instability of some of these compounds has further hindered progress.

1. The Pyridine Nucleotide Coenzymes

The pyridine nucleotides are coenzymes for a group of enzymes known as dehydrogenases, which catalyse oxidation–reduction reactions. The structures of nicotinamide adenine dinucleotide (NAD^+) and nicotinamide adenine dinucleotide phosphate ($NADP^+$) are shown in Fig. III.2. The use of immobilized NAD^+ or $NADP^+$ for the affinity chromatography of the pyridine nucleotide-linked dehydrogenases is substantiated by two major considerations. Firstly, in general the reaction kinetics favoured by the $NAD(P)^+$-dependent dehydrogenases follow an ordered mechanism, involving prior addition of NAD^+ or $NADP^+$ to generate an enzyme–coenzyme binary complex, before binding of the second substrate can occur.[1] This process is followed by interconversion of the ternary complexes, dissociation of the product and then dissociation of the reduced coenzyme. This kinetic sequence is outlined in Fig. III.3. The Michaelis constant for the oxidized coenzyme generally lies within the range 10^{-8}–10^{-3}M, and in general the affinity for the reduced coenzyme is greater than that for the oxidized form.[2] Furthermore, 1 mole of NAD(H) is bound per unit of molecular weight of 35,000–40,000 and in general the binding sites are equal and independent. Competition studies have shown that the reduced and oxidized coenzymes occupy the same site on the protein.[3]

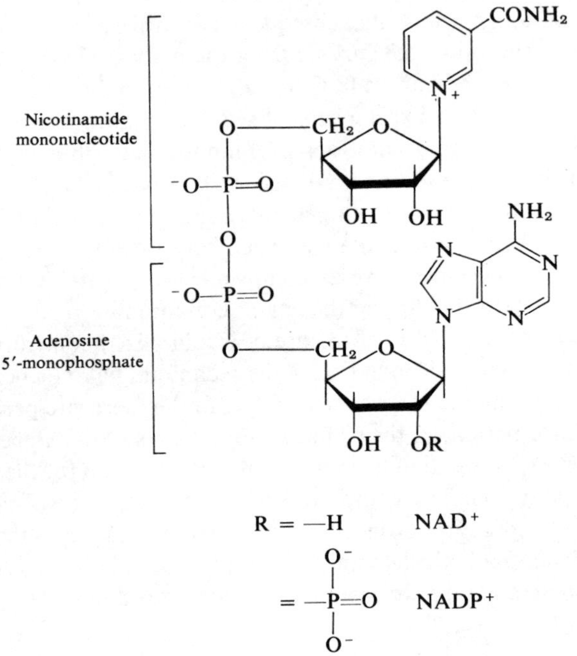

Nicotinamide mononucleotide

Adenosine 5′-monophosphate

R = —H NAD$^+$

= $-PO_3^{2-}$ NADP$^+$

Figure III.2. The structures of nicotinamide adenine dinucleotide (NAD$^+$) and nicotinamide adenine dinucleotide phosphate (NADP$^+$)

The second major contribution to the suitability of NAD(P) as a ligand for the preparation of affinity adsorbents is that the enzymes in general have a greater affinity for their respective coenzymes than for the second substrate. However, several other features almost peculiar to the pyridine nucleotides must be considered in the design of affinity adsorbents. The first concerns the conformation of the coenzyme. Studies by NMR on the geometry of the reduced coenzyme indicate a folded configuration of the molecule in which the adenine moiety and the dihydropyridine ring are in juxtaposition.[4] Studies on the fluorescence changes observed when NADH is bound to enzymes such as lactate, alcohol, glutamate and glycerol-phosphate dehydrogenases suggest that the folded structure of NADH unfolds when it is bound.[5] This is confirmed by the elegant X-ray data on the binding of the coenzyme to dogfish M_4 lactate dehydrogenase.[6]

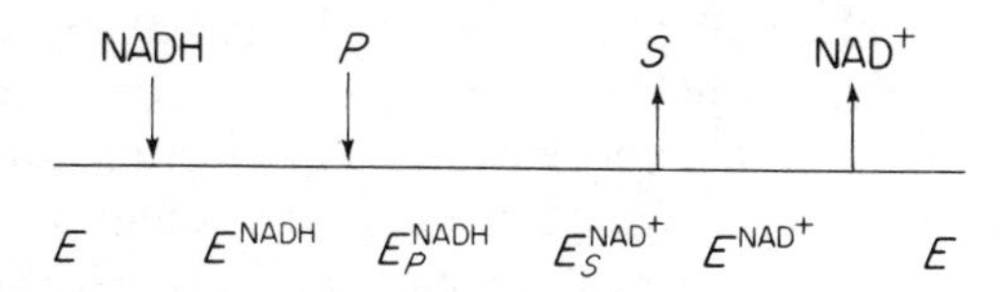

Figure III.3. The ordered kinetic sequence of most pyridine nucleotide-dependent dehydrogenases. *E* represents enzyme, NAD^+ and NADH, oxidized and reduced pyridine nucleotide respectively, *P* product and *S* substrate

The existence of the reduced coenzyme in two different conformational forms[7] (Fig. III.4) can account for the stereospecificity observed in the classic studies of Vennesland *et al.*[8] The reduced pyridine ring is believed to undergo a change in conformation from one boat form to another. These differences

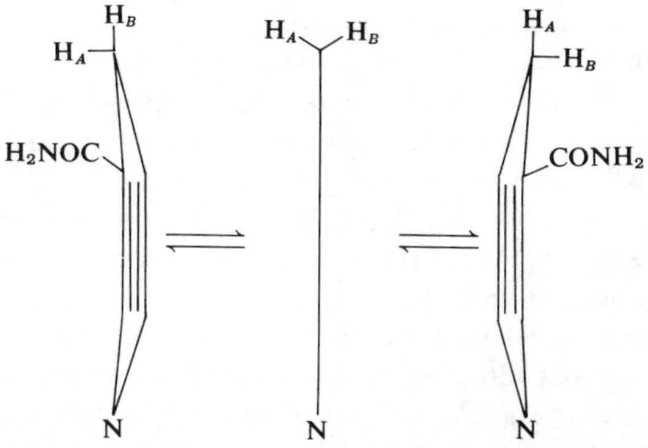

Figure III.4. Proposed shift in conformation of NADH accompanying interaction with dehydrogenases of opposite stereospecificities. Reproduced with permission from H. R. Levy, P. Talalay and B. Vennesland, *Progress in Stereochemistry* (Eds. P. B. de la Mare and W. Klyne), Butterworth, London, 1962

in the relative geometries around the plane of the reduced pyridine ring make the two protons at position 4 sufficiently distinct as to be recognized by two types of dehydrogenase, the so-called *A* and *B* types. The contribution of these conformational and stereospecific effects to the affinity chromatography of the pyridine nucleotide-dependent dehydrogenases is unknown.

The chemistry of the nicotinamide nucleotides is not well documented and the search for methods for the covalent attachment of these coenzymes to insoluble matrices is hindered by their relative instability and lack of reactivity.[9] Of the functional groups of the adenine moiety of NAD^+ the amine at position 6 is potentially available for linkage by CNBr-activated polysaccharide polymers[10] or the epichlorohydrin method.[11] Early studies of affinity chromatography using cofactors insolubilized on cellulose[12] have shown that these polymers possess interesting properties. Thus L-threonine dehydrogenase can be considerably enriched when a crude extract of *Pseudomonas oxalaticus* is chromatographed on NAD^+-cellulose. However, the results obtained with commercial preparations of dehydrogenases were difficult to interpret in terms of the known properties of the enzymes under study. Ion-exchange phenomena resulting from the introduction of charged groups by the chemical procedures employed were subsequently evoked to explain the anomalies.

Methods of immobilization based on attacking the purine ring system with electrophiles such as diazonium ions lead to adsorbents carrying NAD^+ substituted at position 8. These were effective in retaining lactate dehydrogenase[12] and glyceraldehyde 3-phosphate dehydrogenase,[13] whereas methods based on the reaction of the vicinal hydroxyl groups of the ribose rings by periodate oxidation,[14] the titanous chloride method of Novais[15] or borate complexes, produced ineffective adsorbents.

When NAD(H) is coupled to ω-carboxyl-alkyl-Sepharose by a carbodiimide-promoted reaction[12,16] the resultant polymer displays interesting properties. The immobilized NAD(H) was apparently coenzymically active, albeit to a small extent (0·2%). Enzymatically active NAD^+ has also been coupled to the surface of glass beads by a diazonium coupling procedure.[17] The insolubilized NAD^+ was shown to function as a coenzyme in the yeast alcohol dehydrogenase reaction.

Some parameters relevant to the affinity chromatography of pyridine nucleotide-linked dehydrogenases on immobilized nucleotides have been reviewed by Lowe *et al.*[18] Affinity adsorbents prepared by coupling NAD to ω-carboxyl-alkyl-Sepharose with the carbodiimide-promoted reaction of Larsson and Mosbach[16] have been found to contain 2 μmoles NAD^+/g wet weight gel, corresponding to 38·5 μmoles/g dry weight. This is in good agreement with the figure quoted by Larsson and Mosbach[16] for a similar NAD^+ polymer. However, frontal analysis chromatography using commercially available heart muscle lactate dehydrogenase (LDH) and bovine serum albumin (BSA) has shown that these adsorbents contain a heterogeneous population of bound nucleotides. When a constant concentration of BSA and LDH was applied to a column of immobilized NAD^+, about 12·5 I.U. of added LDH were adsorbed before enzyme appeared in the eluate (Fig. III.5). When this capacity was surpassed a further 23 I.U. LDH were adsorbed before the concentration of the enzyme in the eluate approached that of the applied solution. Prior to saturation a shoulder was apparent in the LDH elution profile and could indicate the presence of more than one species of NAD^+ bound to the matrix. The application of a BSA- and enzyme-free buffer to these columns resulted in an initial rapid decrease in the concentration of the enzyme in the eluate followed by a gradual decline to zero, with some 25 I.U LDH being eluted from the column. Subsequent addition of 1M KCl eluted a further 12·6 I.U. LDH. The elution profile for BSA showed a rapid rise in the concentration of the applied solution which remained constant until

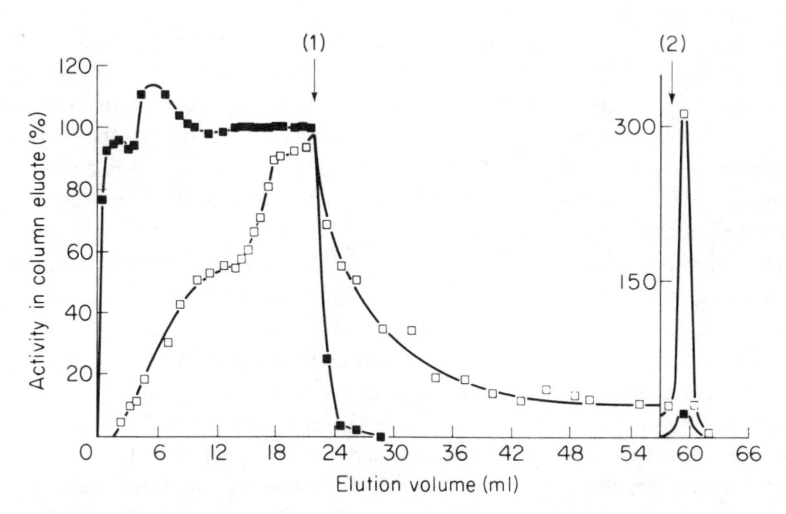

Figure III.5. Frontal analysis chromatography of lactate dehydrogenase on ε-aminohexanoyl-NAD^+-Sepharose. A column (2 mm × 20 mm) containing ε-aminohexanoyl-NAD^+-Sepharose (125 mg moist wt.) was equilibrated with 10 mM KH_2PO_4-KOH, pH 7·5. A mixture containing bovine serum albumin (E_{280}, 0·340) and lactate dehydrogenase (3 units/ml) in the same buffer was applied until the concentration of lactate dehydrogenase in the eluate reached that of the applied solution. Subsequent elution was carried out with (1) 10 mM KH_2PO_4-KOH, pH 7·5, and (2) 1M KCl, 10 mM KH_2PO_4-KOH, pH 7·5. ■ Bovine serum albumin; □ lactate dehydrogenase. Reproduced with permission from C. R. Lowe *et al.*, *Biochem. J.*, **133**, 499 (1973)

application ceased when there was a rapid fall to zero, indicating no interaction with the polymer.

These studies indicate that there is more than one binding site on the NAD^+ polymer and that only one of these sites is associated with the tight binding observed under normal experimental conditions, requiring a high salt concentration for enzyme elution. It can be calculated from a knowledge of the capacities of the tight binding site (12·5 I.U. LDH), together with the nucleotide content of the adsorbent, that of the immobilized cofactor only

0·1% NAD^+ was utilized to bind LDH tightly. Lowe *et al.*[19] have shown that NAD^+-linked dehydrogenases bind strongly to N^6-(6-aminohexyl)-AMP-Sepharose and that the binding pattern resembles that observed with ω-amino-hexanoyl-NAD^+-Sepharose. Thus it is suspected that the tight binding site observed for the NAD^+ polymer corresponds to attachment through an amide linkage between the carboxyl group of the spacer molecule and the 6-amino group of the adenine moiety of the nucleotide.[18] The small proportion of the total nucleotide linked in this fashion could explain the failure of some investigators to observe a change in UV absorption associated with N^6-substitution.[16]

The small proportion of the total nucleotide involved in the tight binding of lactate dehydrogenase to an immobilized NAD^+ adsorbent parallels the observations of Larsson and Mosbach[16] on the percentage coenzymic function that these polymers display. These results could reflect the stringent requirements for interaction between the enzyme and the nucleotide and suggest that most of the immobilized nucleotide is bound in a form unacceptable to the active site of the enzyme.

The binding of lactate dehydrogenase to immobilized NAD^+ was shown to be independent of enzyme and inert protein concentration, equilibration time and flow rate.[18] However, one feature of the system of prime importance was the concentration of bound ligand: thus increasing the concentration of NAD^+ available for the carbodiimide-promoted coupling to ω-amino-hexanoyl-Sepharose led to an increase in the amount of NAD^+ immobilized. The strength of binding of LDH increased with the concentration of NAD^+ immobilized, reaching a maximum at a concentration equivalent to 40 μmoles NAD^+/g dry weight gel. A similar effect was observed when Sepharose-bound nucleotide was diluted with unmodified or control ω-aminohexanoyl-Sepharose. The binding of LDH to the affinity columns decreased exponentially over the 21-fold dilution range used. Clearly, the concentration of bound nucleotide is a dominant factor in determining the strength of binding. Similar results have been observed with other affinity adsorbents, e.g., in the purification of 17β-oestradiol dehydrogenase.[20]

The important foundation of affinity chromatography on immobilized nucleotides is that the enzyme interacts specifically with its complementary nucleotide. The selectivity of this interaction is amply demonstrated by the following observations: first, LDH is only bound tightly to the polymer prepared by coupling NAD^+ to ω-aminohexanoyl-Sepharose in the presence of dicyclohexyl-carbodiimide. Insignificant binding is observed when any of the reaction components are omitted.[18] Second, the strength of the interaction increases with the amount of NAD^+ coupled to the gel. Third, LDH binding is not influenced by increased concentrations of inert protein and, finally, the specificity of the interaction has been shown by frontal analysis chromatography.

Despite the heterogeneous nature of polymers prepared by coupling pyridine nucleotides to ε-aminohexanoyl-Sepharose with a carbodiimide-promoted reaction, these adsorbents have been used effectively for the purification of dehydrogenases. Lowe *et al.*[21] describe how D-3-hydroxybutyrate dehydrogenase may be considerably enriched from a crude extract of *Rhodopseudomonas spheroides* by affinity chromatography on ε-aminohexanoyl-NAD^+-Sepharose (Fig. III.6). This chromatography resulted in a

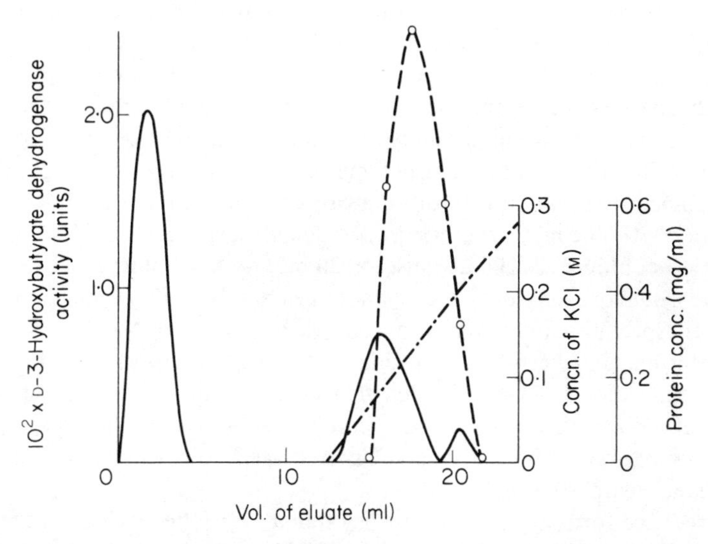

Figure III.6. Affinity chromatography of D-3-hydroxybutyrate dehydrogenase from a crude extract of *Rh. spheroides* on ε-aminohexanoyl-NAD^+-Sepharose. A sample (100 μl) of the dialysed bacterial extract was applied to a 5 mm × 20 mm column of ε-aminohexanoyl-NAD^+-Sepharose equilibrated with 0·05M tris-HCl buffer, pH 8·0. Non-adsorbed protein was washed off with 10 ml of the same buffer and the enzyme eluted with a KCl gradient (0–0·5M; 20 ml total volume) in 0·05M tris-HCl buffer, pH 8·0. Fractions (1·6 ml) were collected at a flow rate of 8 ml/hour and D-3-hydroxybutyrate dehydrogenase activity (○) assayed in the effluent. Protein concentration (——) and KCl concentration (–·–·) were also measured. Reproduced with permission from C. R. Lowe *et al.*, *Biochem. J.*, **133**, 507 (1973)

100-fold increase in specific activity and almost quantitative recovery of activity. In contrast, however, no increase in specific activity was obtained by chromatography of the crude bacterial extract on ε-aminohexanoyl-$NADP^+$-Sepharose since the enzyme was eluted in the void volume.

The general applicability of these adsorbents for the purification of $NAD(P)^+$-linked dehydrogenases has been qualified by O'Carra and Barry,[22] who claimed that some loss of specificity would be experienced. Lowe *et al.*[19]

describe how immobilized pyridine nucleotides can be successfully employed for the separation and purification of several pyridine nucleotide-dependent dehydrogenases from crude systems. Methods for increasing the selectivity and hence the versatility of this approach are suggested, as follows.

(i) *Selection of the appropriate nucleotide:* thus, for NAD^+-dependent dehydrogenases a matrix containing covalently attached NAD^+ yielded satisfactory results, whilst the enzymes were eluted in the void volume when the adsorbent comprised immobilized $NADP^+$. The $NADP^+$-linked dehydrogenases have been purified on matrices containing either NAD^+ or $NADP^+$ although, in general, polymers containing covalently attached $NADP^+$ yielded more satisfactory separations.[19] Thus, D-glucose 6-phosphate dehydrogenase could be enriched 30-fold from a crude extract of *Saccharomyces cerevisiae* by chromatography on ε-aminohexanoyl-$NADP^+$-Sepharose but only 6-fold on an equivalent polymer containing immobilized NAD^+.

(ii) *Choice of eluant:* pulse or stepwise elution with the complementary nucleotide can facilitate the release of dehydrogenases from immobilized nucleotide columns. A 200 μl pulse of 20 mM NAD^+ eluted D-3-hydroxybutyrate dehydrogenase from ε-aminohexanoyl-NAD^+-Sepharose with enhanced specific activity whilst an equivalent pulse of $NADP^+$ failed to effect elution. Furthermore, Hocking and Harris[13] report the elution of D-glyceraldehyde 3-phosphate dehydrogenase from a partially purified extract of *Thermus aquaticus* from a column of NAD^+-Sepharose prepared by a combination of the methods of Cuatrecasas,[23] Weibel *et al.*[17] and Lowe and Dean,[12] with 10 mM NAD^+.

The reduced form of the coenzyme is often a more effective eluant than the oxidized form: a 200 μl pulse of 5 mM NADH was sufficient to elute *Pseudomonas oxalaticus* L-threonine dehydrogenase from immobilized NAD^+. Furthermore, a linear gradient of the appropriate nucleotide will selectively elute enzymes from columns of immobilized cofactors. Thus, L-threonine dehydrogenase from a crude extract of *Ps. oxalaticus* can be quantitatively eluted from ε-aminohexanoyl-NAD^+-Sepharose by a linear gradient of NAD^+ (0–50 mM) (Fig. III.7(a)).

(iii) *Other considerations:* exploitation of the reaction catalysed by the complementary enzyme by incorporating the second substrate in the effluent can selectively alter the chromatographic behaviour of the required enzyme. Fig. III.8(a, b) illustrates the effect of the second substrate, L-threonine, on the chromatographic behaviour of L-threonine dehydrogenase on ε-aminohexanoyl-NAD^+-Sepharose when eluted with a linear gradient of salt. Binding of the enzyme was increased in the presence of 10 mM L-threonine and the resolution from the inert protein was improved with a concomitant increase in specific activity of the enzyme. Fig. III.7(a, b) demonstrates how the behaviour of L-threonine dehydrogenase was influenced by the presence of L-threonine when the column was developed with a linear NAD^+ gradient;

binding was decreased when the eluant was supplemented with 10 mM L-threonine. Elution of the enzyme could not be effected with a linear gradient of L-threonine (0–0·1M) alone.

The apparent difference in the effect of L-threonine on the elution of L-threonine dehydrogenase by ionic and nucleotide gradients can be reconciled on the following basis. In both cases the formation of a ternary complex between the enzyme, immobilized ligand and threonine was promoted. The increased binding observed when the column was developed in a KCl gradient

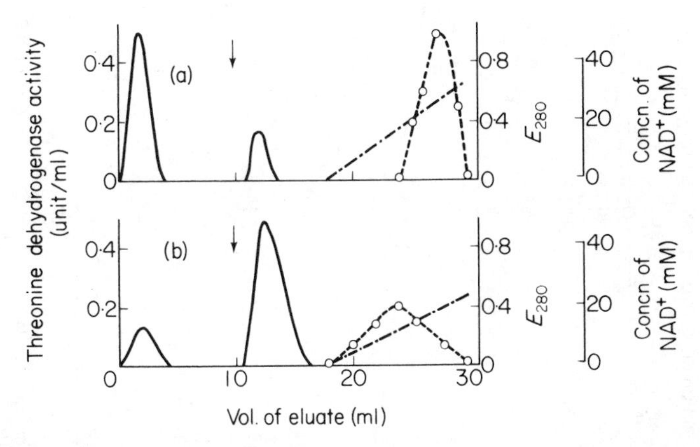

Figure III.7. Effect of L-threonine on the binding of L-threonine dehydrogenase to ε-aminohexanoyl-NAD$^+$-Sepharose. A sample (100 μl) of the crude extract was applied to a 5 mm × 20 mm column of the gel equilibrated with 10 mM potassium phosphate buffer, pH 7·5. Non-adsorbed protein was washed off with 10 ml of the same buffer and inert protein was eluted by increasing the KCl concentration to 100 mM (arrow). L-Threonine dehydrogenase was eluted by a linear gradient of NAD$^+$ (0–50 mM; 20 ml total volume) in 10 mM phosphate buffer, pH 7·5 containing 100 mM KCl. Enzyme (○), inert protein (——) and NAD$^+$ concentration (–·–·) were assayed in the effluent. In (b) conditions were as in (a) except the 10 mM L-threonine were included in all buffers. Reproduced with permission from C. R. Lowe *et al.*, *Biochem. J.*, **133**, 507 (1973)

reflects the greater stability of the ternary complex compared with the binary complex in the absence of threonine. When the ternary complex was subjected to the linear NAD$^+$ gradient supplemented with threonine, release of the enzyme from the matrix would be followed by formation of a tighter ternary complex in free solution. This would prevent subsequent partition of the enzyme through the column.

Any ligand that interacts with the macromolecule in such a way as to alter its chromatographic behaviour can be utilized to increase the selectivity of the desorption phase. However, it is a formidable task to predict the behaviour of pyridine–nucleotide-linked dehydrogenases when crude extracts are chromatographed on adsorbents containing covalently attached nucleotides.

Fig. III.9 demonstrates the behaviour of several $NAD(P)^+$-dependent dehydrogenases when a crude extract of yeast is chromatographed on a

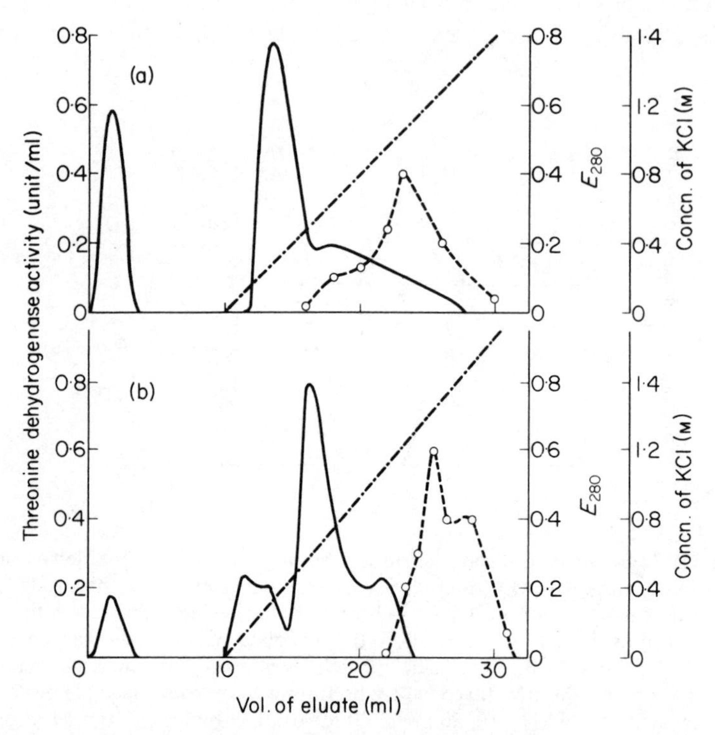

Figure III.8. Effect of L-threonine on the binding of L-threonine dehydrogenase to ε-aminohexanoyl-NAD^+-Sepharose. Chromatographic analyses were performed as described for Fig. III.7 except that the columns were developed with a linear KCl gradient (0–1·0M; 20 ml total volume) in 10 mM potassium phosphate buffer, pH 7·5. L-Threonine dehydrogenase activity (○), protein (——) and KCl concentration (–·–·) were assayed in the effluent. (a) No threonine added; (b) plus threonine (10 mM) in all buffers. Reproduced with permission from C. R. Lowe *et al.*, *Biochem. J.*, **133**, 507 (1973)

column of ε-aminohexanoyl-$NADP^+$-Sepharose and developed with a linear salt gradient. Considerable purification of D-glucose 6-phosphate dehydrogenase (EC 1.1.1.49) was achieved; Glutathione reductase (EC 1.6.4.2) was strongly retained, but was not resolved from alcohol dehydrogenase (EC 1.1.1.1).

The presence of nucleic acid components in crude extracts could also decrease the effectiveness of these adsorbents by interacting with the bound nucleotides. Total or partial removal of nucleic acid contaminants by methods such as salmine sulphate precipitation may thus be desirable. Preliminary passage of an extract through a column of salmine sulphate immobilized on Sepharose has been found satisfactory for the removal of such contaminants (Norris, Lowe and Dean, unpublished observations). The

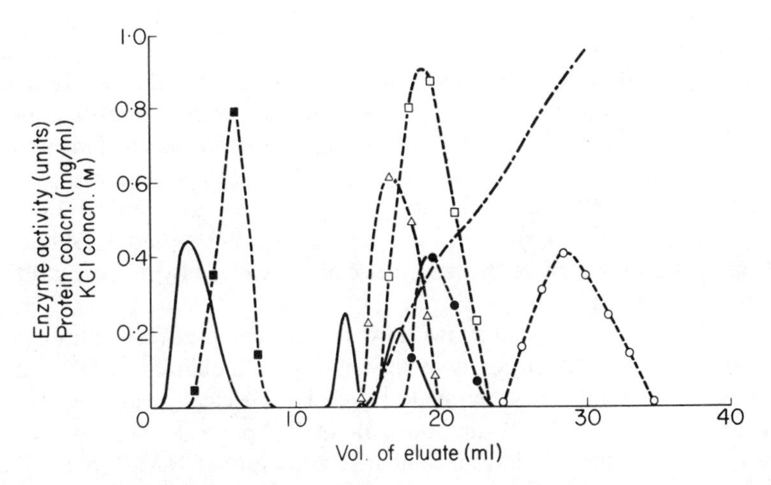

Figure III.9. Affinity chromatography of a crude yeast extract on ε-aminohexanoyl-NADP⁺-Sepharose. A sample (100 μl) of a dialysed yeast extract was applied to a 5 mm × 20 mm column of ε-aminohexanoyl-NADP⁺-Sepharose equilibrated with 10 mM potassium phosphate buffer, pH 7·5. Non-adsorbed protein was washed off with 10 ml of the same buffer and the column was developed with a KCl gradient (0–1·0M; 20 ml total volume) in 10 mM potassium phosphate buffer, pH 7·5. D-Glucose 6-phosphate dehydrogenase (○), L-malate dehydrogenase (△), EC 1.1.1.37, glutathione reductase (●), D-glyceraldehyde 3-phosphate dehydrogenase (■), EC 1.2.1.12 and alcohol dehydrogenase (□) were assayed in the effluent. Protein (——) and the concentration of KCl (–·–·) were also measured. Reproduced with permission from C. R. Lowe *et al.*, *Biochem. J.*, **133**, 507 (1973)

use of a preliminary purification step prior to affinity chromatography has been widely used, particularly where immobilized cofactors have been exploited.[24–27]

The general utility of these adsorbents for the purification of groups of cofactor-dependent enzymes has been demonstrated.[19,21] It is hoped that work currently in progress with chemically defined nucleotide adsorbents will resolve some of the difficulties experienced with the carbodiimide-linked nucleotide matrices.

Mosbach *et al.*[28] reported the preparation of an immobilized 5′-AMP analogue, N^6-(6-aminohexyl)-AMP-Sepharose. D-Glyceraldehyde 3-phosphate dehydrogenase could be eluted from a column of N^6-(6-aminohexyl)-AMP-Sepharose by pulse elution with 50 μl of 0·15 mM NAD^+. Mixtures of D-glyceraldehyde 3-phosphate dehydrogenase and lactate dehydrogenase could also be resolved. Thus, when both were applied to an immobilized AMP column, 0·15 mM NAD^+ afforded elution of the former, the latter being eluted by 10 mM NAD^+. Alternatively, lactate dehydrogenase could be eluted by a pulse of 0·15 mM NADH with 70–80% recovery.

Elution of a mixture of these enzymes could not be effected by a linear gradient of 0·0–0·5M KCl, although the enzymes could subsequently be eluted with low concentrations of cofactor as described above. These results suggest that the binding cannot be attributed to an ion-exchange phenomenon. A linear gradient of 0·0–0·15M salicylate, an inhibitor competitive with NAD^+ for a number of NAD^+-dependent dehydrogenases,[29] was found to elute and resolve the same two dehydrogenases (Fig. III.10). Furthermore, Ohlsson *et al.*[30] have shown that these two enzymes can be resolved by a linear gradient of 0·0–0·5M NADH.

Ohlsson *et al.*[30] have also shown how ternary complex formation can be utilized to increase the selectivity of elution from immobilized AMP. Using a model system, NAD^+ plus pyruvate eluted biospecifically-adsorbed lactate dehydrogenase from N^6-(6-aminohexyl)-AMP-Sepharose whereas neither substance alone effected elution. Similarly, a mixture of NAD^+ plus lactate was also effective, albeit less so. A closely related eluant with efficiency approaching that of NADH is the adduct of NAD^+ and pyruvate synthesized by base-catalysed condensation. Ternary complex formation was subsequently used for resolution of model enzyme mixtures. Thus, yeast alcohol dehydrogenase could be eluted by a solution of NAD^+ and hydroxylamine, and lactate dehydrogenase subsequently by NAD^+ together with pyruvate. Neither enzyme was eluted by these substances separately.

The experiments so far described with chemically defined affinity matrices have all involved simple mixtures of commercially available enzymes. The practical possibilities of group specific adsorbents are demonstrated by the purification of lactate dehydrogenase (LDH) from a crude ox heart extract. Protein lacking LDH activity was eluted in the void volume of the immobilized AMP column and when a pulse of 0·5 mM NAD^+ was applied. A pulse of NAD^+ and pyruvate together yielded 95–100% of the LDH activity applied. The enzyme was purified by a factor of about 35 and was homogeneous by analytical ultracentrifugation and by electrophoresis on polyacrylamide gel. Chromatography of a crude dialysed yeast extract on N^6-(6-aminohexyl)-AMP-Sepharose[32] further demonstrates the usefulness of this immobilized ligand, in that alcohol dehydrogenase can be purified to homogeneity (Fig. III.11). However, the chemically defined affinity adsorbent comprising N^6-(6-aminohexyl)-AMP-Sepharose independently developed by Guilford *et al.*[31] and Craven *et al.*[32] suffers two major disadvantages over polymers prepared by the carbodiimide-promoted condensation to a hexa-carbon spacer attached to Sepharose. Firstly, the immobilized AMP polymer is only the value for NAD^+-dependent dehydrogenases and is of little value for $NADP^+$-dependent dehydrogenases[19] or kinases.[32] For these groups of

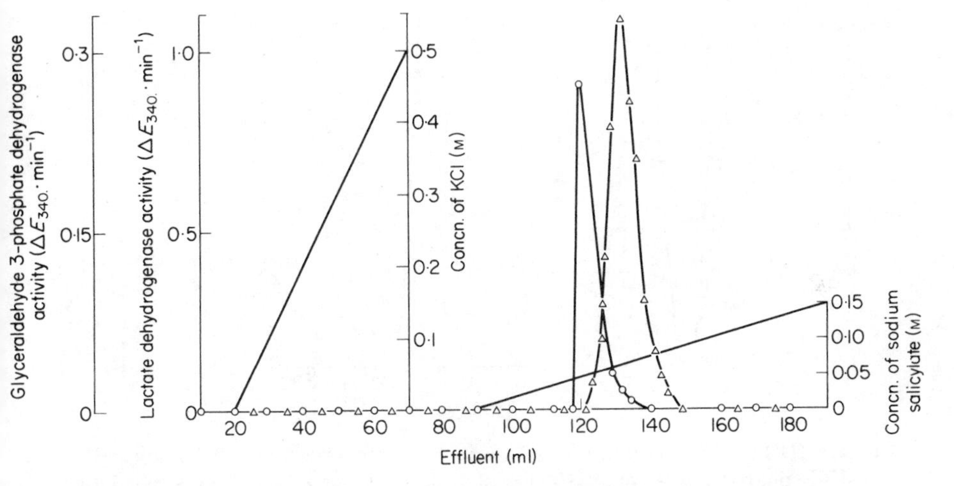

Figure III.10. The resolution of a mixture of two dehydrogenases with a linear gradient of sodium salicylate. The AMP-Sepharose column (0·5 cm × 11 cm) was equilibrated with 0·05M sodium phosphate buffer, pH 7·5. A 1 ml sample of the above buffer containing glyceraldehyde 3-phosphate dehydrogenase (2·8 units) and lactate dehydrogenase (35 units) was added to the column and buffer (20 ml) was applied. Subsequently a linear gradient of 0·0–0·5M KCl in buffer (total volume 50 ml) was applied; 2 ml fractions were collected at a flow rate of 1 ml/10 minutes. The column was washed with buffer (20 ml) and a linear gradient of 0·0–0·15M sodium salicylate in the above buffer (total volume 100 ml) was applied. Fractions (1 ml) were collected at the same flow rate. Enzymic activities were determined as described in the experimental section. ○, Glyceraldehyde 3-phosphate dehydrogenase activity; △, lactate dehydrogenase activity; ——, gradient concentration. Reproduced with permission from R. Ohlsson, P. Brodelius and K. Mosbach, *FEBS Lett.*, **25**, 234 (1972)

enzymes, $NADP^+$ and ATP respectively immobilized on Sepharose by the carbodiimide technique of Larsson and Mosbach[16] have proved satisfactory. Secondly, the undefined polymers are simple to prepare and do not require any extensive chemical expertise which could seriously limit their general utility. However, the results clearly illustrate that the usefulness of group specific affinity adsorbents, despite recently expressed scepticism,[22] is not hampered by the lack of specific eluants. Rapid and efficient purification of a large number of enzymes using only one immobilized ligand is now feasible. Furthermore, the exploration of multi-substrate mechanisms and the nature of hitherto undetected abortive complexes can be encouraged.

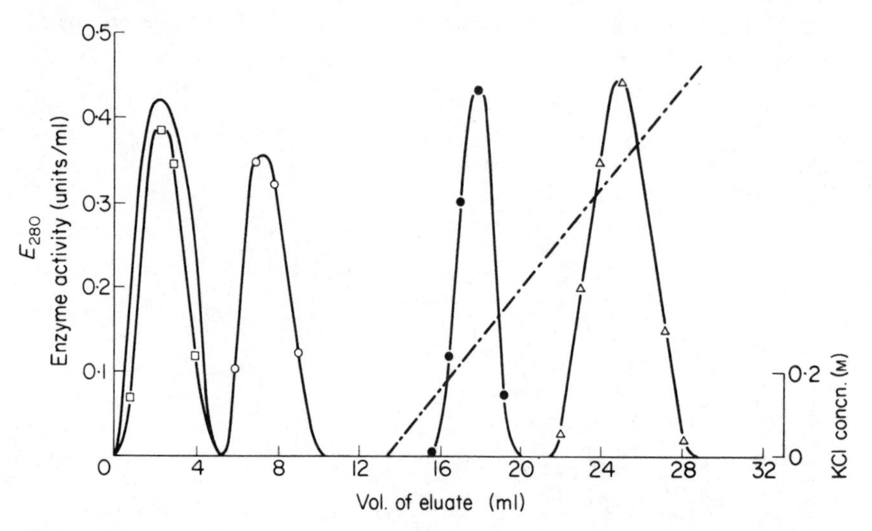

Figure III.11. Chromatography of a crude yeast extract on N^6-(6-aminohexyl)-AMP-Sepharose. A sample (100 μl) of a crude yeast extract was applied to a column (5 mm × 20 mm) containing 0·5 g N^6-(6-aminohexyl)-AMP-Sepharose equilibrated with 10 mM KH_2PO_4-KOH buffer, pH 7·5. After washing through non-adsorbed proteins, enzymes were eluted with a linear salt gradient (0–1 M KCl; 20 ml total volume) at a flow rate of 8 ml/hour. Inert protein (——), glucose 6-phosphate dehydrogenase (□), glutathione reductase (○), malate dehydrogenase (●) and yeast alcohol dehydrogenase (△) were assayed in the effluent

2. Adenine Nucleotides

The number of enzymes that require ATP as substrate or energy source is very large. ATP-dependent reactions may be divided into two main types: those involving transfer of some part of the ATP molecule, the phosphoryl, pyrophorphoryl, adenyl or adenosinyl moieties, and those in which ATP functions as an energy source, such as the ligases or synthetases. The structure of ATP is shown in Fig. III.12.

Several methods have been reported for the immobilization of nucleotides to cellulose. Gilham[33] has reported the coupling of adenosine 5′-phosphate to aminoethyl-cellulose by periodate oxidation of the ribose hydroxyls, followed by Schiff's base formation with the amine and subsequent reduction with borohydride. Furthermore, the terminal phosphate group of the nucleotide can be linked to sterically favoured hydroxyl groups of the cellulose by a water-soluble carbodiimide reaction. Such approaches, however, yield adsorbents that do not retard phosphotransferases or pyridine nucleotide-linked dehydrogenases.

Early studies by Corry, Lowe and Dean (unpublished) on the interaction of phosphokinases with ATP- and ADP-ceullose, prepared by coupling ATP and ADP directly to CNBr-activated cellulose, showed little selective binding. This may be expected since it is the adenine moiety of ATP that is important in its binding to kinases.

ATP-Sepharose prepared in an analogous manner retards highly purified succinyl-CoA synthetase compared with unmodified Sepharose.[34] Recovery of the enzyme was quantitative and the specific activity following elution from the adsorbent was slightly enhanced. The immobilized ATP column was found

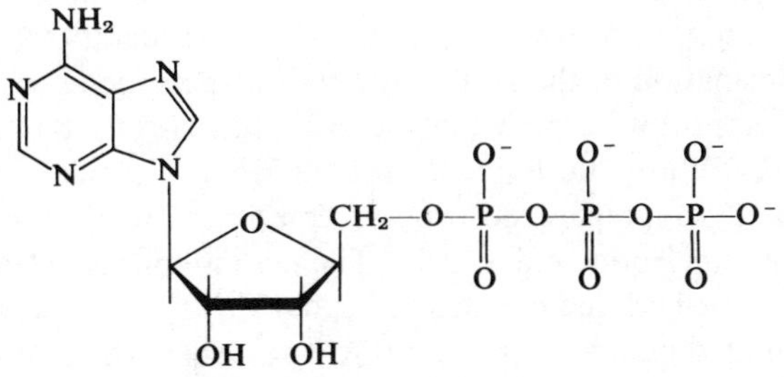

Figure III.12. The structure of adenosine 5′-triphosphate (ATP)

to be remarkably stable and was catalytically active in converting succinyl-CoA synthetase into a phosphorylated intermediate. The phosphorylated enzyme was shown to be an intermediate in the catalytic action since succinyl-CoA was formed both in the absence and presence of ATP.

$$E + \mathrm{ATP} \xrightleftharpoons{\mathrm{Mg}^{2+}} E - P + \mathrm{ADP}$$

$$E - P + \mathrm{Succinate} + \mathrm{CoA} \xrightleftharpoons{\mathrm{Mg}^{2+}} E + P_i + \mathrm{Succinyl\text{-}CoA}$$

Since ATP can be efficiently coupled to Sepharose by this method, the preparation of other phosphorylated enzyme intermediates similar to succinyl-CoA synthetase should be possible.

ATP, ADP or AMP can be coupled to ϵ-aminohexanoyl-Sepharose by a carbodiimide-promoted reaction.[35] Adsorbents prepared in this fashion are effective in retarding the pyridine nucleotide-dependent dehydrogenases but are relatively ineffective for affinity chromatography of the phosphokinases. Of the six enzymes tested, creatine kinase, glycerokinase, hexokinase, phosphofructokinase, pyruvate kinase and myokinase, only myokinase showed any significant retardation on ε-aminohexanoyl-ATP-Sepharose. In fact, myokinase was retarded to such an extent that it could not be eluted with salt concentrations up to 1M KCl and required a 200 μl pulse of 50 mM ATP for quantitative release. In contrast, the enzyme was not retarded by ε-aminohexanoyl-ADP or AMP-Sepharose.

Studies on the interaction of phosphokinases with N^6-(6-aminohexyl)-AMP-Sepharose have confirmed the above observations on chemically undefined polymers[36] of phosphoglycerokinase, pyruvate kinase, hexokinase, creatine kinase, myokinase and glycerokinase, of which only glycerokinase was bound to any significant extent. It should be noted, however, that a ligand concentration of 2 μmoles AMP/g wet weight of Sepharose was employed in this study, and that increased binding could be achieved by increasing the ligand concentration. Observations by Mosbach *et al.*,[28] however, have shown that pyruvate kinase is unretarded when adsorbents containing very high concentrations of 5′-AMP are used. These considerations would suggest that the conformation of the bound nucleotide or its mode of attachment to the insoluble support were important factors influencing the binding of kinases.

ATP and *d*ATP are allosteric effectors of ribonucleotide reductase[37] and have been used to prepare adsorbents competent for the purification of T_4-induced ribonucleotide reductase.[38] The *p*-aminophenylesters of ATP and *d*ATP were synthesized and coupled to Sepharose 4B activated with cyanogen bromide. A limited number of proteins were adsorbed when a crude extract of T_4-infected *E. coli* cells was chromatographed on ATP- or *d*ATP-Sepharose. The T_4-ribonucleotide reductase could be eluted with a linear gradient of *d*ATP (0–1 mM), ATP (0–10 mM) or *d*AMP (0–20 mM) in 0·1M phosphate buffer, pH 7·0. The enzyme was not released by 5 mM *d*CTP, 5 mM *d*TTP, 1 mM CDP or 5 mM *d*CDP.

Using this procedure, T_4-ribonucleotide reductase was purified on *d*ATP-Sepharose to about 70% purity from a crude extract, and to homogeneity from a crude ammonium sulphate fraction.

The nucleotide, cyclic 3′,5′-adenosine monophosphate (*c*AMP), acts as a trigger for many biologically important reactions.[39] One of its principal functions is to activate a protein kinase[40,41] by binding to regulatory proteins (R) of the inactive enzyme, with the concomitant release of enzymatically active subunits (C),

$$\mathrm{RC} + c\mathrm{AMP} \rightleftharpoons \mathrm{R}\cdot c\mathrm{AMP} + \mathrm{C}$$

The affinity of *c*AMP for the 'binding proteins' is high ($K_{diss} \simeq 2 \times 10^{-9}$M) and consequently the isolation by affinity chromatography of such proteins and regulatory subunits from solutions containing inactive kinases was envisaged. Tesser *et al.*[42] prepared derivatives of 8-bromo-*c*AMP although their instability seriously compromised their usefulness as affinity adsorbents.

Wilcheck *et al.*[43] have chromatographed the protein kinases of rat parotid and rabbit skeletal muscle on a preparation of Sepharose-ε-aminohexanoyl-*c*AMP. Fully active enzymes which had lost their ability to bind *c*AMP were eluted. It is suggested that *c*AMP-Sepharose activates the protein kinase by removal of a regulatory subunit that binds *c*AMP, although as yet no method for the recovery of the *c*AMP-binding unit from the matrix has been devised.

3. Other Nucleotides

Specific adsorbents comprising immobilized nucleotides of, for example, adenine, guanine or uridine may be applied generally to the purification of a wide variety of enzymes. In principle, for example, any enzyme that interacts with uridine phosphate derivatives as substrate or product could be purified with the aid of UDP-Sepharose. Many enzymes of this type have K_m or K_i values for uridine derivatives in the order of 10^{-4} to 10^{-5}M, and hence especially low K_m or K_i values are not necessary for effective binding.

A UDP-derivative was synthesized by a method which in principle could also be applied to the preparation of GDP- and ADP-derivatives.[44] The effectiveness of the UDP-adsorbent was tested primarily with the galactosyl-transferase of lactose synthetase from bovine milk.[45] It was found that the transferase from milk whey was weakly bound at pH 7·4 but that maximum adsorption could be achieved in the presence of manganous ions. The capacity of the UDP-Sepharose was not altered by high concentrations of sodium chloride (0·5 to 1·0M) indicating the absence of non-specific ion-exchange effects. Furthermore, enzyme was not eluted by washing the adsorbent with large volumes of manganese-ion-containing buffers.

The transferase bound to UDP-Sepharose could be eluted by several means. Dilute solutions of urea (1·5M) eluted the enzyme in a small volume but with only half the expected yield. Elution with borate buffers pH 8·5 resulted in improved yields but a more dilute preparation.

The columns of UDP-Sepharose could be used routinely to purify the enzyme from whey under the conditions given in Fig. III.13 with buffers containing manganous ions. The majority of the whey proteins passed through unretarded. The enzyme was subsequently eluted with a buffer containing EDTA and *N*-acetylglucosamine. A 120- to 160-fold enrichment of specific activity of the enzyme with 80% yield was obtained by this procedure.

In order to test the effectiveness of UDP-Sepharose for the purification of other UDP-glycosyltransferases, its interaction with glycogen synthetase was studied.[44] A partially purified preparation of the rabbit muscle enzyme was applied to a column of UDP-Sepharose. A small amount of the synthetase emerged unretarded with the inert proteins but the majority was eluted when glycogen was added to the irrigating buffer. The eluted protein appeared homogeneous by gel electrophoresis.

Danenberg *et al.*[46] have synthesized an affinity adsorbent by coupling 2′-deoxyuridylic acid through its 5′-phosphate group to Sepharose by a 6-*p*-aminobenzamidohexyl chain. This resin quantitatively adsorbed thymidylate synthetase from a partially purified extract of lysed *Lactobacillus casei*.

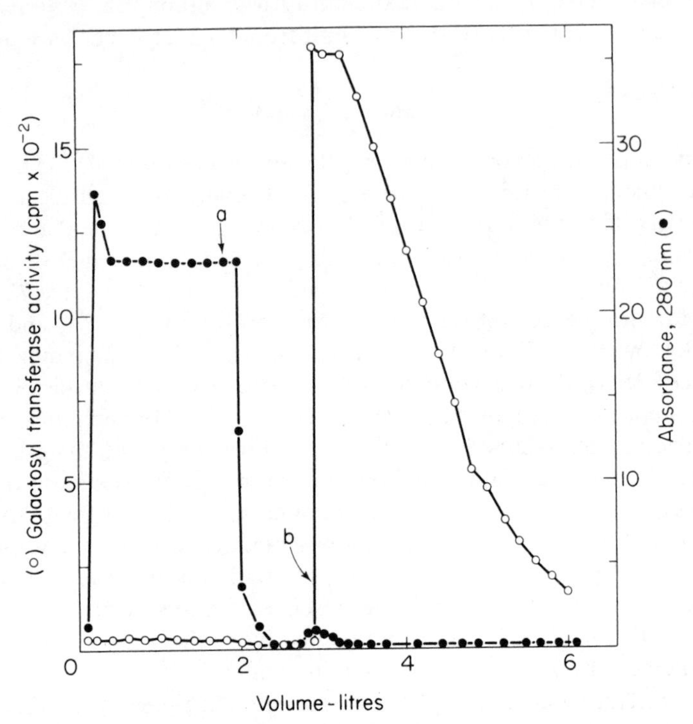

Figure III.13. The purification of the galactosyltransferase from whey with UDP-Sepharose. A column (3 cm × 15 cm) of UDP-Sepharose (4 to 5 μM of UDP per ml) was equilibrated at 5° with 0·025M sodium cacodylate buffer, pH 7·4, containing 0·025M $MnCl_2$ and 0·01M mercaptoethanol. Whey (1·8 litres) was applied to the column at a flow rate of about 100 ml per hour at 5°C. The column was then washed at *A* (arrow) with 1 litre of the equilibration buffer. The enzyme was then eluted by washing the column at *B* (arrow) with 0·025M sodium cacodylate, pH 7·4, containing 0·025M EDTA, 0·01M mercaptoethanol and 0·005M *N*-acetylglucosamine. The column was washed and eluted at about 100 ml per hour. Fractions (25 ml) were collected automatically. Absorbance at 280 nm (●). Galactosyltransferase activity (○). Reproduced with permission from R. Barker, K. W. Olsen, J. H. Shaper and R. L. Hill, *J. Biol. Chem.*, **247**, 7135 (1972)

The enzyme was subsequently eluted in a homogeneous state by raising the ionic strength of the buffer. Agarose adsorbents containing RNA and UMP have also been prepared.[47] Periodate oxidized RNA or UMP was reacted with the hydrazide of Sepharose-ε-aminocaproic acid methyl ester.

Jackson *et al.*[48] have prepared immobilized derivatives of GTP, GDP and guanosine by a similar procedure and have used them for the purification of D-*erythro*-dihydroneopterin triphosphate synthetase from *Lactobacillus plantarum*. The GTP-substituted Sepharose bound the synthetase from a partially purified extract. The enzyme was eluted with a pulse of GTP with a recovery of 60–70% and a 20-fold increase in specific activity. Partially purified extracts were used in most cases since the presence of phosphatases in crude extracts could destroy the bound ligand. However, when dialysed crude extracts were applied directly to the adsorbent, purifications of 350-fold and yields of approximately 60% were obtained. The immobilized GDP and guanosine derivatives were ineffective for the purification of the enzyme.

4. Flavin Nucleotides

The known flavoproteins, that utilize either flavin mononucleotide (FMN) or flavin adenine dinucleotide (FAD), are universally distributed in nature as catalysts for certain categories of dehydrogenation reactions. In the course of the oxidation of substrates such as pyridine nucleotides, α-hydroxy-acids, aldehydes, α-amino-acids and others, the isoalloxazine ring system of the coenzyme becomes reduced. The reduced coenzymes subsequently react with other electron acceptors, thus regenerating the oxidized form of the coenzyme. The structures of FMN and FAD are given in Fig. III.14.

Observations on the reconstitution of flavo-apoenzymes with flavo-coenzyme analogues have shown that the specificity for flavin binding is not as high as one might expect. Flavin analogues may in certain flavoproteins replace the natural coenzymes forming stable and in some cases even active holoenzymes.

Studies on the substrate specificity of rat liver flavokinase indicate that the enzyme will bind to several flavin substrates and competitive inhibitors.[49–51] Modifications of the side-chain in position 9 of the isoalloxazine system (Fig. III.14) with a 1′-D-ribityl or methyl substituent do not interfere with binding to the enzyme. Considerable tolerance in size and inductive effect of the 6- and 7-substituents is also observed. Thus, 9-(1′-D-ribityl) and 9-methyl-flavin when covalently linked to a cellulose matrix by an acetamido group in position 6 or 7 should generate a biochemically specific matrix for the selective purification of liver flavokinase (Fig. III.15). The chromatography of flavokinase on 6-cellulose-acetamido-9-(1′-D-ribityl)-isoalloxazine shows that the enzyme precedes most of the protein eluted from the column.[52] The interaction of flavokinase with relatively few flavin moieties is weak compared with the electrostatic interactions of most of the inert protein with the large number of residual carboxymethyl groups remaining on the original CM-cellulose. Nevertheless, greater retention of flavokinase is effected with the flavin-cellulose adsorbent than with control CM-cellulose.

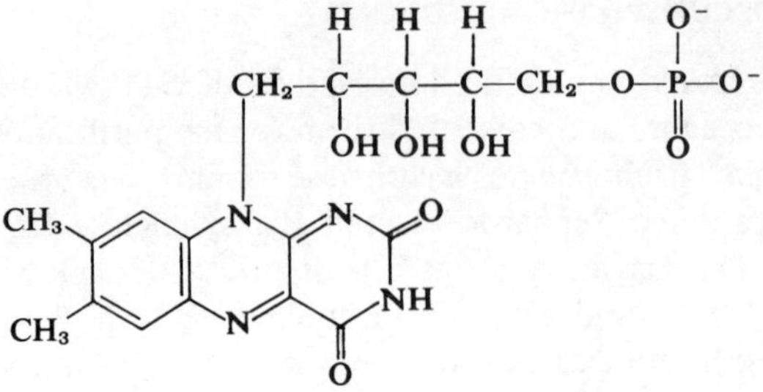

FMN

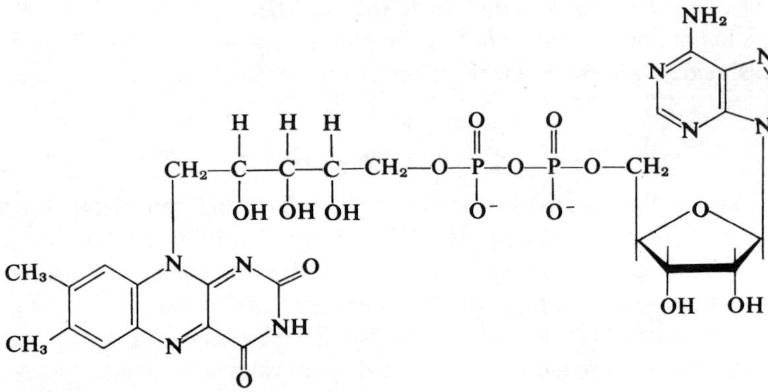

FAD

Figure III.14. The structures of flavin mononucleotide (FMN) and flavin adenine dinucleotide (FAD)

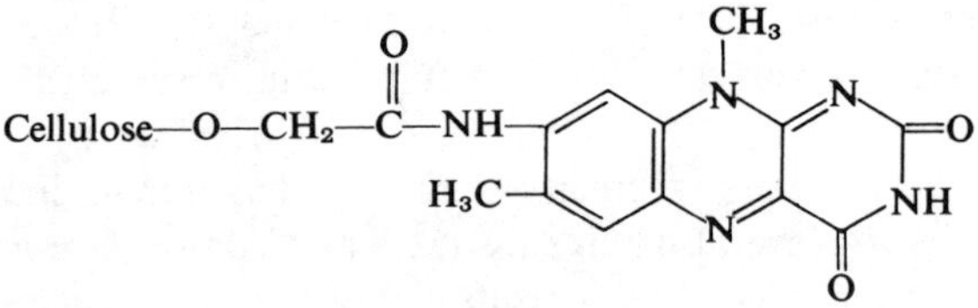

I

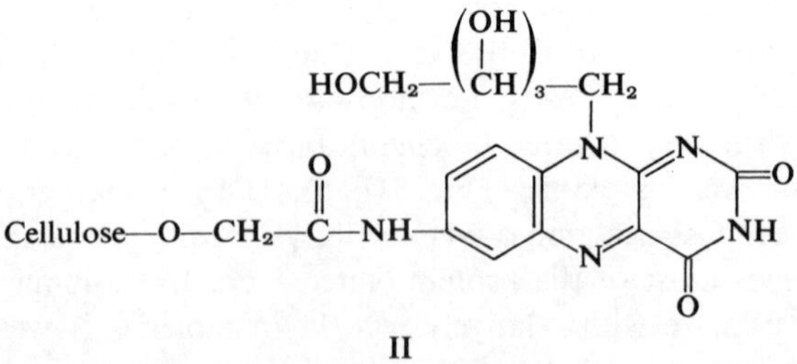

II

Figure III.15. Affinity adsorbents for the selective purification of liver flavokinase: (I) 7-celluloseacetamido-6,9-dimethylisoalloxazine, and (II) 6-celluloseacetamido-9-(1′-D-ribityl)isoalloxazine. Reproduced with permission from C. Arsenis and D. B. McCormick, *J. Biol. Chem.*, **239**, 3093 (1964)

The use of a matrix containing a larger amount of lumiflavin attached accentuates the separation of flavokinase from the bulk of the inert protein. The chromatography of flavokinase on 7-celluloseacetamido-6,9-dimethylisoalloxazine is shown in Fig. III.16.

Studies on the coenzyme specificity of yeast NADPH-cytochrome *c* reductase indicate that the enzyme exhibits some binding to flavin phosphate analogues.[53] Furthermore, a monomethylester of FMN functions with NADPH-cytochrome *c* reductase, pyridoxine (or pyridoxamine) phosphate

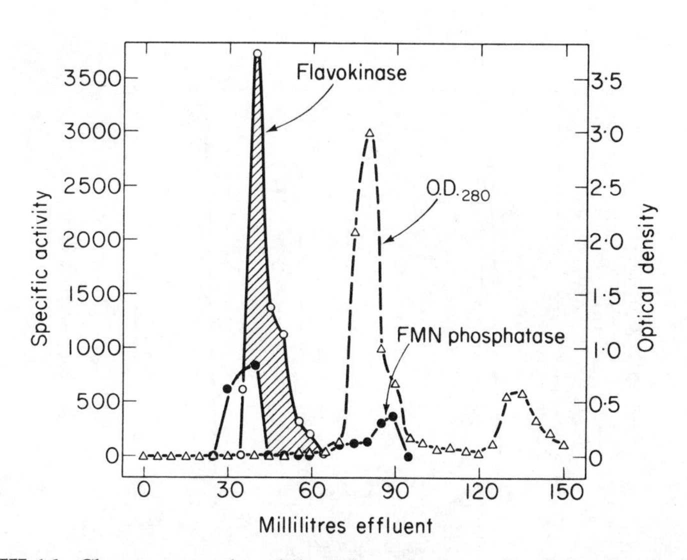

Figure III.16. Chromatography of liver flavokinase on 7-celluloseacetamido-6,9-dimethylisoalloxazine. A partially purified preparation of liver flavokinase was applied to a column of the cellulose derivative (1·4 cm × 65 cm) equilibrated with 0·005M potassium phosphate, pH 7·0. Enzyme and inert protein was eluted with a linear potassium phosphate gradient (0·005M–0·1M; 250 ml total volume). Flavokinase (○), FMN phosphatase (●) and inert protein (△) were assayed in the effluent. Reproduced with permission from C. Arsenis and D. B. McCormick, *J. Biol. Chem.*, **239**, 3093 (1964)

oxidase and glycollate oxidase. The chromatographic use of FMN esters of cellulose as biochemically specific adsorbents for the selective purification of these enzymes has been established.

Unfortunately, ion-exchange phenomena complicate the interpretation of these results, and in each case a cellulose derivative of similar charge type has been used as a control. Nevertheless, over 65-fold enrichment results when crude yeast NADPH-cytochrome *c* apo-reductase is chromatographed on FMN-cellulose phosphate. Similarly, pyridoxine (or pyridoxamine) phosphate apo-oxidase is much more effectively retained by adsorbents bearing FMN moieties than on a control column of DEAE-cellulose esterified with non-flavin phosphate to the same extent as the FMN-derivative of DEAE-cellulose. Furthermore, marked purification of crude glycollate apo-oxidase by chromatography on FMN-cellulose (Fig. III.17) has been achieved. Inert protein is eluted in the void volume, whilst the apo-oxidase is retained and maximally enriched over 400-fold after quantitative elution at higher ionic strength. In each case considerable enrichment and quantitative recoveries were effected.

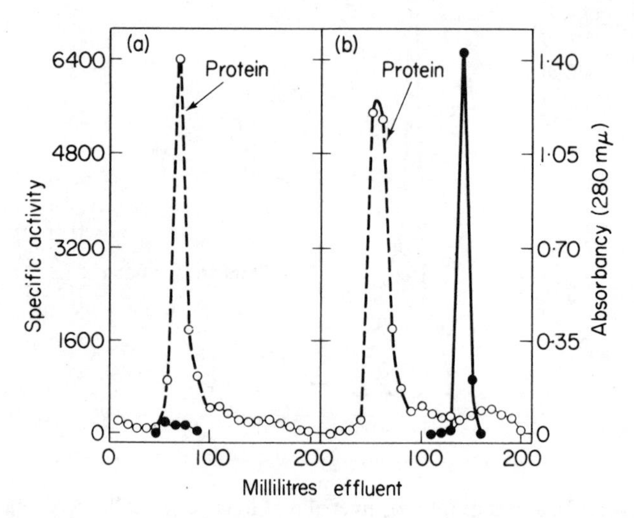

Figure III.17. Chromatography of glycollate apooxidase on cellulose (a) and FMN-cellulose (b). A partially purified glycollate oxidase from spinach leaves was applied to columns (1·5 cm × 30 cm) of (a) and (b) equilibrated with 0·005M potassium phosphate buffer, pH 8·0. Enzyme was eluted with a linear gradient of potassium phosphate buffer, pH 8·0 (0·005M–0·05M; 200 ml total volume). Glycollate apooxidase (●) and inert protein (○) were assayed in the effluent. Reproduced with permission from C. Arsenis and D. B. McCormick, *J. Biol. Chem.*, **241**, 330 (1966)

5. Pyridoxal Coenzymes

Pyridoxal phosphate is a versatile vitamin derivative which participates in the catalysis of several important reactions of amino-acid metabolism, namely transamination, decarboxylation and racemization. The structures of pyridoxal phosphate (PLP) and the corresponding amine, pyridoxamine phosphate (PMP), are shown in Fig. III.18.

It is generally accepted that of the functional groups available on the coenzyme, the ring nitrogen, the phosphate group and probably also the methyl group and the ionized phenolic hydroxyl are involved in binding PLP to its apoenzyme.[54] Furthermore, with most PLP-dependent enzymes the aldehyde group is known to participate in a transaldimination reaction generating a Schiff's base either with the amino group of a potential substrate or with the ε-amino group of a lysine residue of the protein. This suggests that adequate binding to the enzyme could be achieved with a coenzyme bearing a modified aldehyde or amino group, but that the binding would be loose enough to permit dissociation under relatively mild conditions. Thus, an *N'*-derivative of PMP has been used for the affinity chromatography of transaminases.[55]

Pig heart glutamic-oxaloacetic transaminase is a typical, well-characterized PLP-enzyme whose apoenzyme can be readily obtained from commercial preparations. The enzyme was subjected to affinity chromatography on

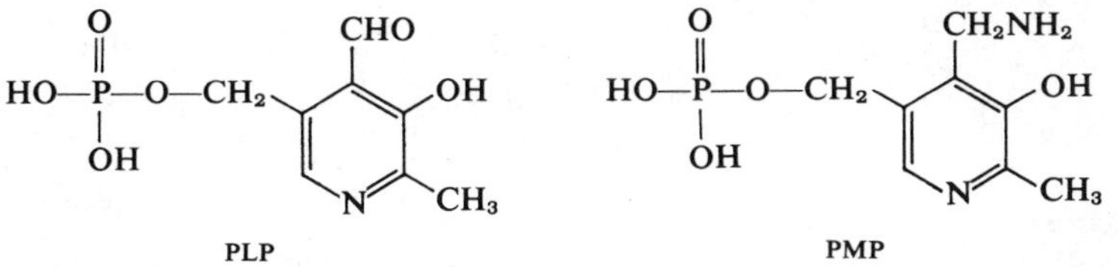

PLP PMP

Figure III.18. The structures of pyridoxal phosphate (PLP) and pyridoxamine phosphate (PMP)

PMP-Sepharose, *N'*-(ω-aminohexyl)-PMP-Sepharose and *N'*-(ω-amino-dodecyl)-PMP-Sepharose. Whilst the holoenzyme was eluted in the void volume with proteins displaying no affinity for PLP, the apoenzyme was bound to the matrix. Quantitative elution of the enzyme was achieved by raising the ionic strength and lowering the pH. Changing the pH or ionic strength alone did not effect elution of the enzyme. When the apoenzyme was applied to PMP-Sepharose, the enzyme was eluted in the void volume, emphasizing the importance of an extension arm interposed between the ligand and the matrix backbone.

Similar selective adsorbents have been utilized to partially purify tyrosine-aminotransferase (TAT) from rat liver.[56] Pyridoxamine phosphate was covalently linked via intermediate groups of three different lengths to agarose. Adsorbents prepared from these ligands bound TAT with increasing affinity as the length of the spacer group was increased. The bound enzyme was eluted with a buffer containing 0·01M pyridoxal phosphate and 0·5M NaCl at pH 9·0 with a resultant purification of some 120-fold. An additional 5-fold purification could be achieved by chromatography on Sephadex G-200. Some evidence suggests that the apoenzyme and the holoenzyme in the pyridoxamine form bind more readily than does the holoenzyme in the pyridoxal form.

However, the authors point out that this type of affinity chromatography may not be generally applicable to other pyridoxal enzymes. Rat liver

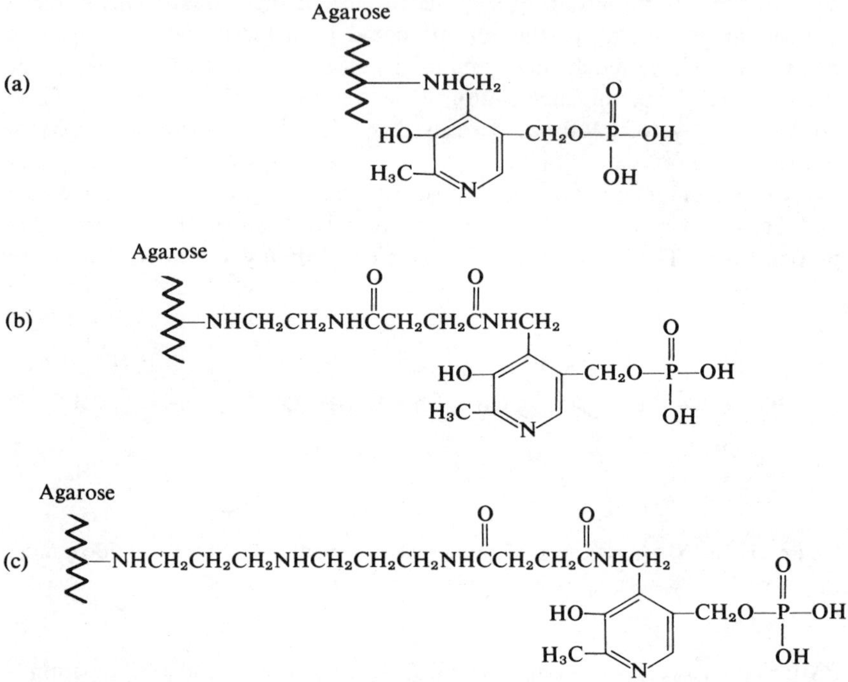

(a) Agarose—$NHCH_2$–pyridine ring (HO, H_3C, N, $CH_2O-P(=O)(OH)-OH$)

(b) Agarose—$NHCH_2CH_2NHC(=O)CH_2CH_2C(=O)NHCH_2$–pyridine ring (HO, H_3C, N, $CH_2O-P(=O)(OH)-OH$)

(c) Agarose—$NHCH_2CH_2CH_2NHCH_2CH_2CH_2NHC(=O)CH_2CH_2C(=O)NHCH_2$–pyridine ring (HO, H_3C, N, $CH_2O-P(=O)(OH)-OH$)

Figure III.19. Affinity adsorbents prepared by linking pyridoxamine phosphate to Sepharose 4B through extension arms of increasing length. (a) Pyridoxamine phosphate coupled directly to CNBr-activated agarose. (b) Pyridoxamine phosphate coupled to succinylated aminoethyl-Sepharose with a water-soluble carbodiimide. (c) Pyridoxamine phosphate coupled to succinylated 3,3′-diaminodipropylamine-Sepharose as described for (b). Reproduced with permission from J. V. Miller, P. Cuatrecasas and E. B. Thompson, *Biochim. Biophys. Acta*, **276**, 408 (1972)

aspartate and alanine transaminases in the pyridoxal form and bacterial D-serine dehydratase and tryptophanase in the apoenzyme forms do not bind to these adsorbents under the conditions that TAT was bound.

Agarose columns containing covalently bound PMP will also selectively remove TAT from hepatoma tissue culture cell homogenates.[57] Columns with PMP linked directly to Sepharose (Fig. III.19(a)) did not bind TAT from homogenates of these cells. In contrast, adsorbents (b) and (c) (Fig. III.19) gave substantial binding. The enzyme could be eluted from adsorbent (b) in good yield at pH 4 and from adsorbent (c) with pH values of 9 or above. In the latter cases the enzyme could be purified at least 100-fold. Furthermore, these matrices have been applied to the partial purification of TAT-synthesizing ribosomes from the same source.[58] Since TAT is thought to consist of four identical subunits, each binding one molecule of PLP, it is argued that

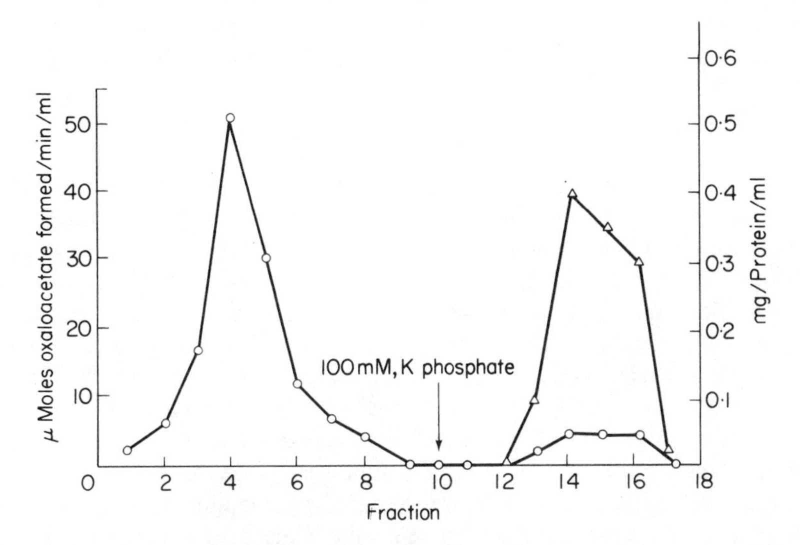

Figure III.20. Affinity chromatography of apoaspartate aminotransferase from *Rhizobium japonicum* on pyridoxal-5′-phosphate-Sepharose. Partially purified apoAAT was applied to a column (1·5 cm × 15 cm) of PLP-Sepharose equilibrated with 0·005M potassium phosphate buffer, pH 5·5. After washing through non-adsorbed protein, ATT apo-enzyme was eluted with 0·01M potassium phosphate buffer, pH 5·5. A_{280} (–○–○–) apoAAT activity (–△–△–△–). Reproduced with permission from E. Ryan and P. F. Fottrell, *FEBS Lett.*, **23**, 73 (1972)

at least one TAT subunit being synthesized on a polysome might be sufficiently complete to bind to the PMP-agarose column. If the entire complex of peptidyl-*t*RNA, *m*RNA and ribosome remained intact and could be eluted, synthesis of TAT might proceed to completion in a cell-free system. Evidence is presented to suggest that this method can indeed be used for the partial purification of ribosomes that synthesize TAT. When ribosomes prepared from induced HTC cells were applied to adsorbent B (Fig. III.19), less than 1% were retained. Subsequent elution at pH 4 and concentration by high-speed centrifugation showed that they were enriched for the ability to synthesize immunologically detectable TAT when compared to unfractionated ribosomes.

Ryan and Fottrell[59] have demonstrated how effectively affinity chromatography on immobilized PLP can be used to purify an apo-transaminase from a crude bacterial extract. Aspartate transaminase (AAT) from *Rhizobium japonicum*, partially purified by heat, ammonium sulphate precipitation and gel filtration steps, was applied to a column of PLP-Sepharose. Elution of the apo-AAT could be achieved by increasing the phosphate concentration in the elution buffer or by inclusion of free PLP. This purification procedure enriched the enzyme 170-fold with 69% recovery as illustrated in Fig. III.20.

6. Folate and Folate Analogues

Immobilized folate and folate analogues have been used as effective adsorbents for the purification of folate binding proteins and enzymes.

Intracellular dihydrofolate reductase (5,6,7,8-tetrahydrofolate: $NADP^+$ oxidoreductase, EC 1.5.1.3) catalyses the NADPH-dependent reduction of dihydrofolate and holds considerable pharmacological interest as the target enzyme for a number of chemotherapeutic agents.[60] The enzyme uses folate or 7,8-dihydrofolate as substrate and is powerfully inhibited by the corresponding 2,4-diamine analogues, e.g., methotrexate. The structures of these compounds are shown in Fig. III.21. Since these 2,4-diamino substituents are primarily responsible for binding of the inhibitors to the enzyme, immobilized derivatives of these compounds are prepared by coupling a distal carboxyl group to aminoalkyl-Sepharose. These insoluble matrices will retard or retain the mammalian and bacterial enzyme according to the K_m or K_i of the analogue.[61]

Although the pteridine moiety of the folate molecule is primarily responsible for the interaction with the active site of the enzyme, the *p*-aminobenzoyl-L-glutamate side-chain also contributes to the binding.[62] Thus, for maximal interaction between the reductase and the matrix there must be an extended spacer molecule between the agarose backbone and the inhibitor. With high affinity compounds ($K_i \sim 10^{-8}$–10^{-9}M) a spacer length of 4–6 methylene groups is required for good recovery of catalytic activity (> 50%).[63]

The influence of the matrix backbone on the interaction of enzymes with immobilized ligands has been discussed in detail elsewhere in this book. The coupling of folate analogues to insoluble matrices demonstrates the importance of the matrix itself in determining the strength of the interaction with the ligand. The attachment of the carboxyl group of methotrexate to aminoethyl-cellulose is readily accomplished as evidenced by the formation of a highly coloured cellulose derivative. However, columns prepared from this material did not adsorb significant amounts of enzyme.[64] Mell *et al.*[65]

attached methotrexate to soluble aminoethyl-starch and were able to resolve the resulting high molecular weight enzyme–methotrexate–starch complex by a series of gel filtrations. Erickson and Mathews[66] have shown that polyacrylamide matrices can be successfully exploited for the purification of dihydrofolate reductase specified by bacteriophage T_4. The enzyme is abundant in cultures of *E. coli* infected by T-even bacteriophages or T_5, due to synthesis of a virus-coded enzyme in infected cells. Dihydrofolate reductase

5,6,7,8-Tetrahydrofolic acid

Aminopterin

Methotrexate (amethopterin)

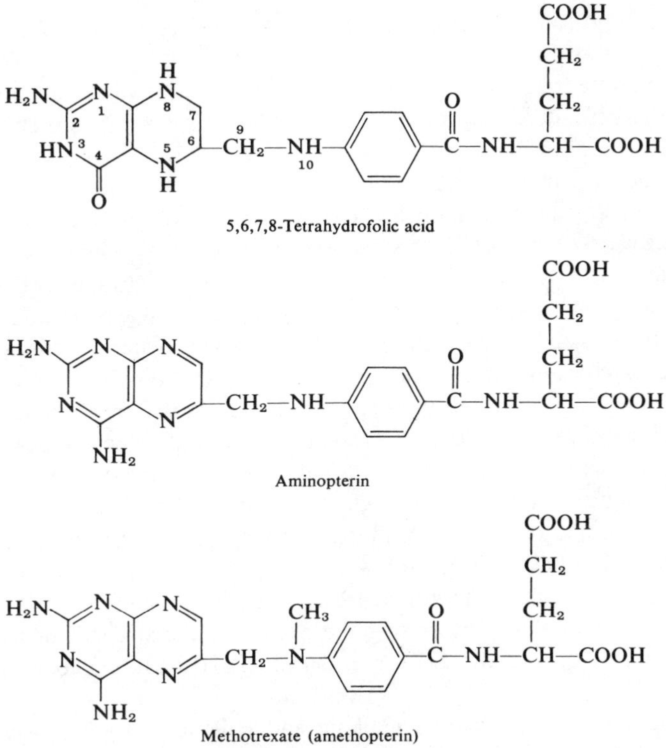

Figure III.21. The structures of 5,6,7,8-tetrahydrofolate, aminopterin and methotrexate

specified by bacteriophage T_4 has been purified 6000-fold on N^{10}-formylaminopterin immobilized by peptide linkage to aminoethyl Bio-gel P-150. About 4·5 mg enzyme could be obtained from 500 g of infected cells (80% recovery) in a homogeneous state.

Newbold and Harding[61] describe a rapid method for the retrieval of small amounts of catalytically active dihydrofolate reductase and other pteridine binding proteins on columns of agarose containing covalently linked folate derivatives. Dihydrofolate reductase was purified from *Lactobacillus casei* MTX/R on immobilized folate and methotrexate. The folate derivatives were covalently attached to 6-aminohexyl-Sepharose by a carbodiimide condensation. Columns containing folic acid retarded but did not retain the enzyme whilst methotrexate at pH 6 was particularly effective for retention of the enzyme. This behaviour reflects the affinity for folate ($K_i \sim 5 \times 10^{-6}$M) and methotrexate ($K_i \sim 10^{-8}$M) that the enzymes display in free solution.

For elution of the active enzyme in high yields, 5-formyltetrahydrofolate, 7,8-dihydrofolate and folate are particularly effective counter-substrates. The buffer composition is important, not only for maintaining the stability of the enzyme but also for elution of reductase-independent proteins. Dissociation of the enzyme from the matrix can lead to inactivation and is apparently due to irreversible deformation of the secondary or tertiary structure of the enzyme. Inactivation occurs particularly when folate and NADPH are simultaneously present in the eluant.

The same authors demonstrated how dihydrofolate reductase could be purified nearly 4000-fold in a single-step procedure by applying a crude rat skin eluate to a methotrexate-agarose column. A similar affinity matrix prepared by coupling methotrexate to Sepharose via a six-carbon chain quantitatively adsorbs the dihydrofolate reductase activity from a partially purified extract of chicken liver.[64] An approximately 250-fold purification can be achieved by elution of the enzyme with dilute potassium phosphate supplemented with dihydrofolate to yield a preparation which appeared to be homogeneous. Furthermore, Chello *et al.*[67] purified to homogeneity the dihydrofolate reductase from a high enzyme mutant of L1210 mouse leukaemia by a simple two-step procedure involving pH 5·1 precipitation of inert protein and affinity chromatography on a matrix prepared by coupling methotrexate to aminoethyl-Sepharose. A peak of pure enzyme was eluted from a loaded column by raising the pH and ionic strength of the eluting buffer from 0·05M citrate pH 6·0 to 0·05M tris-HCl pH 8·5 containing 0·1M KCl.

Using a similar procedure, Gauldie and Hillcoat[68] applied a crude ammonium sulphate precipitate of an extract from the L1210 lymphoma to a column of methotrexate-agarose. The adsorbent was washed with buffer containing NADPH to increase the binding of the enzyme, whence dihydrofolate reductase activity was eluted with buffer at high pH containing dihydrofolate and a high concentration of salt. The elution pattern from a column of methotrexate-agarose is shown in Fig. III.22.

Poe *et al.*[69] describe a convenient large-scale procedure for the isolation of dihydrofolate reductase from *E. coli B* (strain MB 1428) based on batchwise affinity chromatography. The crude extract (1 litre containing 2000 units of enzyme activity) was treated with 100 ml of methotrexate coupled to aminoethyl-Sepharose. Washing with 1M NaCl eluted less than 1% of the bound enzyme and quantitative recovery could be achieved by elution with dihydrofolate in 1M NaCl.

Folate in cow's milk is strongly bound to a minor whey protein and the use of affinity chromatography has resulted in a marked improvement in the purity of the carrier protein compared with previous isolation methods.[70] Folic acid was chosen for immobilization because it binds to the protein more strongly than the natural milk folate, 5-methyl-5,6,7,8-tetrahydrofolate. Agarose-folic acid was added to the protein fraction containing the folate-binding protein from whey and the mixture stirred. The pH was lowered to

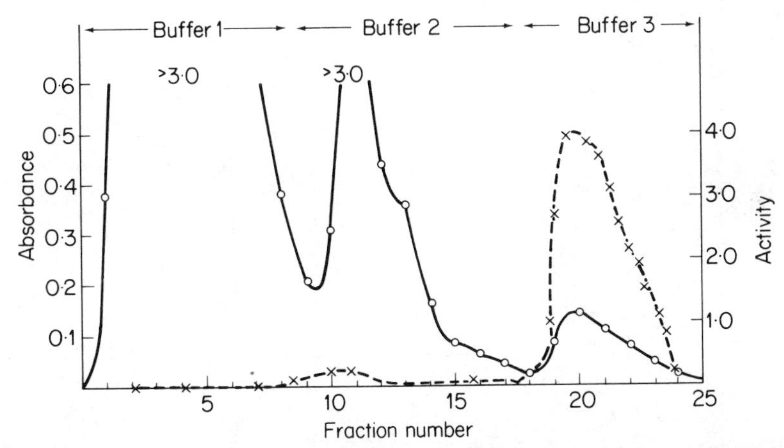

Figure III.22. Affinity chromatography of dihydrofolate reductase from a subline of L1210 lymphoma on methotrexate-agarose. A desalted $(NH_4)_2SO_4$ fraction from a L1210 lymphoma cell extract (26 ml) was applied to a column of methotrexate-agarose and the column washed sequentially with three buffers. Protein absorbance at 280 nm was measured and enzyme activity assayed. Buffer 1, 0·1M tris-HCl (pH 7·5) containing 1×10^{-5}M NADPH; buffer 2, 0·2M tris-glycine (pH 9·5) containing 2·0M KCl and 1×10^{-5}M NADPH; buffer 3, 0·2M tris-glycine (pH 9·5) containing 2·0M KCl and 1×10^{-5} dihydrofolate. O—O, absorbance at 280 nm; × - - - ×, tetrahydrofolate dehydrogenase activity, μmoles/minute per ml. Reproduced with permission from J. Gauldie and B. L. Hillcoat, *Biochim. Biophys. Acta*, **268**, 36 (1972)

3·6 with NaHCl to dissociate the native folate from the binding protein and then raised to 7·0 with NaOH to allow the binding protein to become attached to the immobilized folate. The adsorbent was packed into a column and eluted sequentially with:

(i) 0·05M NaH_2PO_4 pH 7·2 containing 1M NaCl;
(ii) 0·05M NaH_2PO_4 pH 6·0;
(iii) 0·1M sodium acetate pH 5·0 containing 8M urea;
(iv) 1N HCl containing 8M urea.

The folate binding protein was eluted with 8M urea at pH 5·0 (Fig. III.23).

In common with other adsorbents based on immobilized coenzymes significant binding of non-specific inert protein can be experienced. Thus, Newbold and Harding[61] observed that other proteins were eluted from methotrexate-agarose and distinguished from dihydrofolate reductase on polyacrylamide gel electrophoresis. Similar behaviour was noted by Gauldie and Hillcoat[68] and Kaufmann and Pierce.[64] The strong attachment of the

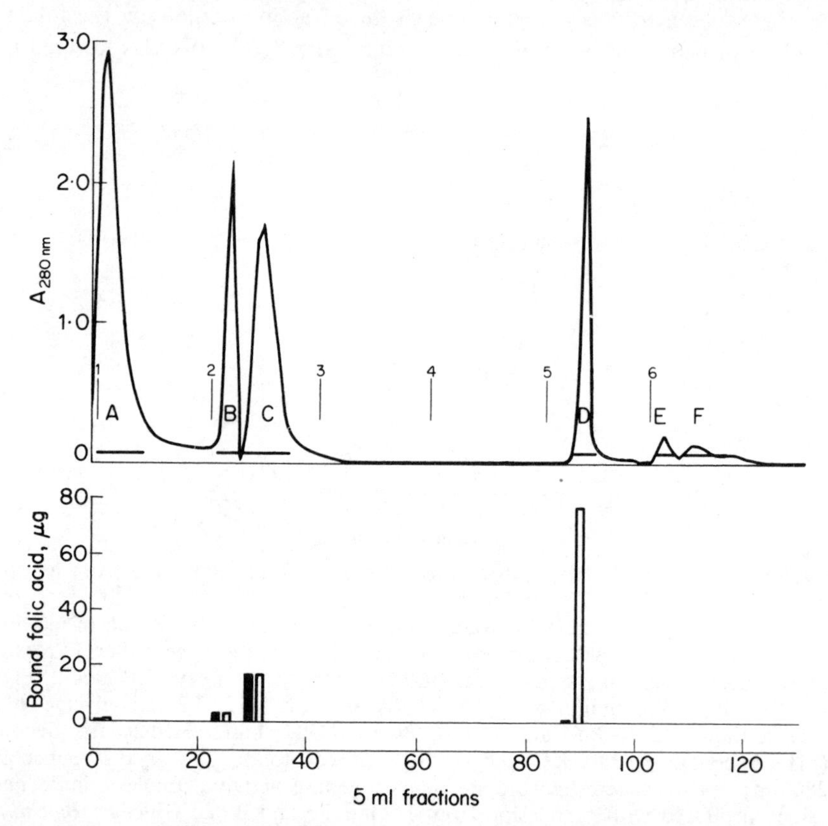

Figure III.23. Affinity chromatography of partially purified folate-binding protein from whey on a column (1·2 cm × 19 cm) of agarose-folic acid. Proteins were first adsorbed to the gel as described in the text. The column was eluted with the following buffers, all containing 3 mM 2-mercaptoethanol, the point of application of each buffer being as indicated: (1) 0·05M sodium phosphate, pH 7·2; (2) 0·05M sodium phosphate, pH 7·2, containing 1M NaCl; (3) 0·05M sodium phosphate, pH 6·0; (4) 0·1M sodium acetate, pH 5·0; (5) 0·1M sodium acetate, pH 5·0, containing 8M urea; (6) NHCl containing 8M urea. (——): A_{280}. Fractions were pooled as indicated by the bars, dialysed against buffer A (see text) and the bound folate content of the resulting solutions measured before (■) and after (□) the addition of an excess of folic acid. Reproduced with permission from D. N. Salter, J. E. Ford, K. J. Scott and P. Andrews, *FEBS Lett.*, **20**, 302 (1972)

folate binding protein to the affinity adsorbent proved an advantage in separating it from other whey proteins which had also been adsorbed. This non-specific adsorption of proteins could have arisen by attraction to basic groups on the gel or to the remaining free carboxyl of the immobilized folic acid itself. It is thus recommended that a preliminary purification step be included to reduce these effects.

Another problem, also experienced with affinity chromatography on immobilized adenine and pyridine nucleotides, is the binding of nucleic acid contaminants by these columns. Gauldie and Hillcoat[68] showed how nucleic acid components were adsorbed onto methotrexate-agarose and were leached off during the development of the column. They were readily removed from the enzyme preparation by a column of Sephadex G-75.

7. Biotin

Affinity chromatography on immobilized biotin has been applied to the purification of avidin, a protein abundant in egg white and the genital tracts of animals. Avidin exhibits a remarkable affinity for biotin with a dissociation constant of 10^{-15}M[71] for the biotin–avidin complex.

Highly purified avidin has been obtained from egg white by classical procedures utilizing three successive CM-cellulose columns.[72] Pioneering work by McCormick led to some degree of purification of 10% pure avidin on columns of cellulose to which biotin had been esterified.[73] Such columns, however, displayed weak affinity for avidin since the applied avidin was incompletely adsorbed, easily eluted and poorly resolved from extraneous proteins also bound.

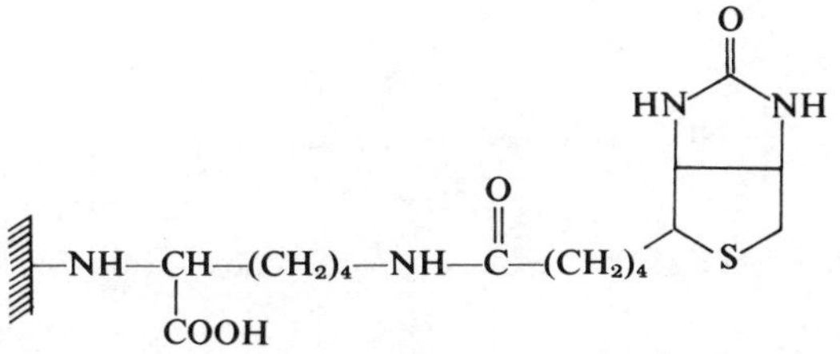

—NH—CH(COOH)—$(CH_2)_4$—NH—C(=O)—$(CH_2)_4$—[biotin ring: HN, NH, C=O, S]

Figure III.24. The structure of the avidin adsorbent: ε-*N*-biotinyl-L-lysyl-Sepharose

Cuatrecasas and Wilchek[74] prepared an avidin-specific adsorbent by coupling biocytin (ε-*N*-biotinyl-L-lysine) to Sepharose. The lysyl substitution on the carboxyl group does not impair binding to avidin and the amino group of this derivative can be conveniently coupled to Sepharose. The structural units essential for avidin binding, the ureido and thiophan rings, are distant from the matrix backbone (Fig. III.24).

Avidin was adsorbed so strongly that elution could not be effected by extremes of pH (1·5 or 12) or with 3M guanidine–HCl. Slow elution of the protein occurred when the buffer was supplemented with biotin in concentrations that approach the limit of solubility. A combination of two extreme conditions, 6M guanidine–HCl and pH 1·5 was the most effective way to elute avidin. The resulting denatured avidin was completely and rapidly reconstituted by 10-fold dilution with buffer or by dialysis to lower the concentration of guanidine.

The biocytin-Sepharose adsorbent can be used to obtain a 4000-fold single-step purification of avidin from crude egg white with 90% recovery. Approximately 1 mg of avidin was obtained from the raw white of a single egg.

Wolpert and Ernst-Fonberg[75] describe the synthesis of *N*-agarose-*N'*-biotin propylamide-1,3-diaminopropane by coupling biotin to 3,3'-diamino-dipropylamine with dicyclohexyl-carbodiimide in *N,N'*-dimethylformamide and subsequently coupling the biotin derivative to Sepharose. A column prepared from this adsorbent selectively retarded acetyl coenzyme A carboxylase, a biotin-containing enzyme.

8. Lipoic Acid

Lipoic acid (1,2-dithiolane-3-valeric acid) is involved in the oxidative decarboxylation of α-keto acids by the multi-enzyme dehydrogenation complexes. Lipoic acid is covalently bound to the complex by a linkage between the carboxyl function of the coenzyme and an ϵ-amino group of a lysine residue. The oxidized form of lipoic acid is regenerated by a specific flavoprotein, lipoyl dehydrogenase.

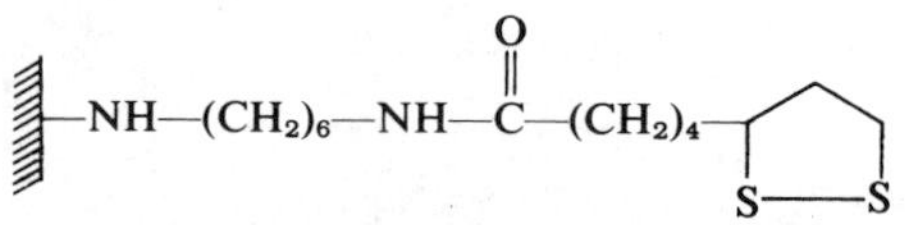

$$\text{—NH—}(CH_2)_6\text{—NH—}\overset{O}{\overset{\|}{C}}\text{—}(CH_2)_4\text{—(dithiolane ring, S—S)}$$

Figure III.25. The structure of N^6-(6-aminohexyl)-lipoyl-Sepharose

Cox and Lowe[76] have prepared an immobilized-lipoate adsorbent by coupling α-lipoic acid to 6-aminohexyl-Sepharose with 1-ethyl-3(3-dimethyl-aminopropyl) carbodiimide hydrochloride in ethanol at pH 4·7. Fig. III.25 shows the structure of the resulting adsorbent, N^6-(6-aminohexyl)-lipoyl-Sepharose, which is similar to the natural substrate for lipoyl dehydrogenase, protein-bound N^{ϵ}-lipoyl-L-lysine. Furthermore, the reactive dithiolane ring is held distant from the matrix backbone and hence this adsorbent should be competent for lipoyl dehydrogenase. However, chromatography of a commercial enzyme preparation from pig heart resolved lipoyl dehydrogenase into two active fractions, one appearing in the void volume of the column and the other eluted on a linear salt gradient. These observations are the subject of further investigation.

Scouten *et al.*[140] have recently reported the purification of lipoamide dehydrogenase from pig heart, yeast and *E. coli* on columns of propyl-lipoamide glass beads. Chromatography of a crude or partially purified lipoamide dehydrogenase resulted in a 100-fold purification.

9. Cobalamins

Vitamin B_{12} derivatives are implicated in many metabolic processes, although B_{12} coenzymes have been established as required for only five enzymatic reactions. These enzymes include methylmalonyl-CoA mutase (methylmalonyl-CoA-CoA carbonyl mutase), glutamate mutase (L-*threo*-3-methylaspartate carboxyaminomethyl mutase), dioldehydrase, ribonucleotide reductase, and the enzyme responsible for the biosynthesis of methionine from homocysteine.

Vitamin B_{12} is composed of two parts: the corrin ring system and the nucleotide. The corrin ring contains tervalent cobalt chelated to the 4 nitrogen atoms of the ring, the nucleotide and a small ligand such as cyanide ion, hydroxide ion or water (Fig. III.26). The entire structure is termed cobalamin and a series of analogues exists depending on the small ligand and the nucleotide moiety.

Significant advances in the study of vitamin B_{12} binding proteins have been lessened by difficulties encountered in their isolation. The two major cobalamin binding proteins in human serum, transcobalamin I and II, are present in minute concentrations and have only recently been isolated in pure form.[77]

Recently, Olesen *et al.*[78] have described methods for the covalent attachment of cobalamins to several support materials. Hydroxocobalamin was coupled to poly-L-lysine or albumin with a water-soluble carbodiimide and the resulting hydroxocobalamin–albumin conjugate reacted with bromoacetyl-cellulose to yield an insoluble support. This matrix adsorbed cobalamin-binding proteins from serum and gastric juice which could be recovered by specific elution with hydroxocobalamin.

Olesen *et al.*[78] also coupled 5′-deoxy-adenosyl-cobalamin to poly-D-glutamic acid or succinylated γ G-globulin. However, the cobalamin moiety of 5′-deoxy-adenosyl-cobalamin-succinylated γ-globulin complex could readily be released by treatment with cerous hydroxide or by exposure to light. This would suggest that the cobalamin was linked to the matrix through its axial ligands. However, small alterations in the nucleoside or nucleotide moiety of 5′-adenosylcobalamin greatly influence the biochemical activity of the coenzyme. Yamada and Hogenkamp[79] have suggested an alternative approach. They reported the synthesis of a 5′-deoxy-adenosyl-cobalamin-agarose for the purification of enzymes requiring 5′-deoxy-adenosyl-cobalamin as coenzyme. Cyanocobalamin ϵ-amino-carboxylic acid was coupled to diaminododecane with a water-soluble carbodiimide and the resulting extended ligand reacted with agarose previously activated with CNBr. The complex was reduced with chromous chloride and reacted with 5′-*O*-*p*-tolylsulphonyl-adenosine to yield the desired adsorbent.

The adsorbent was competent for the purification of ribonucleotide reductase from a partially purified extract of *Lactobacillus leichmanii*. Applica-

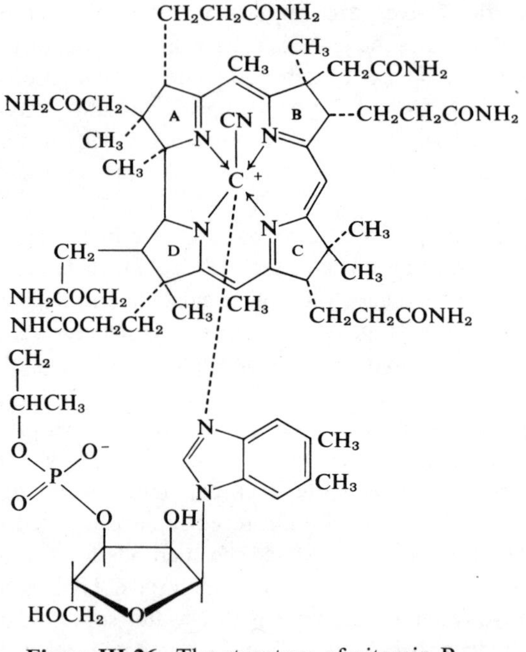

Figure III.26. The structure of vitamin B_{12}

tion of the enzyme to the adsorbent equilibrated with an appropriate buffer, in the presence of substrate, allosteric effector or dithiol, resulted in some retention of the enzyme. When the enzyme extract was applied in the presence of both dithiol and an allosteric effector, almost quantitative adsorption ensued. Furthermore, enzyme activity could be recovered by removing the allosteric effector from the irrigating buffer. Ribonucleotide reductase activity was enriched 6-fold with a recovery of 53%, as shown in Fig. III.27. Interestingly, only the deoxyribonucleotide allosteric effectors, such as *d*GTP, and not the ribonucleotide substrates, were able to retard the enzyme.

This matrix could be used as the starting point for the synthesis of affinity adsorbents competent for other B_{12} requiring systems. Thus, the cyanocobalamin-agarose can be reduced with borohydride and oxidized in the absence of cyanide to yield hydroxo-cobalamin-agarose; such an adsorbent could be used for the purification of intrinsic factor from gastric juice or the transcobalamins from serum. Methyl-cobalamin-agarose may be useful in the purification of methyl-tetrahydrofolate-homocysteine transmethylase apoenzyme. Furthermore, other considerations would suggest that the 5′-deoxyadenosyl-cobalamin-agarose would purify dioldehydrase.

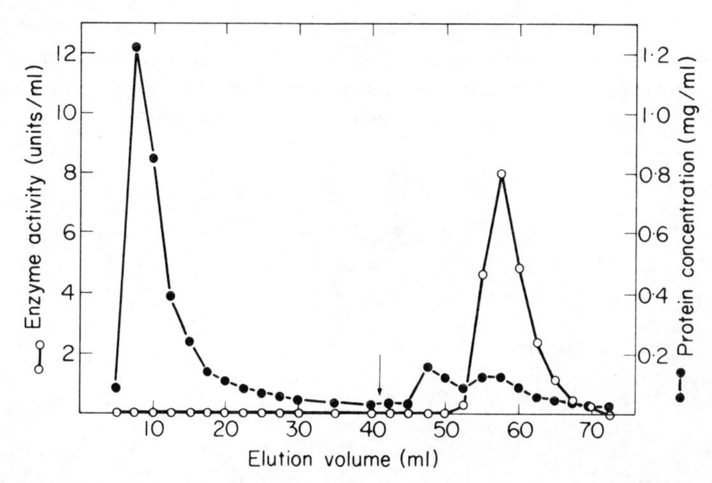

Figure III.27. Affinity chromatography of partially purified ribonucleotide reductase. The deoxyadenosylcobalamin-agarose column (0·4 cm × 35 cm) was equilibrated with 0·05M tris-phosphate buffer, pH 7·2, containing 10 mM dithiothreitol and 0·1 mM dGTP. A protein solution (specific activity 9·2 units per mg) containing 10·25 mg in 1 ml of the same buffer was applied to the column. The column was run at 4°C in the dark; the flow rate was 3·5 ml per hour; 2·5 ml fractions were collected. Elution was started (arrow) with 0·1M tris-phosphate buffer, pH 7·2, containing 10 mM dithiothreitol. Reproduced with permission from R.-H. Yamada and H. P. C. Hogenkamp, *J. Biol. Chem.*, **247**, 6266 (1972)

Allen and Majerus[80] have developed a technique of affinity chromatography which is effective for the isolation of traces of the vitamin B_{12} binding proteins. The affinity ligand was prepared by mild acid hydrolysis of the unsubstituted propionamide side-chains of the corrin ring of vitamin B_{12}. The hydrolysis products, a mixture of mono-, di- and tri-carboxylic vitamin B_{12} derivatives, were separated by QAE-Sephadex and the mono-carboxyl derivatives coupled to the free amino group of 3,3′-diaminodipropylamine-substituted Sepharose using a water-soluble carbodiimide. The resultant polymer contained 0·7 μmoles vitamin B_{12} per ml of packed Sepharose.

The vitamin B_{12} binding proteins of human granulocytes, human plasma, human gastric juice, an extract of hog gastric mucosa and a partially purified preparation of human transcobalamin II were almost quantitatively (>90%) bound to the vitamin B_{12}-Sepharose.[77,81] Utilizing affinity chromatography as the sole purification step, the vitamin B_{12} binding protein from human granulocytes has been purified nearly 10,000-fold to homogeneity with a recovery of over 90%.[81]

10. Haem Coenzymes

Coenzymes derived from a metal ion and the porphyrin nucleus play a central role in the metabolism of organisms. Porphyrins are derived from the parent tetrapyrrole porphin by substitution of methyl, ethyl, vinyl or propionic acid functions for the hydrogen atoms at positions 1 to 8. The proto-

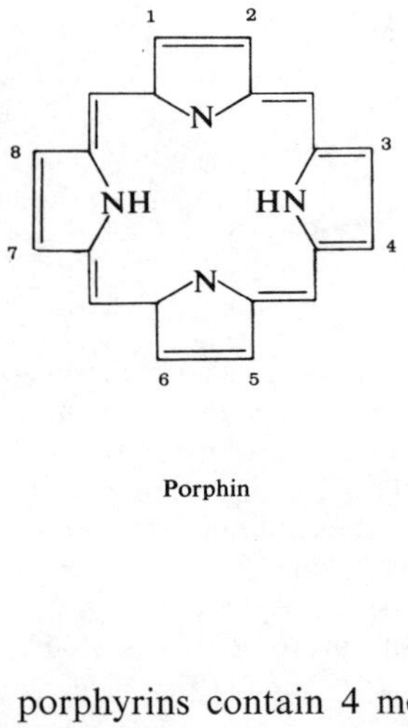

Porphin

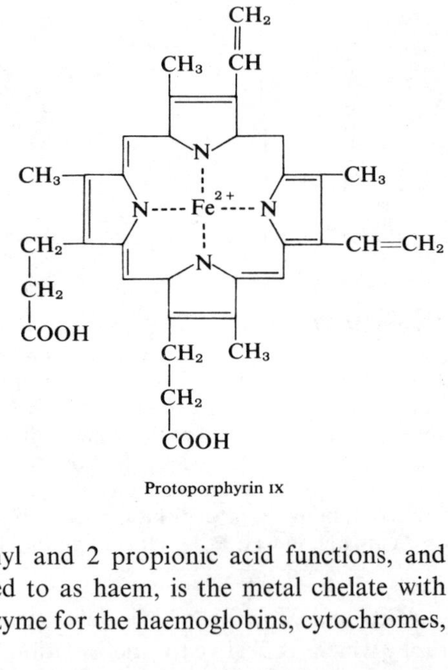

Protoporphyrin IX

porphyrins contain 4 methyl, 2 vinyl and 2 propionic acid functions, and protoporphyrin IX, generally referred to as haem, is the metal chelate with ferrous iron. Haem serves as a coenzyme for the haemoglobins, cytochromes, catalases and peroxidases.

Conway and Muller-Eberhard[82] have coupled haemin and protoporphyrin IX to aminoethyl-Sepharose in dimethylformamide. The resulting resin had a relatively low capacity for the two major haem-binding proteins in serum, albumin and haemopexin. In contrast, when the water soluble 2,4-disulphonic acid deuteroporphyrin was attached to Sepharose, 10^{-6} moles of immobilized porphyrin bound 10^{-7} moles of iodinated (^{125}I) haemopexin. The adsorption was unaffected by 0·1M acetate pH 4·5, 0·1M borate pH 9·5 or 1M NaCl.

Apo-haemopexin could be eluted in good yield (90%) by 2·5M guanidine–HCl and, following dialysis, was electrophoretically and immunologically indistinguishable from native haemopexin.

B. NUCLEIC ACID ENZYMES

Affinity chromatography was first introduced to nucleic acid research by Alberts *et al.*[83] and Litman,[84] who developed DNA-cellulose as a chromatographic adsorbent. DNA-cellulose has subsequently been used for the purification of DNA-specific proteins from several organisms.[85–88]

Deoxyribonucleic acid polymerase from a partially purified extract of *Micrococcus luteus* has been isolated on DNA-cellulose and eluted by high salt concentrations.[84] Almost all of the proteins and 30% of the enzyme activity were eluted in the void volume; fractions eluted with 0·4M NaCl contained 6·5% of the protein and 10% of the polymerase activity while the 0·7M NaCl fraction contained 2% of the protein and 15–20% of the activity. Protein recovery was quantitative whilst polymerase recovery was 50–60%. The same DNA-cellulose could be used for repeated isolations with identical results and the purified polymerase preparation was entirely free of *endo*- and *exo*-nucleases.

The DNA has previously been immobilized by drying onto cellulose[83] but it is not possible to insolubilize low molecular weight DNA by this method and even the DNA that is bound is slowly leached out of the cellulose. Entrapment of DNA in polyacrylamide-agarose[89,90] and covalent attachment to Sephadex G-200[91] or agarose beads has also been employed to immobilize DNA for affinity chromatography.[89,90] In most cases the physiological double-stranded DNA was the preferred ligand.

Schaller *et al.*[92] have utilized single-stranded DNA-agarose for affinity chromatography of DNA-specific enzymes. This adsorbent has a high binding capacity for many DNA-specific enzymes and can be used for the analysis and preparation of these proteins. The success of the adsorbent can be attributed to the following:

(i) Single-stranded DNA is a substrate or substrate analogue for many DNA-specific enzymes.

(ii) A high concentration of denatured DNA can readily be immobilized on the agarose matrix.

(iii) Commercial preparations of DNA can be employed.

(iv) The resultant matrix contains little solid material and non-specific adsorption can be minimized.

Single-stranded circular DNA from phage fd or denatured calf thymus DNA was immobilized on agarose. The DNA content of this gel was stable under the conditions of enzyme purification. Single-stranded DNA agarose gels have a high capacity to bind DNA-specific proteins and were used to differentiate between similar enzymatic activities in DNA-free extracts of *E. coli*. Endogenous DNA competes with bound DNA and therefore prevents quantitative binding of proteins to the column material. Removal of DNA can be achieved satisfactorily by digestion with DNAse, by precipitation with polyethylene glycol 6000 in concentrated salt or by treatment with DEAE-cellulose. Other methods such as streptomycin, protamine or polyethylene-imine precipitation are of limited use since, in addition to DNA, some other proteins are also removed.

Application of a DNA-free crude extract of *E. coli* to a DNA-agarose column and subsequent gradient elution demonstrated how DNA-dependent RNA-polymerase, DNA-polymerase, nuclease and polynucleotide ligase activity could be resolved (Fig. III.28). Furthermore, the characteristic elution profile was reproduced when the DNA-specific fraction was rechromatographed.

Nucleases were weakly bound; the DNAse fraction could be resolved into at least two activities and the two main DNA polymerases were clearly separated. The majority of the ligase was unretarded by the column, although 10% of the activity was retained specifically.

On a preparative scale, crude protein fractions from large amounts of *E. coli* cells were purified on relatively small DNA-agarose columns. Typical recoveries were between 75% and 100% with purifications up to 200-fold and although the enzyme fractions obtained were purified to some extent (5–80%), a further classical purification step was necessary to yield a homogeneous preparation.

RNA-polymerase from *E. coli* could be purified on single-stranded DNA-agarose columns.[93] The polymerase was separated into two fractions, one containing core enzyme and another which was shown by template specificity and subunit composition to be RNA polymerase holoenzyme. Fig. III.29 demonstrates the chromatography of RNA polymerase on DNA-agarose. The first peak contained polymerase with template specificity of the core enzyme and was eluted around 0·5M KCl. The second enzyme peak, eluted between 0·6M and 1·0M KCl, was probably the RNA polymerase holoenzyme. Rechromatography of each of the two peaks showed that both fractions eluted homogeneously at their characteristic salt concentrations.

The single-stranded DNA-agarose columns can also be used to separate *E. coli* ribonucleases from the bulk of the cellular proteins.[94] By successive chromatography on DNA-agarose, hydroxylapatite and DEAE-cellulose it is possible to resolve the enzymes ribonuclease II, III and H, and polynucleotide phosphorylase. The DNA-agarose columns were loaded with a 42–60% ammonium sulphate fraction of a cellular extract. Almost all the inert protein was eluted in the void volume, whence the mixture of adsorbed ribonucleases were eluted with a linear KCl gradient. The enzymes could subsequently be resolved by chromatography on hydroxylapatite and

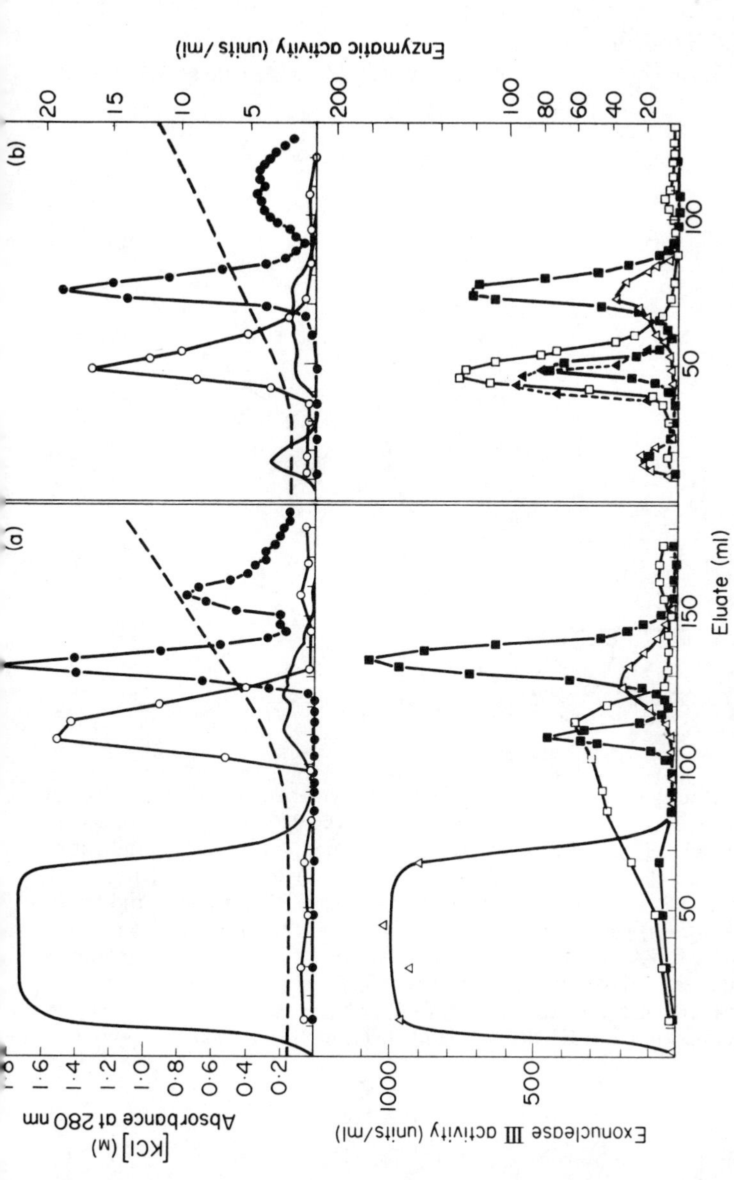

Figure III.28. Chromatography of *E. coli* proteins on calf thymus DNA-agarose. Protein from 3·3 g *E. coli* H 560 was used as a starting material. (A) The crude DNA-free protein fraction, (B) the DNA-specific protein fraction. Both fractions were complemented with 0·3 μg purified DNA polymerase I, loaded onto the columns (0·8 cm² × 15 cm) at 0·15M salt, and eluted with a 120 ml linear gradient of 0·15–1·2M KCl in standard buffer. Columns were operated from bottom to top at a flow rate of 10 ml/hour, 1·2 ml fractions were collected and assayed for ionic strength (- - - -), absorbance at 280 nm (——) and enzymatic activities; RNA polymerase (●——●); RNAase (○——○); DNA polymerase (■——■); DNAase (exonuclease III assay with poly[d(A-T)], □——□) (exonuclease I assay with denatured *E. coli* DNA, ▲- - - -▲); polynucleotide ligase (△——△). Reproduced with permission from H. Schaller *et al.*, *Eur. J. Biochem.*, **26**, 474 (1972)

DEAE-cellulose. Thus, DNA-agarose seems to be a useful step in the purification of *E. coli* ribonucleases. Furthermore, the existence in *E. coli* of an enzyme that can degrade RNA in RNA–DNA hybrids, RNAse H, has been confirmed by this method.

The purification of several amino-acyl-*t*RNA synthetases by affinity

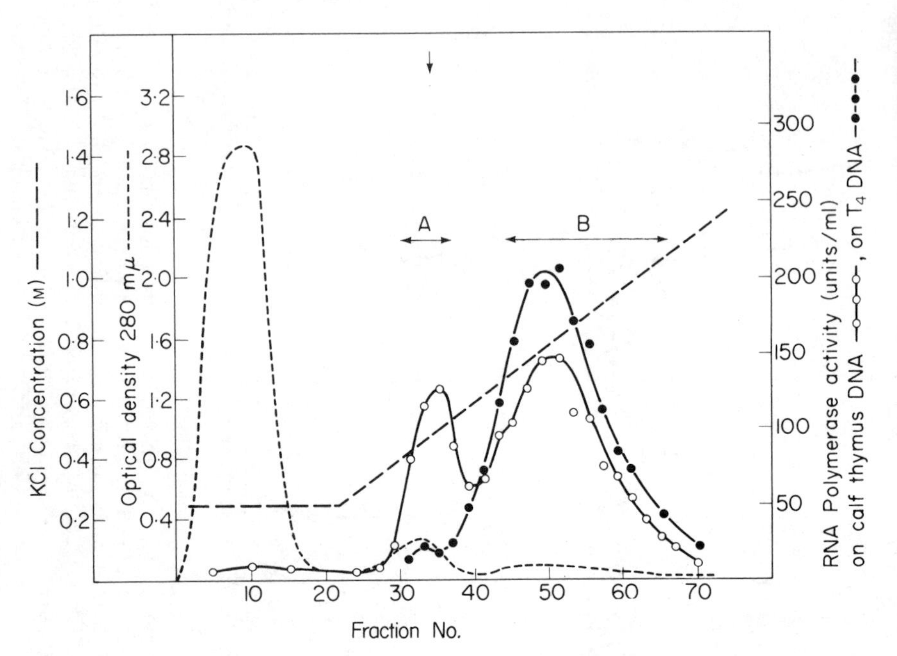

Figure III.29. Affinity chromatography of RNA polymerase on DNA-agarose. A partially purified preparation of RNA polymerase (45,000 units) was applied to a 3·1 cm² × 15 cm column of DNA-agarose and eluted with a linear gradient of potassium chloride (0·25–1·25M; 600 ml total volume) in a 0·01M tris pH 8·0 buffer containing 10^{-3}M EDTA, 10^{-4}M dithioerythritol, 5% glycerol and 0·25M KCl. Fractions (10 ml) were collected and assayed for RNA polymerase activity, using calf thymus DNA and T_4 DNA as template. Peak fractions with relatively low (A) and high (B) activity on T_4 DNA were pooled and concentrated. Fraction A contains core enzyme and fraction B RNA polymerase holoenzyme. Reproduced with permission from C. Nüsslein and B. Heyden, *Biochem. Biophys. Res. Commun.*, **47**, 282 (1972)

chromatography on immobilized *t*RNA has been described.[95,96] Of the reported procedures, all suffer the disadvantage of retaining contaminating proteins either by specific interactions with the ligand or by non-specific ion-exchange. Consequently, the isolation to homogeneity of an aminoacyl-*t*RNA synthetase has been difficult to achieve. However, Remy *et al.*[97]

describe the complete purification of yeast phenylalanine-*t*RNA synthetase on an adsorbent comprising *t*RNA^{Phe} coupled to a hydrazinyl-Sepharose matrix.

A crude yeast extract, partially purified by ammonium sulphate fractionation, was applied to the *t*RNA column, whence the bulk of the inert proteins emerged in the void volume. The phenylalanine-*t*RNA synthetase was eluted by altering the pH and ionic strength under the conditions required to dissociate the *t*RNA–enzyme complex. The enzyme eluted by this technique was not homogeneous but contained a small number of contaminants. These contaminating proteins could either be those that recognize general *t*RNA features, such as the methylases or *t*RNA–ATP–CTP nucleotidyltransferase, or those that are specific for a general RNA character, for example, the RNAses. It should be possible to eliminate these specifically adsorbed proteins by passage of the enzyme extract through a column containing a mixture of covalently attached *t*RNAs lacking *t*RNA^{Phe}. Thus, use of a generalized column charged with *t*RNAs without *t*RNA^{Phe} in conjunction with the specific *t*RNA^{Phe} column resulted in the isolation of pure phenylalanyl-*t*RNA synthetase.

Nelidova *et al.*[95] used a mixture of *t*RNAs bound to a polyacrylhydrazid-agar gel to obtain some enrichment of valyl- and lysyl-*t*RNA synthetases from a rat liver extract. Similarly, Bartkowiak *et al.*[96] reported a 27·5-fold purification of *E. coli* isoleucyl-*t*RNA synthetase on isoleucyl-*t*RNA^{Ile} bound to bromoacetamidobutyl-Sepharose.

C. SULPHYDRYL ENZYMES

An agarose adsorbent with a high capacity for sulphydryl proteins should be very useful for a variety of separative procedures. The preparation of an organomercurial derivative of agarose has been described by Cuatrecasas.[98] Sodium *p*-chloromercuribenzoate was coupled to aminoethyl-agarose with a water-soluble carbodiimide in 40% *N,N'*-dimethylformamide. This organomercurial adsorbent (Fig. III.30) had a high capacity for horse haemoglobin,

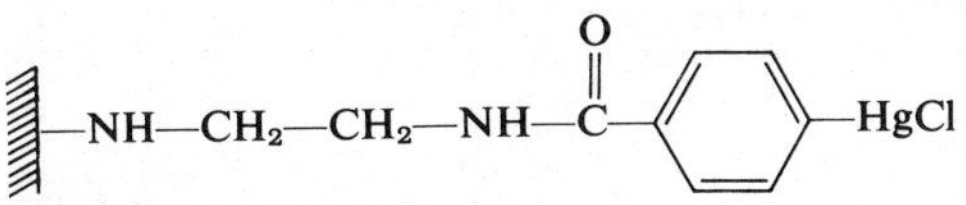

$$\text{—NH—CH}_2\text{—CH}_2\text{—NH—C(=O)—C}_6\text{H}_4\text{—HgCl}$$

Figure III.30. An organomercurial derivative of Sepharose

which could subsequently be eluted by passing a solution of 0·5M cysteine into the column, arresting the flow for 1 to 2 hours, and then collecting the effluent. Solutions of low pH or of complexing agents such as EDTA released the bound protein only slowly. The adsorbent was also effective in resolving thyroglobulin molecules from purified preparations that have no free sulphydryl groups from molecules that have a single free thiol group.[99] Elution could be effected with buffers containing chelating or reducing agents.

Ruiz-Carrillo and Allfrey[100] have purified calf thymus histone F3 in high yield by affinity chromatography on an identical organomercurial-Sepharose adsorbent. Of the five major classes of histone that occur in mammalian cell nuclei only fraction F3 contains cysteine.[101,102] The presence of reactive

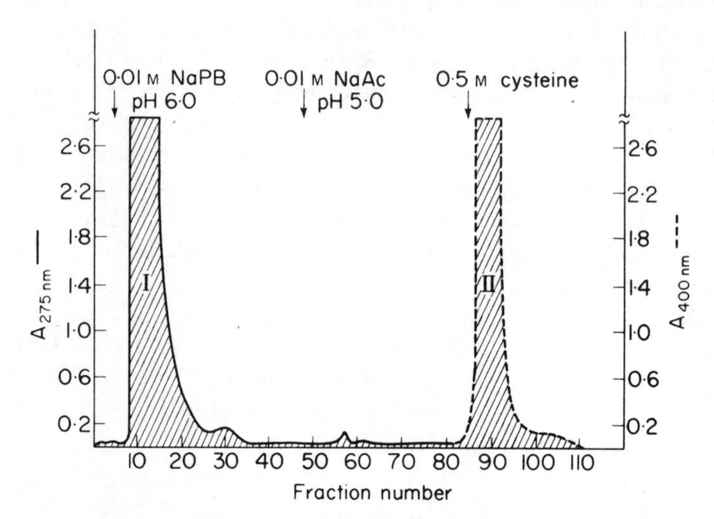

Figure III.31. Purification of calf thymus histone F3 by affinity chromatography. A mixture of histones F3, F2A2 and F1 was applied to an organomercurial-Sepharose 4B column (30 cm × 1·5 cm) and eluted stepwise with 0·01M Na phosphate buffer, pH 6·0; 0·01M Na acetate buffer, pH 5·0; and 0·5M cysteine. Fractions of 7 ml were collected. The absorbancy at 275 nm is plotted against fraction number as a solid line. The protein content of later fractions was determined by turbidity assay, adding 250 μl of 25% trichloroacetic acid to 50 μl of each fraction. After 70 minutes the turbidity was measured at 400 nm. Peak I contains histone F2A2 and F1. Peak II is pure histone F3. Reproduced with permission from A. Ruiz-Carrello and V. G. Allfrey, *Arch. Biochem. Biophys.*, **154**, 185 (1973)

sulphydryl groups in fraction F3 suggests its resolution from fractions F1 and F2A2 might be achieved by affinity chromatography on an organomercurial adsorbent. Fig. III.31 shows the elution profile of a calf thymus histone fraction on the column. The first peak, consisting mainly of histones F2A2 and F1, was eluted with equilibration buffer. A small peak of non-specifically adsorbed protein was eluted with 0·01M sodium acetate buffer, pH 5·0, whence elution of the F3 histone was effected with a freshly prepared solution of 0·5M cysteine. Elution with 0·5M solutions of β-mercaptoethanol or dithiothreitol was equally effective.

Sluyterman and Wijdenes[103] coupled *p*-aminophenylmercuric acetate to agarose by the cyanogen bromide method and found the adsorbent effective for the resolution of enzymically active and inactive papain. Active papain was bound strongly and elution could be achieved with solutions containing 0·5 mM $HgCl_2$ or 0·5 mM β-mercaptoethanol. The procedure could be scaled up to encompass preparative separations.

Sulphydryl derivatives of cellulose[104] and cross-linked dextran[105] have also been prepared, although the procedures were complicated and the adsorbents less effective than the agarose derivatives described above. Mercurated dextran was found to fractionate mononucleotides according to their affinity for organomercurial Hg^+.[106] Elution with 0·1M borate buffer of increasing pH released the nucleotides from the adsorbent in the sequence cytidine 5′-monophosphate, adenosine 3′-monophosphate, guanosine 5′-monophosphate and thymidine 5′-monophosphate.

D. THE HYDROLASES (PROTEASES)

In recent years, the proteolytic enzymes have received considerable attention, partly because of their availability from commercial sources and their great utility as tools for the study of protein structure, and partly because of the interest in their chemical nature and mechanism of action.

The catalytic action of the proteolytic enzymes was demonstrated with synthetic peptides and their derivatives, which proved to be suitable models for their more complex naturally occurring substrates. The reaction catalysed is the hydrolysis of a peptide, amide or ester bond. All proteolytic enzymes attack peptide bonds, although their specificity depends not on the chain length of the substrate, but on the nature of the amino acid side-chains and the presence or absence of nearby ionic groups. Furthermore, although most of the naturally occurring substrates of the proteolytic enzymes are proteins or polypeptides in which all the peptide bonds involve α-carboxyl or α-amino groups and in which the optically active amino acids are of the so-called natural or L-configuration, the action of many of the proteases is not restricted to such bonds.

Exopeptidases hydrolyse peptide bonds adjacent to terminal amino or carboxyl groups and include the carboxypeptidases, aminopeptidases and the dipeptidases. Conversely, the endopeptidases can attack centrally located bonds in addition to the hydrolysis of terminal peptide bonds. This class includes pepsin, trypsin, chymotrypsin and papain.

The isolation and purification of proteases is usually a long and complicated process. Autodigestion during classical isolation procedures can result in low yields. A simpler and more rapid procedure should greatly enhance the probability of obtaining a preparation free of degradation products and aggregated species. Thus, a single passage through a suitable affinity adsorbent may greatly facilitate the isolation of proteolytic enzymes from crude systems.

In general, the simplest approach to the group separation of proteolytic enzymes by affinity chromatography would be to use proteins as the insolubilized ligands. Thus, although the affinity adsorbent would be relatively unspecific and bind nearly all classes of protease, the method may be particularly useful for endopeptidases. A rapid chromatographic procedure for the isolation and purification of proteases from malted wheat flour has been

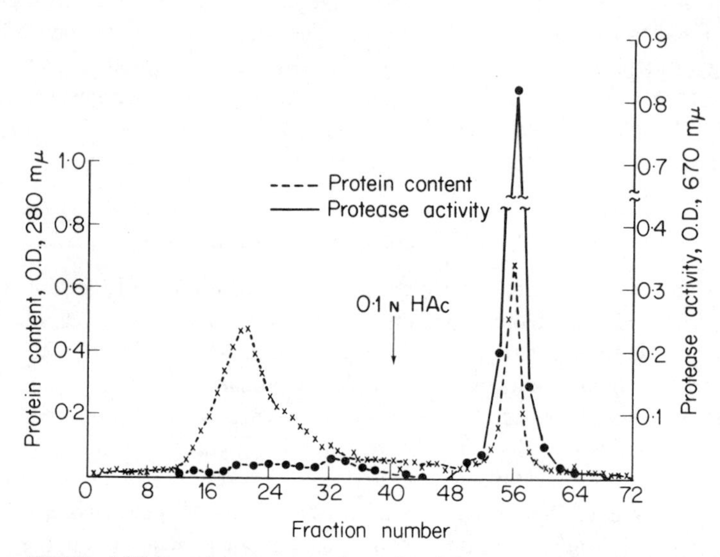

Figure III.32. Affinity chromatography of wheat proteases on haemoglobin-Sepharose. A crude extract of malted wheat flour was applied to a column of haemoglobin-Sepharose equilibrated with 0·05M acetate buffer, pH 5·5. The adsorbed proteases were eluted with 0·1N acetic acid. Reproduced with permission from G. K. Chua and W. Bushuk, *Biochem. Biophys. Res. Commun.*, **37**, 545 (1969)

described.[107] The separation was achieved by passing the crude extract through a column of haemoglobin-Sepharose. Non-adsorbed proteins were washed off with the starting buffer and the adsorbed proteins were eluted with 0·1M acetic acid (Fig. III.32). The recovery of protein and proteolytic activity was over 90% and a two-fold increase in specific activity was achieved. Analysis of the eluants by disc electrophoresis showed that all of the non-proteolytic components were separated from the proteolytically active proteins, and that the active peak comprised three major and one minor protein with similar mobilities. The efficiency of this adsorbent in the purification of proteases was particularly striking in view of the very low proteolytic activity of wheat and the tendency of its proteases to aggregate with other proteins.

Another approach to the isolation of proteases by affinity chromatography exploits the specific inhibitory action of the naturally occurring protein inhibitors. They form stable complexes with the corresponding proteases at neutral pH but are dissociable at low pH values. These properties of the natural proteinase inhibitors, combined with the ease of coupling proteins to Sepharose, suggests their use for the isolation and purification of proteases.

Chicken ovomucoid binds bovine trypsin in a 1:1 molar ratio to form a very stable complex which dissociates only when the pH is lowered from neutral to acidic pH. Ovomucoid covalently linked to Sepharose exhibits the same specificity as soluble ovomucoid and will bind trypsin but not α-chymotrypsin.[108] This technique was found to be particularly useful for removing contaminants from commercial trypsin preparations. An affinity adsorbent for trypsin is especially desirable since this enzyme exhibits a tendency to autodigestion. The content of active enzyme in commercial preparations, determined by active site titration, lies within the range 60–80%.[109] The specific activity of trypsin purified by affinity chromatography on ovomucoid-Sepharose was about 10% higher than the specific activity of the initial crystalline trypsin from the commercial source.

This adsorbent has also been successfully applied to the isolation of bovine trypsin, porcine trypsin and trypsin from pancreatic extracts of the spiny Pacific dogfish.[110] Furthermore, ovomucoid-Sepharose was capable of resolving α- and β-trypsin, the two predominant species found in conventional preparation of bovine trypsin.[111] However, for this method to be generally applicable to the purification of trypsins, the adsorbent must not bind other proteins usually present in crude preparations of trypsin. Neither bovine trypsinogen nor chymotrypsin were retarded by this adsorbent at pH 7·5 and, in fact, none of the proteins found in acidified activated pancreatic juice, other than trypsin, were adsorbed under these conditions (Fig. III.33). The technique represents a facile approach to the large-scale purification of trypsins by affinity chromatography.

Soybean trypsin inhibitor-Sepharose (STI-Sepharose) has been successfully employed for the purification of several proteases from activated bovine pancreatic juice.[112] Equilibration of the STI-Sepharose with tris-HCl buffer containing 0·5M NaCl results in retardation of carboxypeptidases A and B from the bulk of the inert protein, whilst 0·1M sodium acetate pH 3·0 eluted trypsin and chymotrypsin. No trypsin activity could be detected in the carboxypeptidase fraction and only 0·4% contamination of carboxypeptidase B activity with chymotrypsin was found. Using STI-Sepharose one may obtain approximately 200 mg carboxypeptidase A with a single preparation. Furthermore, the STI column can be developed in a few hours and the resulting enzyme fraction requires only concentrating by ultrafiltration. This

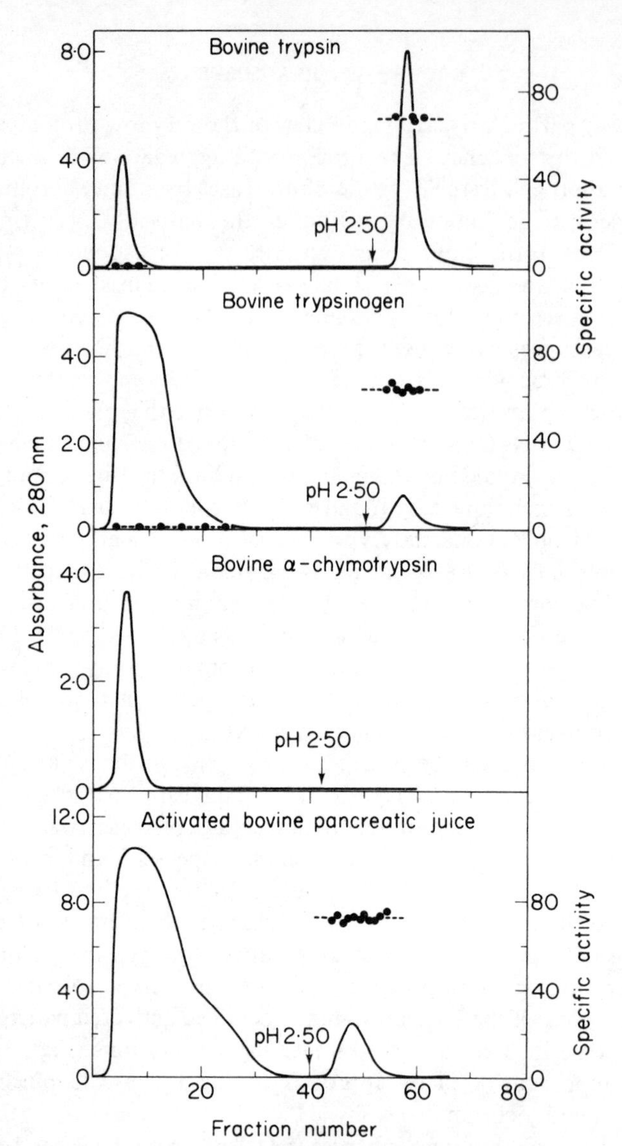

Figure III.33. Chromatography of bovine pancreatic enzymes on a column (1·5 cm × 25 cm) of chicken ovomucoid-Sepharose. The column was equilibrated with 0·10M tris-chloride, 0·05M calcium chloride, 0·50M potassium chloride (pH 7·50) at 60 ml/hour. The enzymes were dissolved in 10 ml of buffer, applied to the column, and washed until no further protein was eluted. Active trypsin was then eluted with 0·10M potassium formate, 0·50M potassium chloride (pH 2·50) and 10 ml fractions were collected. The following amounts of enzyme were applied: bovine trypsin, 250 mg; bovine trypsinogen, 60 mg, containing 2–5% trypsin contamination; bovine α-chymotrypsin, 50 mg; activated bovine pancreatic juice, 20 ml. The specific activity towards α-*N*-benzoyl-L-arginine ethyl ester (μmole min^{-1} mg^{-1}) is indicated by the black dots. Reproduced with permission from N. C. Robinson, R. W. Tye, H. Neurath and K. A. Walsh, *Biochemistry*, **10**, 2744 (1971)

purification scheme could also be extended to produce α- and β-trypsins from the pancreatic juice used for isolation of the carboxypeptidases. The chromatographic system developed by Robinson *et al.*,[110] for the separation of these forms of bovine trypsin using chicken ovomucoid covalently attached to Sepharose, could be utilized to resolve the trypsin and chymotrypsin eluted by 0·1M sodium acetate from the STI-Sepharose column. The trypsin activity would be fractionated into the α- and β-forms, while the chymotrypsin activity would appear in the void volume and would thus be unfractionated.

Porath and Sundberg[113] have shown that the proteolytic enzymes in a pancreas extract can also be fractionated on STI-Sepharose. The pancreas extract was applied to the adsorbent in tris-HCl buffer, pH 7·5, whence elution of chymotrypsin followed by trypsin was accomplished with a pH gradient of increasing acidity. Alternatively, chymotrypsin may be displaced by tryptamine, and trypsin by benzamidine. Sundberg and Christiansen[114] have isolated elastin from a pancreas extract by single-step affinity chromatography on potato inhibitor-Sepharose.

Fritz *et al.*[115] have fractionated kallikrein and plasmin by chromatography on a resin of kallikrein inhibitor cross-linked to a copolymer of ethylenemaleic acid. Furthermore, Fritz *et al.*[116] have isolated human and porcine kallikrein by affinity chromatography on Kunitz Soybean Inhibitor-cellulose. A partially purified extract of kallikrein was applied to a column of the SBI-cellulose. Non-adsorbed proteins were washed off, whence the kallikreins were eluted with 2M urea or 0·05M benzamidine, a competitive inhibitor. Depending on the specific activity of the starting material, a 100 to 350-fold purification was achieved with a yield of 60–91%. The use of competitive, low molecular weight trypsin–kallikrein inhibitors to elute the kallikreins from the SBI-cellulose has the advantage that they stabilize the eluted kallikreins, which are unstable in highly purified form.

Fritz *et al.*[117] have demonstrated the usefulness of affinity chromatography in the isolation of the natural inhibitors of some proteases. Thus, they cross-linked trypsin, α-chymotrypsin and kallikrein to ethylenemaleic acid and resolved their complementary natural inhibitors. Feinstein[118] has purified chicken ovoinhibitor from ovomucoid by single-step affinity chromatography on bovine chymotrypsin-Sepharose, and Stewart and Doherty[119] have exploited to advantage the interaction of trypsin inhibitors with trypsinogen and trypsin. They prepared trypsinogen-agarose and trypsin-agarose columns and have compared the chromatographic behaviour of crude peanut trypsin inhibitor (PTI) and Kunitz soybean trypsin inhibitor (SBTI) on these columns. Only the PTI was bound to the trypsinogen-agarose columns while both PTI and SBTI were bound to the trypsin-agarose columns. The PTI could be eluted from the trypsinogen-agarose columns under milder conditions than from the trypsin-agarose columns. The trypsin-agarose adsorbent has also been used successfully for the isolation of a basic trypsin-inhibitor (Kunitz inhibitor) from bovine pancreas[120] and ovary.[121]

One serious disadvantage of the use of immobilized trypsin has been highlighted by Hixon and Laskowski;[122] some plant protease inhibitors were extensively modified when isolated by affinity chromatography on immobilized trypsin.[123] Stewart and Doherty[119] suggest that these proteases could be isolated in unmodified form by affinity chromatography on insolubilized trypsinogen. Such adsorbents should also be useful for the purification of those protease inhibitors which bind to trypsinogen, such as peanut trypsin inhibitor, basic pancreatic trypsin inhibitor and the cow colostrum inhibitor.[124]

Immobilized proteins can thus be utilized as very effective affinity adsorbents with some specificity being achieved by the use of the natural protein inhibitors. Greater selectivity of the adsorbents can be accomplished by careful selection of proteins or peptides with amino-acid sequences relevant to the specificity of the protease to be isolated. The use of dipeptides as insolubilized ligands permits considerable latitude in specificity, whilst also affording the advantages of a chemically defined polymer. In practical terms, these peptides can be coupled to Sepharose directly or by an extension arm of some kind. The choice of the peptide will reflect the bond specificity of the protease to be isolated and hence it is necessary to consider this aspect of the problem before an approach to affinity chromatography can be made. Carboxypeptidase A, for example, shows preferential action on peptides which contain carboxyl-terminal aromatic amino acids, phenylalanine, tyrosine or tryptophan or on branched-chain aliphatic amino acids such as leucine and isoleucine. The terminal carboxyl group must be unsubstituted. The enzyme is completely inactive towards substrates containing carboxyl-terminal arginine and lysine and towards any peptides containing proline or hydroxyproline as terminal or penultimate. Conversely, carboxypeptidase B acts uniquely on peptides containing arginine, lysine or ornithine as carboxyl-terminal group and is inhibited by the corresponding structural analogues δ-amino-valeric acid and ε-amino-caproic acid.

These considerations allow a first approximation to the affinity chromatography of carboxypeptidases to be attempted. Some selectivity can be achieved by linking a single amino-acid residue relevant to the bond specificity of the enzyme to Sepharose via a hydrocarbon extension arm. An arginine-based adsorbent, prepared by coupling the azide of CM-Sephadex to ε-amino-caproyl-D-arginine was shown to be a competent matrix for the affinity chromatography of bovine carboxypeptidase B.[125] The enzyme was eluted specifically by a linear gradient of the structural analogue, ε-amino-caproic acid (Fig. III.34). Furthermore, ε-amino-caproyl-D-tryptophan-Sepharose has been used to resolve carboxypeptidases A and B from a partially purified preparation of bovine pancreatic juice.[126] Both carboxypeptidases were retarded by the modified Sepharose, although carboxypeptidase B was eluted by 0·15M NaCl and carboxypeptidase A required a higher salt concentration. The purity of carboxypeptidase B thus prepared varied from 70–90% as judged by its specific activity towards benzoyl-glycine-L-arginine.

A more general approach has been to link dipeptide inhibitors to an insoluble support. Uren[127] has described the behaviour of one such high-capacity adsorbent, *N*(2-ethyl-cellulose)-glycyl-D-phenylalanine, for the group separation of the proteolytic enzymes, trypsin, chymotrypsin A and B, and carboxypeptidase A and B, by affinity chromatography. This adsorbent will bind carboxypeptidase A by virtue of its enzymatic specificity, for the enzyme is not retarded by unsubstituted aminoethyl-cellulose nor by cellulose derivatives which do not conform to the specificity requirements, such as glycyl-D-arginine. The homologous carboxypeptidase B, on the other hand, is bound

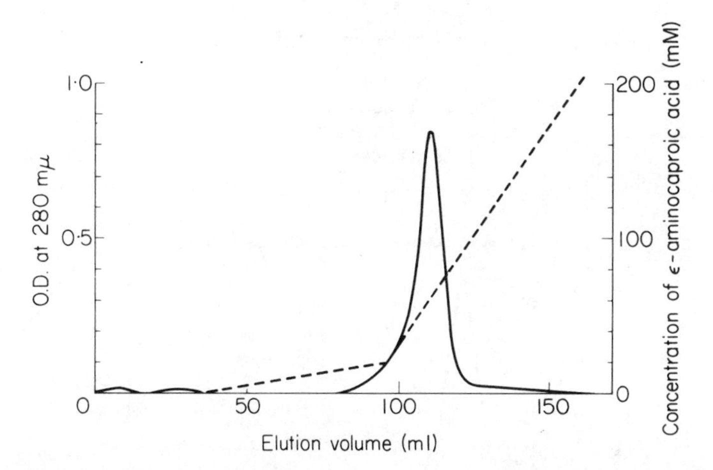

Figure III.34. Elution of bovine carboxypeptidase B from an ε-aminocaproyl-D-arginine-coupled CM-Sephadex column by ε-aminocaproic acid. Amount of enzyme applied: 10 mg, column size: 1·0 cm × 11·2 cm, solid line: optical density at 280 mμ, eluant: 10 mM sodium phosphate buffer containing 100 mM NaCl, a gradient from 0 to 200 mM ε-aminocaproic acid was applied as indicated by the dotted line, pH 7·05 throughout. Reproduced with permission from H. Akanuma, A. Kasuga, T. Akanuma and M. Yamasaki, *Biochem. Biophys. Res. Commun.*, **45**, 27 (1971)

by the glycyl-D-arginine derivative but not by unsubstituted aminoethyl-cellulose or by a glycylglycine-derivative.

Further examination of the specificity requirements of the carboxypeptidases shows that careful selection of the component amino acids of the dipeptide can lead to adsorbents of even higher specificity. Thus, an effective adsorbent for carboxypeptidase A was prepared by coupling the dipeptide L-tyrosyl-D-tryptophan directly to Sepharose.[128] The column did not retain carboxypeptidase B, whereas carboxypeptidase A was only eluted when 0·1M acetic acid was applied (Fig. III.35). These facts are in direct concordance with the known enzymatic specificity of the enzymes. Furthermore, specific adsorbents for bovine carboxypeptidase B and porcine carboxypeptidase B have been obtained by coupling the dipeptides L-leucyl-D-arginine and D-alanyl-L-arginine respectively to Sepharose.[129]

Kasai and Ishii[109] describe a simple procedure for the preparation of an adsorbent competent for trypsin. A mixture of di- and tri-peptides containing

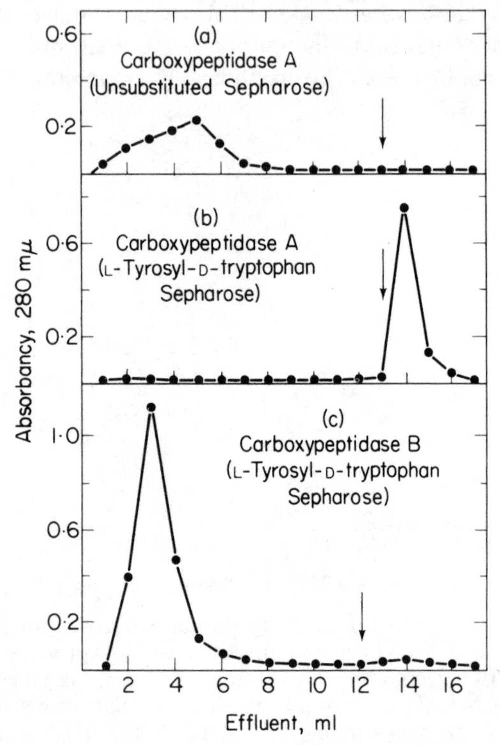

Figure III.35. Affinity chromatography of carboxypeptidase A on a column (0.5 cm × 6 cm) of Sepharose coupled with L-tyrosine-D-tryptophan. The buffer used was 0·05M tris-HCl, pH 8·0, containing 0·3M NaCl. About 1 mg of pure carboxypeptidase A [a, b] and 1·8 mg of carboxypeptidase B (c), in 1 ml of the same buffer, were applied to the columns. Elution was accomplished with 0·1M acetic acid (arrow). Reproduced with permission from P. Cuatrecasas, M. Wilchek and C. B. Anfinsen, *Proc. Nat. Acad. Sci., U.S.A.*, **61**, 636 (1968)

only L-arginine as carboxyl-terminal (Arg-peptides) were obtained by digestion of salmine sulphate with trypsin. The Arg-peptides were subsequently separated on an ion-exchange column and coupled to CNBr-activated Sepharose.

Bovine trypsin was strongly retained on the adsorbent at pH 7·3 and could be eluted with over 90% recovery by lowering the pH. The adsorbent was able to remove enzymatically inactive protein from a commercial preparation of trypsin. Chymotrypsin, trypsinogen and ribonuclease I were not retained by the matrix. Furthermore, adsorbed trypsin could be eluted with 0·01 M benzamidine-HCl, a competitive inhibitor, whilst it was not eluted by an NaCl solution of similar ionic strength. On the other hand, L-arginyl-Sepharose, prepared by coupling L-arginine directly to Sepharose, did not retard trypsin.

A similar approach has been employed for the design of an adsorbent for papain, an enzyme with catalytic activity towards basic bonds but without the restricted action shown by trypsin. Blumberg *et al.*[130] have attached the inhibitory peptide glycylglycyl(*O*-benzoyl)-L-tyrosyl-L-arginine to Sepharose and shown that the enzyme binds in the presence of 20 mM EDTA and can be eluted with buffers of low ionic strength. The matrix can be employed for resolving catalytically active from inactive enzyme, since half of the protein content of a commercially available preparation was bound whilst the inactive half was unretarded. The specific activity of the enzyme was doubled and the purified enzyme had a thiol content of 0·94 groups per molecule. L-Lysyl-Sepharose is a useful adsorbent for plasminogen, a protease with preference for peptide bonds bounded by arginine or lysine residues,[131] and poly-L-lysine is a novel matrix for the isolation of bovine pepsinogen and pepsin.[132]

Considerable specificity in the affinity adsorbent can be introduced by coupling chemically synthesized small molecular weight inhibitors and substrate analogues to Sepharose via an extended hydrocarbon spacer. A number of proteolytic enzymes, such as α-chymotrypsin and carboxypeptidase A, are capable of binding but not hydrolysing, the enantiomeric substrate analogue. D-Tryptophan methyl ester is thus an ideal ligand for the affinity chromatography of α-chymotrypsin. When this inhibitory analogue is coupled directly to Sepharose, incomplete and ineffective resolution of the enzyme occurs.[128] However, the interposition of a hexacarbon spacer between the ligand and the matrix backbone dramatically improved the strength of binding (Fig. III.36). The adsorbent was particularly useful for removing impurities from commercial preparations of α-chymotrypsin and for the resolution of native and DFP-treated enzyme. The relatively unfavourable affinity constant of the *N*-ϵ-aminocaproyl-D-tryptophan methyl ester ($K_i \sim 0{\cdot}1$ mM) is compensated for by the covalent attachment of large amounts of the inhibitor to the Sepharose. The resulting ligand concentration of 10 μmoles/ml Sepharose produced an effective ligand concentration of 10 mM in the column.

The adsorbent was almost completely specific for chymotryptic proteases, and did not retard pancreatic ribonuclease, subtilisin or DFP-trypsin. The technique could thus be a useful way of removing chymotryptic impurities from other proteases used in structural studies, where even small traces can lead to unpredicted cleavages. It is interesting to note that chymotrypsinogen A is slightly but significantly retained by the chymotrypsin-specific column, suggesting that this precursor is capable of a weak interaction with the analogue. The Sepharose-ε-aminocaproyl-D-tryptophan methyl ester adsorbent also binds pineapple stem bromelain strongly and provides a simple and efficient means for its isolation from crude extracts.[133]

Another approach has been developed for the isolation of mammalian chymotrypsin-like enzymes.[134] The phenyl moiety of 4-phenyl-butylamine is

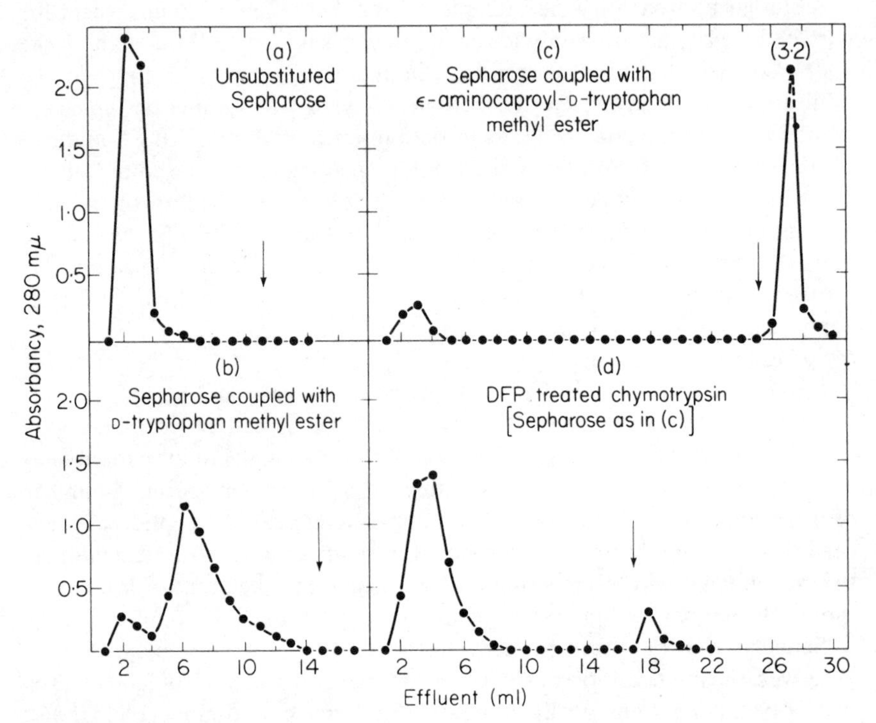

Figure III.36. Affinity chromatography of α-chymotrypsin on inhibitor Sepharose columns. The columns (0·5 cm × 5 cm) were equilibrated and run with 0·05M tris-Cl buffer, pH 8·0, and each sample (2·5 mg) was applied in 0·5 ml of the same buffer. One-millilitre fractions were collected, the flow rate was about 40 ml per hour, and the experiments were performed at room temperature. α-Chymotrypsin was eluted with 0·1M acetic acid, pH 3·0 (arrows). Peaks preceding the arrows in *B, C, D* were devoid of enzyme activity. Reproduced with permission from P. Cuatrecasas, M. Wilchek and C. B. Anfinsen, *Proc. Nat. Acad. Sci., U.S.A.*, **61**, 636 (1968)

known to bind to the active site of α-chymotrypsin and act as a weak competitive inhibitor with a K_i of 2·8 mM. Sepharose, to which 4-phenylbutylamine has been covalently attached, has been found selectively to adsorb α-chymotrypsin, chymotrypsin B and chymotrypsin-like enzymes isolated from the pancreas of the moose (*Alces alces*) while rejecting porcine trypsin, trypsinogen, subtilisin and chemically modified α-chymotrypsin. Chymotrypsinogen A and bovine trypsin were slightly retarded but not retained by the column.

Jameson and Elmore[135] have separated bovine α- and β-trypsin by affinity chromatography on Sepharose 4B or cellulose to which the potent competitive inhibitor, *p*-(*p*-aminophenoxypropoxy)benzamidine, has been coupled by its terminal amino group. Furthermore, bovine thrombin of high purity has been isolated on *p*-chloro-benzylamido-ϵ-aminocaproyl-Sepharose.[136] One disadvantage of this method is the low solubility of *p*-chlorobenzylamine in aqueous solutions, which complicates the synthesis of the agarose adsorbents. Schmer[137] has circumvented the problem by coupling *p*-aminobenzamidine, which is readily soluble in water, to ϵ-aminohexanoyl-Sepharose with a water-soluble carbodiimide. Although this inhibitor is not as specific for thrombin as *p*-chlorobenzylamine,[138] the resulting benzamidine-Sepharose is a potent adsorbent for thrombin. The specific activity of crude bovine thrombin was increased 16-fold in a single step when eluted with 0·2M benzamidine. However, plasmin is also retained and co-eluted with thrombin by the benzamidine, which has about the same K_i for both enzymes. This contaminant can be removed by chromatography on SE-Sephadex either before or after the affinity chromatography.

In summary, affinity adsorbents have found many applications in the field of protease enzymology. Besides the obvious use in the isolation and purification of proteases, affinity adsorbents have been employed for the resolution of active and inactive enzymes, and the separation of native and chemically modified forms. The removal of contaminants from commercial preparations is important for analytical and structural studies where defined cleavages of proteins are required. Likewise, the separation of isoenzymic forms, e.g., α- and β-trypsins, is important. Furthermore, a column of Butesin-Sepharose has been used as the basis for a new affinity chromatography assay for human plasminogen in plasma.[139] Plasminogen is bound to the adsorbent whilst unretarded proteins, including plasmin, are washed off with phosphate-saline buffer, pH 7·4. Adsorbed plasminogen is eluted with 0·1M lysine and determined by its absorbance at 280 nm.

REFERENCES

1. Dalziel, K. (1970): in *Pyridine Nucleotide Dependent Dehydrogenases* (Ed. H. Sund), Springer-Verlag, Heidelberg, p. 3.
2. Holbrook, J. J., and Gutfreund, H. (1973): *FEBS Lett.*, **31**, 157.
3. Stinson, R. A., and Holbrook, J. J. (1973): *Biochem. J.*, **131**, 719.
4. Sarma, R. H., Ross, V., and Kaplan, N. O. (1968): *Biochemistry*, **7**, 3052.
5. Velick, S. F. (1970): in *Light and Life* (Eds. W. D. McElroy and B. Glass), Johns Hopkins Press, Baltimore, p. 108.
6. Adams, M. J., McPherson, A., Rossmann, M. G., Shevitz, R. W., Smiley,

I. E., and Wonacott, A. J. (1970): in *Pyridine Nucleotide Dependent Dehydrogenases* (Ed. H. Sund), Springer-Verlag, Heidelberg, p. 157.

7. Levy, H. R., Talalay, P., and Vennesland, B. (1962): in *Progress in Stereochemistry*, **3** (Eds. P. B. D. de la Mare and W. Klyne), Butterworth, London, p. 239.
8. Vennesland, B., and Westheimer, F. H. (1954): in *The Mechanism of Enzyme Action* (Eds. W. D. McElroy and B. Glass), Johns Hopkins Press, Baltimore, p. 357.
9. Dean, P. D. G., and Lowe, C. R. (1972): *Biochem. J.*, **127**, 11P.
10. Axen, R., Ernbäck, S., and Porath, J. (1967): *Nature* (*London*), **215**, 491.
11. Porath, J., and Fornstedt, N. (1970): *J. Chromatogr.*, **51**, 479.
12. Lowe, C. R., and Dean, P. D. G. (1971): *FEBS Lett.*, **14**, 313.
13. Hocking, J. D., and Harris, J. I. (1973): *FEBS Lett.*, **34**, 280.
14. Gilham, P. T. (1971): *Methods Enzymol.*, **21**, 191.
15. Novais, F. M. (1971): Ph.D. Thesis, University of Birmingham.
16. Larsson, P.-O., and Mosbach, K. (1971): *Biotechnol. Bioeng.*, **13**, 393.
17. Weibel, M. K., Weetall, H. H., and Bright, H. J. (1971): *Biochem. Biophys. Res. Commun.*, **44**, 347.
18. Lowe, C. R., Harvey, M. J., Craven, D. B., and Dean, P. D. G. (1973): *Biochem. J.*, **133**, 499.
19. Lowe, C. R., Mosbach, K., and Dean, P. D. G. (1972): *Biochem. Biophys. Res. Commun.*, **48**, 1004.
20. Nicolas, J. C., Pons, M., Descomps, B., and Crastes de Paulet, A. (1972): *FEBS Lett.*, **23**, 175.
21. Lowe, C. R., Harvey, M. J., Craven, D. B., Kerfoot, M. A., Hollows, M. E., and Dean, P. D. G. (1973): *Biochem. J.*, **133**, 507.
22. O'Carra, P., and Barry, S. (1972): *FEBS Lett.*, **21**, 281.
23. Cuatrecasas, P. (1970): *J. Biol. Chem.*, **245**, 3059.
24. Arsenis, C., and McCormick, D. B. (1966): *J. Biol. Chem.*, **241**, 330.
25. Collier, R., and Kohlhaw, G. (1971): *Anal. Biochem.*, **42**, 48.
26. Gauldie, J., and Hillcoat, B. L. (1972): *Biochim. Biophys. Acta*, **268**, 35.
27. Ryan, E., and Fottrell, P. F. (1972): *FEBS Lett.*, **23**, 73.
28. Mosbach, K., Guilford, H., Ohlsson, R., and Scott, M. (1972): *Biochem. J.*, **127**, 625.
29. Hines, W. J. W., and Smith, M. J. H. (1964): *Nature* (*London*), **201**, 192.
30. Ohlsson, R., Brodelius, P., and Mosbach, K. (1972): *FEBS Lett.*, **25**, 234.
31. Guilford, H., Larsson, P.-O., and Mosbach, K. (1972): *Chem. Scripta*, **2**, 165.
32. Craven, D. B., Harvey, M. J., Lowe, C. R., and Dean, P. D. G. (1973): *Eur. J. Biochem.*, submitted for publication.
33. Gilham, P. T. (1971): *Methods Enzymol.*, **21**, 191.
34. Wider de Xifra, E. A., Mendiara, S., and Battle, A. M. del C. (1972): *FEBS Lett.*, **27**, 275.
35. Lowe, C. R., Harvey, M. J., Craven, D. B., and Dean, P. D. G. (1973): *Biochem. J.*, **133**, 499.
36. Harvey, M. J., Lowe, C. R., and Dean, P. D. G. (1973): *Eur. J. Biochem.*, submitted for publication.
37. Berglund, O., Karlström, O., and Reichard, P. (1969): *Proc. Nat. Acad. Sci. U.S.A.*, **62**, 829.
38. Berglund, O., and Eckstein, F. (1972): *Eur. J. Biochem.*, **28**, 492.
39. Robison, G. A., Butcher, R. W., and Sutherland, E. W. (1971): in *Cyclic AMP*, Academic Press, New York, London.

40. Walsh, D. A., Perkins, J. P., and Krebs, E. G. (1968): *J. Biol. Chem.*, **243**, 3763.
41. Corbin, J. D., Reimann, E. M., Walsh, D. A., and Krebs, E. G. (1970): *J. Biol. Chem.*, **245**, 4849.
42. Tesser, G. I., Fisch, H.-U., and Schwyzer, R. (1972): *FEBS Lett.*, **23**, 56.
43. Wilchek, M., Salomon, Y., Lowe, M., and Selinger, Z. (1971): *Biochem. Biophys. Res. Commun.*, **45**, 1177.
44. Barker, R., Olsen, K. W., Shaper, J. H., and Hill, R. L. (1972): *J. Biol. Chem.*, **247**, 7135.
45. Trayer, I. P., and Hill, R. L. (1971): *J. Biol. Chem.*, **246**, 6666.
46. Danenberg, P. V., Langenbach, R. J., and Heidelberger, C. (1972): *Biochem. Biophys. Res. Commun.*, **49**, 1029.
47. Robberson, D. L., and Davidson, N. (1972): *Biochemistry*, **11**, 553.
48. Jackson, R. J., Wolcott, R. M., and Shiota, T. (1973): *Biochem. Biophys. Res. Commun.*, **51**, 428.
49. McCormick, D. B. (1962): *J. Biol. Chem.*, **237**, 959.
50. McCormick, D. B., and Butler, R. C. (1962): *Biochim. Biophys. Acta*, **65**, 326.
51. McCormick, D. B., Arsenis, C., and Hemmerich, P. (1963): *J. Biol. Chem.*, **238**, 3095.
52. Arsenis, C., and McCormick, D. B. (1964): *J. Biol. Chem.*, **239**, 3093.
53. Arsenis, C., and McCormick, D. B. (1966): *J. Biol. Chem.*, **241**, 330.
54. Braunstein, A. E. (1964): in *Vitamins and Hormones*, Vol. 22 (Eds. R. S. Harris, I. G. Wool and J. A. Loraine), Academic Press, New York, p. 451.
55. Collier, R., and Kohlhaw, G. (1971): *Anal. Biochem.*, **42**, 48.
56. Miller, J. V., and Thompson, E. B. (1971): *Fed. Proc.*, **30**, 516.
57. Miller, J. V., Cuatrecasas, P., and Thompson, E. B. (1972): *Biochim. Biophys. Acta*, **276**, 407.
58. Miller, J. V., Cuatrecasas, P., and Thompson, E. B. (1971): *Proc. Nat. Acad. Sci. U.S.A.*, **68**, 1014.
59. Ryan, E., and Fottrell, P. F. (1972): *FEBS Lett.*, **23**, 73.
60. Hitchings, G. H., and Burchall, J. J. (1965): *Adv. Enzymol.*, **27**, 417.
61. Newbold, P. C. H., and Harding, N. G. L. (1971): *Biochem. J.*, **124**, 1.
62. Baker, B. R. (1967): in *The Design of Active Site Directed Irreversible Enzyme Inhibitors*, John Wiley, New York.
63. Dann, J. G., Harding, N. G. L., Newbold, P. C. H., and Whiteley, J. M. (1972): *Biochem. J.*, **127**, 28P.
64. Kaufman, B. T., and Pierce, J. V. (1971): *Biochem. Biophys. Res. Commun.*, **44**, 608.
65. Mell, G. P., Whiteley, J. M., and Huennekens, F. M. (1968): *J. Biol. Chem.*, **243**, 6074.
66. Erickson, J. S., and Mathews, C. K. (1971): *Biochem. Biophys. Res. Commun.*, **43**, 1164.
67. Chello, P. L., Cashmore, A. R., Jacobs, S. A., and Bertino, J. R. (1972): *Biochim. Biophys. Acta*, **268**, 30.
68. Gauldie, J., and Hillcoat, B. L. (1972): *Biochim. Biophys. Acta*, **268**, 35.
69. Poe, M., Greenfield, N. J., Hirshfield, J. M., Williams, M. N., and Hoogsteen, K. (1972): *Biochemistry*, **11**, 1024.
70. Salter, D. N., Ford, J. E., Scott, K. J., and Andrews, P. (1972): *FEBS Lett.*, **20**, 302.
71. Green, N. M. (1963): *Biochem. J.*, **89**, 585.
72. Melamed, M. D., and Green, N. M. (1963): *Biochem. J.*, **89**, 591.
73. McCormick, D. B. (1965): *Anal. Biochem.*, **13**, 194.

74. Cuatrecasas, P., and Wilchek, M. (1968): *Biochem. Biophys. Res. Commun.*, **33**, 235.
75. Wolpert, J. S., and Ernst-Fonberg, M. L. (1973): *Anal. Biochem.*, **52**, 111.
76. Cox, E. A., and Lowe, C. R. (1973): *FEBS Lett.*, in preparation.
77. Allen, R. H., and Majerus, P. W. (1972): *J. Biol. Chem.*, **247**, 7709.
78. Olesen, H., Hippe, E., and Haber, E. (1971): *Biochim. Biophys. Acta*, **243**, 66.
79. Yamada, R. H., and Hogenkamp, H. P. C. (1972): *J. Biol. Chem.*, **247**, 6266.
80. Allen, R. H., and Majerus, P. W. (1972): *J. Biol. Chem.*, **247**, 7695.
81. Allen, R. H., and Majerus, P. W. (1972): *J. Biol. Chem.*, **247**, 7702.
82. Conway, T. P., and Muller-Eberhard, U. (1973): *Fed. Proc.*, **32**, 1382.
83. Alberts, B. M., Amodio, F. J., Jenkins, M., Gutman, E. D., and Ferris, F. L. (1968): *Cold Spring Harbour Symp. Quant. Biol.*, **33**, 289.
84. Litman, R. M. (1968): *J. Biol. Chem.*, **243**, 6222.
85. Alberts, B. M., and Frey, L. M. (1970): *Nature* (*London*), **227**, 1313.
86. Bautz, E. K. F., and Dunn, J. J. (1972): in *Procedures in Nucleic Acid Research*, Vol. II (Eds. G. L. Cantoni and D. R. Davies), Harper and Row, New York.
87. Alberts, B. M., and Herrick, G. (1972): *Methods Enzymol.*, **21**, 198.
88. Salas, J., and Green, H. (1971): *Nature* (*London*), **229**, 165.
89. Cavalieri, L. F., and Carroll, E. (1970): *Proc. Nat. Acad. Sci. U.S.A.*, **67**, 807.
90. Poonian, M. S., Schlabach, A. J., and Weissbach, A. (1971): *Biochemistry*, **10**, 424.
91. Rickwood, D. (1972): *Biochim. Biophys. Acta*, **269**, 47.
92. Schaller, H., Nüsslein, C., Bonhoeffer, F. J., Kurz, C., and Nietzschmann, I. (1972): *Eur. J. Biochem.*, **26**, 474.
93. Nüsslein, C., and Heyden, B. (1972): *Biochem. Biophys. Res. Commun.*, **47**, 282.
94. Weatherford, S. C., Weisberg, L. S., Achord, D. T., and Apirion, D. (1972): *Biochem. Biophys. Res. Commun.*, **49**, 1307.
95. Nelidova, O. D., and Kiselev, L. L. (1968): *Molekul. Biol.*, **2**, 60.
96. Bartkowiak, S., and Pawelkiewicz, J. (1972): *Biochim. Biophys. Acta*, **272**, 137.
97. Remy, P., Birmele, C., and Ebel, J. P. (1972): *FEBS Lett.*, **27**, 134.
98. Cuatrecasas, P. (1970): *J. Biol. Chem.*, **245**, 3059.
99. Cuatrecasas, P. (1972): *Advan. Enzymol.*, **36**, 29.
100. Ruiz-Carrillo, A., and Allfrey, V. G. (1973): *Arch. Biochem. Biophys.*, **154**, 185.
101. Phillips, D. M. P. (1965): *Biochem. J.*, **97**, 669.
102. Fambrough, D. M., and Bonner, J. (1968): *J. Biol. Chem.*, **243**, 4434.
103. Sluyterman, L. A. A. E., and Wijdenes, J. (1970): *Biochim. Biophys. Acta*, **200**, 595.
104. Shainoff, J. R. (1968): *J. Immunol.*, **100**, 187.
105. Eldjarn, L., and Jellum, E. (1963): *Acta Chem. Scand.*, **17**, 2610.
106. Gruenwedel, D. W., and Fu, J. C. C. (1971): *Proc. Nat. Acad. Sci. U.S.A.*, **68**, 2002.
107. Chua, G. K., and Bushuk, W. (1969): *Biochem. Biophys. Res. Commun.*, **37**, 545.
108. Feinstein, G. (1970): *FEBS Lett.*, **7**, 353.
109. Kasai, K., and Ishii, S. (1972): *J. Biochem.* (*Tokyo*), **71**, 363.
110. Robinson, N. C., Tye, R. W., Neurath, H., and Walsh, K. A. (1971): *Biochemistry*, **10**, 2743.
111. Schroeder, D. D., and Shaw, E. (1968): *J. Biol. Chem.*, **243**, 2943.

112. Reeck, G. R., Walsh, K. A., and Neurath, H. (1971): *Biochemistry*, **10**, 4690.
113. Porath, J., and Sundberg, L. (1971): in *Protides of the Biological Fluids* (Ed. H. Peeters), Pergamon Press, New York, p. 401.
114. Sundberg, L., and Christiansen, T. (1972): *Biotechnol. Bioeng. Symp.*, **3**, 165.
115. Fritz, H., Brey, B., Schmal, S., and Werle, E. (1969): *Hoppe-Seyler's Z. Physiol. Chem.*, **350**, 617.
116. Fritz, H., Wunderer, G., and Dittmann, B. (1972): *Hoppe-Seyler's Z. Physiol. Chem.*, **353**, 893.
117. Fritz, H., Schult, H., Hutzel, M., Wiedemann, M., and Werle, E. (1967): *Hoppe-Seyler's Z. Physiol. Chem.*, **348**, 308.
118. Feinstein, G. (1971): *Biochim. Biophys. Acta*, **236**, 73.
119. Stewart, K. K., and Doherty, R. F. (1971): *FEBS Lett.*, **16**, 226.
120. Kassel, B., and Marciniszyn, M. B. (1971): in *Proc. Int. Res. Conf. on Proteinase Inhibitors*, de Gruyter, Berlin, p. 43.
121. Chauvet, J., and Acher, R. (1972): *FEBS Lett.*, **23**, 317.
122. Hixson, H. F., and Laskowski, M. (1970): *J. Biol. Chem.*, **245**, 2027.
123. Hochstrasser, K., Werle, E., Siegleman, R., and Schwarz, S. (1969): *Hoppe-Seyler's Z. Physiol. Chem.*, **350**, 897.
124. Keil, B. (1970): in *Structure and Function Relationships of Proteolytic Enzymes* (Eds. P. Desnuelle, H. Neurath and M. Ottesen), Academic Press, New York, p. 102.
125. Akanuma, H., Kasuga, A., Akanuma, T., and Yamasaki, M. (1971): *Biochem. Biophys. Res. Commun.*, **45**, 27.
126. Reeck, G. R., Walsh, K. A., Hermodson, M. A., and Neurath, H. (1971): *Proc. Nat. Acad. Sci. U.S.A.*, **68**, 1226.
127. Uren, J. R. (1971): *Biochim. Biophys. Acta*, **236**, 67.
128. Cuatrecasas, P., Wilchek, M., and Anfinsen, C. B. (1968): *Proc. Nat. Acad. Sci. U.S.A.*, **61**, 636.
129. Sokolovsk, M., and Zisapel, N. (1971): *Biochim. Biophys. Acta*, **250**, 203.
130. Blumberg, S., Schechter, I., and Berger, A. (1969): *Israel J. Chem.*, **7**, 125.
131. Deutsch, D. G., and Mertz, E. T. (1970): *Science*, **170**, 1095.
132. Nevaldine, B., and Kassell, B. (1971): *Biochim. Biophys. Acta*, **250**, 207.
133. Bobb, D. (1972): *Prep. Biochem.*, **2**, 347.
134. Stevenson, K. J., and Landman, A. (1971): *Can. J. Biochem.*, **49**, 119.
135. Jameson, G. W., and Elmore, D. T. (1971): *Biochem. J.*, **124**, 66P.
136. Thompson, A. R., and Davie, E. W. (1971): *Biochim. Biophys. Acta*, **250**, 210.
137. Schmer, G. (1972): *Hoppe-Seyler's Z. Physiol. Chem.*, **353**, 810.
138. Markwardt, F., Landmann, H., and Walsmann, P. (1968): *Eur. J. Biochem.*, **6**, 502.
139. Zolton, R. P., and Mertz, E. T. (1972): *Canad. J. Biochem.*, **50**, 529.
140. Scouten, W. H., Torok, F., and Gitomer, W. (1973): *Biochim. Biophys. Acta*, **309**, 521.

Chapter IV

Some Applications and Special Techniques of Affinity Chromatography

The previous sections have dealt almost exclusively with applications of affinity chromatography to the preparation and isolation of enzymes from biological sources. The adsorbents have comprised a wide range of immobilized ligands that have been selective either for individual enzymes or for groups of enzymes. This chapter extends the technique to the purification of other biologically important components and to several analytical and special applications.

A. APPLICATIONS TO THE PURIFICATION OF REGULATORY MACROMOLECULES AND COMPLEX BIOLOGICAL STRUCTURES

Many of the functions of the cell now recognized, such as the interaction with hormones, drugs, viruses and toxins, discrimination of self from non-self, cell–cell recognition and information transfer, presumably involve specific macromolecular and supramolecular structures. These biological components carry specific binding or receptor sites for biochemically important ligands which are not only an integral part, but also characterize their function, the transport and reception of compounds of biochemical importance.

These important regulatory macromolecules in general share the common features of being present in trace amounts, of having a high affinity and selectivity for their complementary ligand, and of often being associated with other supramolecular structures. Furthermore, they are often difficult to assay on the basis of a discrete 'enzymic' activity. These features imply that the purification of such macromolecules is almost unapproachable by conventional techniques. Thus, selective adsorbents with biological specificity are well suited to the resolution and isolation of complex biological structures, and important regulatory macromolecules. Polypeptide and hormone receptors, drug receptors, transport proteins and repressor molecules are amenable to this approach because they display specific binding functions with a high degree of affinity.

1. Antibodies and Antigens

The introduction of alien substances, termed *antigens*, into vertebrates may induce the formation of certain proteins, termed *antibodies*, which have the property of combining very selectively with the homologous antigen. During the past 50 years there have been innumerable ingenious attempts to prepare homogeneous antibodies. The purpose of this section is to indicate a number of the major problems encountered in antibody isolation and to describe how some of them may be circumvented.

Antibodies are globulins with a wide range of electrophoretic mobilities that are classified as IgG, IgM or IgA depending on their antigenic properties. The IgG immunoglobulins have a molecular weight of 150,000–160,000 and are composed of two identical halves each of which contain two polypeptide chains: a heavy chain of molecular weight 55,000–60,000 and a light chain of molecular weight 20,000–24,000 which are held together by disulphide bridges and non-covalent interactions. Each complete immunoglobulin molecule contains two identical sites capable of combining with the antigenic group (Fig. IV.1).

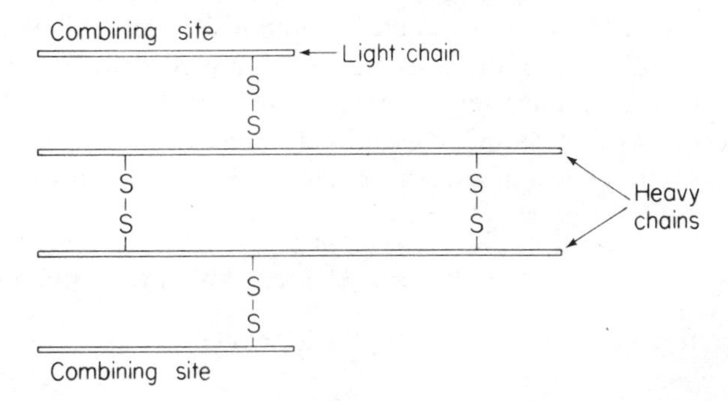

Figure IV.1. Schematic structure of the γG-globulin molecule

When an antibody combines with its complementary antigen the binding energy is derived from the relatively weak non-specific attractive forces which exist between atoms, ions and molecules. This is supplemented by the effects due to the interaction between solvent molecules which push the antibody and antigen together. These attractive forces are effective over short distances only and require that the combining region of the antibody fit closely over the antigen to produce sufficient binding force to hold the molecules together. Thus, the selectivity or specificity of the antigen–antibody interaction is an important prerequisite for a strong interaction. Furthermore, for the same reason, the distribution of groups on the antigen must be matched by complementary groups on the antibody.

Obermeyer and Pick[1] and Landsteiner and Lampl[2] coupled chemically defined groups, *haptens*, to proteins and used the conjugated proteins as immunogens for antibodies. Their work highlighted a major complication in that antihapten antibodies were heterogeneous. The heterogeneity may arise through the following causes.

(1) Haptens being attached to different parts of the carrier molecule and residing in different micro-environments. The problem may be partly circumvented by using carrier proteins containing only one amino acid residue available for the coupling reaction or by using a synthetic polypeptide containing only one type of amino acid.[3]

(2) The hapten may be oriented in different ways with respect to the surface of the antigen molecule.

(3) The antibody may be directed against different parts of the hapten molecule.

With antiprotein antibodies the situation is more complex since different parts of the protein are involved and the antigenic groupings are not as well defined as in the case of simple haptens. Furthermore, in an antihapten serum there are several classes of protein with antihapten antibody activity, e.g., the IgG, IgM and IgA immunoglobulins. The antibody response is conditioned by the immunization schedule, route of administration and nature of the antigen. Early collection of the antiserum yields a predominance of IgM antibodies whilst hyperimmune serum generates IgG antibodies.

The heterogeneity of antihapten antibody binding sites generates a spectrum of association constants for the combination of hapten with antibody. For the following reaction,

$$\text{Hapten (H)} + \text{antibody (Ab)} \rightleftharpoons \text{Hapten–antibody complex (HAb)}$$

The dissociation constant of the complex is defined by

$$K_d = \frac{(\text{H})(\text{Ab})}{(\text{HAb})}$$

and generally lies within the range of 10^{-6}–10^{-12}M.

Antibody purification depends on the separation of antibodies from other serum proteins by the formation of an insoluble antibody–antigen complex, the removal of non-antibody protein by thorough washing of the precipitate, dissociation of the complex and subsequent separation of the antibody from the antigen. The method of choice is to use an insoluble antigen as a specific immunoadsorbent. Such immunoadsorbents permit a more facile and complete separation of antibody from serum than methods based on precipitation with soluble antigen. The approach is indispensable in the comparatively rare cases of non-precipitating antibodies or in cases where the antibody is present in such low concentrations that isolation by conventional methods is difficult.

Furthermore, the use of immobilized haptens or antigens as immunoadsorbents together with suitable elution procedures can be useful in the resolution of heterogeneous antibodies.

A useful immunoadsorbent should possess the following features. It should:

(1) Specifically adsorb the complementary antibody from a mixture of components.

(2) Release the adsorbate quantitatively and under conditions innocuous to its specific antibody activity.

(3) Have a high capacity for adsorbing the specific antibody.

(4) Retain biological activity on repeated use and storage.

(5) Have adequate mechanical properties to allow centrifugation, filtration or use in column form.

Specific immunoadsorbents have been prepared by coupling the hapten or protein to a variety of inert matrices.[4–7] The rather low capacity of these adsorbents for antibody has been largely overcome by the use of agarose derivatives. The advantages of agarose as an inert supporting matrix have been delineated elsewhere (Chapter II) and for antigen–antibody interactions the greater adsorptive capacity for proteins and the increased strength of the interaction is particularly noticeable.[7] Where the antigen itself is a protein, the severe restrictions caused by immobilization can alter the allowable three-dimensional conformations that the protein can adopt and thus compromise the specificity of the interaction with the complementary macromolecule. Overimmobilized antigen can also fail to release antibodies because the necessary conformational change cannot occur. Thus, as described in a previous section (Chapter II, Section C.2), superior protein-agarose adsorbents can be prepared by linkage of the protein by the fewest possible points. Nevertheless, high-density protein matrices containing up to 25 mg antibody per ml wet weight Sepharose[8] can be readily achieved by activation of Sepharose by cyanogen halides.[9]

One other problem associated with the preparation of immunoadsorbents relates to the occlusion of the antigen binding site by immobilization on an inert matrix. This can be overcome to a certain extent by including the antigen or hapten in the coupling reaction and covalent linkage of the antigen–antibody complex *in toto*. The hapten affords some protection of the binding site and stabilizes the tertiary structure of the globulin during the immobilization process. Subsequent release of the antigen should yield an adsorbent competent for binding the original antigen.

The low dissociation constants of many antibody–antigen interactions (10^{-6}–10^{-12}M) coupled with the remarkable specificity mean that dissociation of the adsorbed antigen or antibody can be a formidable task. In general, it is important to release the adsorbed entity as soon as possible after the contaminating material has been washed off the column. Prolonged contact of the adsorbate with the immobilized hapten can lead to a drastic drop in the recovery of the active material, possibly by partial denaturation of one or both of the components or by an adverse conformational change in the antibody.

The dissociation of the antigen–antibody complex into its two components can generally be achieved by lowering the pH to less than 3, using either a glycine–HCl buffer,[10,11] hydrochloric acid,[8] acetic acid,[12] 20% formic acid[13] or 1M propionic acid.[14] Kleinschmidt and Boyer[15,16] reported that the formation of protein–antibody immune precipitates was inhibited by some concentrated electrolyte solutions, by detergents, denaturing agents and other substances of low molecular weight. Other electrolytes such as sodium thiocyanate,[17] sodium iodide[18] and magnesium chloride[19] have also been used for the elution of antibodies from immunoadsorbents. The dissolving power of bivalent cations is greater than that of the monovalent ones with $Mg^{2+} \geqslant Ba^{2+} \geqslant Ca^{2+} \geqslant Sr^{2+}$.[20] Nevertheless, many immunoadsorbents will not

Table IV.1. Some eluants for immunoadsorbents

Hapten	Eluant	Reference
	(1) *Acids* (pH 3)	
Insulin	0·01–1M HCl	8
Insulin	1M Acetic acid, pH 2·0	12
Anti-DNP-amino acids	20% Formic acid	13
	Glycine–HCl, pH 2·8	20
Ragweed allergens	Glycine–H_2SO_4, pH 2·7	31
	1M Propionic acid	14
	(2) *Salts*	
IgG	Sodium thiocyanate	17
	2·5M NaI, pH 7·5	18
	2·8M $MgCl_2$	19
Semliki Forest virus	Carbonate/bicarbonate, pH 11	32
Serum albumin	1% NaCl (pH 2·0) 37°C	33
	(3) *Protein denaturants*	
Human chorionic somato-mammotropin	6M Guanidine–HCl, pH 3·1	22
β-Lipoprotein	4M Urea	23
Galactosidase	8M Urea	21
	(4) *Haptens*	
p-Aminophenyl-β-lactoside	0·5M Lactose	27
DNP-Lysine	DNP, DNP-lysine	27
	(5) *Others*	
Human complement, Cl	0·2M 1,4-Diaminobutane	30
IgG	Distilled water	20

release their adsorbed protein unless elution is effected with high concentrations of protein denaturants such as urea[21] or guanidine–HCl,[22] particularly if combined with extreme pH values.[23]

For antihapten antibodies, it is usually possible to dissociate the complex at pH 7 with a 0·1M solution of the appropriate hapten.[24,25] Once the complex has been dissociated from the insoluble immunoadsorbent, the antigen can easily be separated from the soluble antibody by centrifugation or filtration.

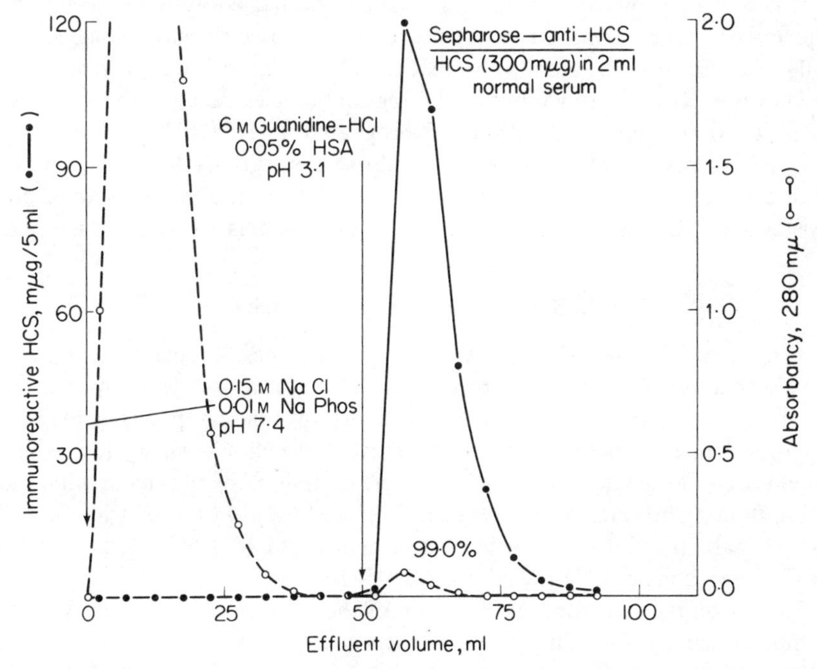

Figure IV.2. The purification of human chorionic somato-mammotropin (HCS) from normal serum on a column of anti-HCS-Sepharose. No hormone was detected in the unretarded fraction by radioimmunoassay, while 99% was recovered in the guanidine eluate. The small eluate peak (about 1% of the applied absorbancy) was not attributable to HCS and represented non-specific adsorption of serum proteins. Reproduced with permission from B. Weintraub, *Biochem. Biophys. Res. Commun.*, **39**, 83 (1970)

The conditions for dissociation of antigen–antibody complexes are summarized in Table IV.1.

No attempt will be made here to review the rapidly expanding field of immunoadsorbents.[25,26] Suffice it to give several representative examples of the application of the technique. Agarose immunoadsorbents have been employed to purify a variety of antihapten antibodies by elution with the appropriate hapten,[27] a mouse myeloma IgA protein which binds nitrophenyl groups,[28] antibodies to Staphylococcal nuclease,[29] the first component of human complement[30] and insulin antibodies.[8]

Antibodies have been covalently attached to agarose to purify various antigens. Thus, an antiinsulin serum globulin immunoadsorbent bound labelled and non-labelled insulin at pH 8·2–8·4.[12] Furthermore, this adsorbent quantitatively removed insulin from a sample of dog pancreatic vein serum and bound insulin was eluted with 1M acetic acid.

Columns of Sepharose containing covalently attached antibody to human chorionic somato-mammotropin (HCS) have been used to purify labelled and unlabelled hormone by affinity chromatography.[22] The technique concentrates the hormone from large volumes of biological tissue extract and yields almost quantitative recoveries on elution with 6M guanidine–HCl, pH 3·1 (Fig. IV.2). This represents a 500-fold increase in ultimate sensitivity for the hormone with a lower limit of as little as 2 pg/ml (1×10^{-13}M) and has enabled the existence of HCS in plasma and some tumour extracts to be demonstrated.

2. Binding and Transport Proteins

The limited supply of vitamins and hormones has led many animals to develop elaborate mechanisms for the adsorption, transport and conservation of these trace substances. Specific transport or binding proteins play an important role in these processes and prevent the rapid urinary loss which would occur if the vitamin or hormone were present in the plasma in unbound form. Binding proteins are thus present in exceedingly low concentrations and exhibit a high avidity for their complementary vitamin or hormone with dissociation constants in the range of 10^{-7}–10^{-16}M.

Studies on the structure and function of these proteins have been severely compromised by the difficulties experienced in their isolation. In human plasma, vitamin B_{12} is tightly bound to at least two proteins, transcobalamin I and transcobalamin II, which must be purified approximately 1 and 2 million-fold respectively to achieve homogeneity.[34] This means that 1000 litres of human plasma contain only about 80 and 20 mg respectively of each of these proteins. Purification to this extent with operational yields is beyond the scope of classical purification techniques.

Affinity chromatography is an attractive method to be exploited in these cases since the interactions are specific and of high affinity and large volumes of starting material can be dealt with. However, like antibody–antigen interactions, the central problem relates to the subsequent dissociation of the protein from the immobilized ligand. This has been achieved under alkaline conditions[35] or with protein denaturants at acid pH values.[36,39] Furthermore, elevated temperatures have been utilized to assist the release of corticosterone-binding globulin from cortisol-Sepharose,[40] and L-fucose binding proteins from a Sepharose derivative of *N*-(ϵ-aminocaproyl)-β-L-fucosamine.[41] In the latter case, however, the affinity of L-fucose for the binding proteins was relatively low ($K_{diss} \simeq 10^{-4}$M) and elution was effected with 40 mM L-fucose. However, in contrast, the effective removal of steroid-binding protein from affinity adsorbents has only been reported for corticosterone-binding globulin.[40] Thus, so far, active steroid-binding proteins have not been effectively recovered from affinity matrices based on deoxycorticosterone- or estriol-Sepharose,[42] estradiol-Sepharose,[43] polyvinyl-estradiol[44] or androgen-agarose.[36] One major problem which has hampered the interpretation of these results is the leakage of free steroid into the milieu.[42,45] The free steroid binds with great avidity to the binding protein and effectively inhibits its capacity to bind the small amount of radio-labelled hormone necessary for assay. It would appear that this disappearance of binding capacity has been erroneously interpreted in terms of adsorption to the support. Steroids and other aromatic compounds are strongly adsorbed to the solid matrix and for quantitative removal it is necessary to wash the adsorbent with large volumes of solutions, including organic solvents, for prolonged periods of time. Cuatrecasas[43] has shown that, if such precautions are taken, an estradiol-Sepharose derivative prepared by coupling 3-*O*-succinyl-^{3}H-estradiol to aminoethyl-Sepharose with a carbodiimide reaction is very effective in extracting estradiol-binding proteins from human serum and calf uterine cytosol. Furthermore, protein denaturants were avoided in the elution of the strongly adsorbed protein, by selectively cleaving the estradiol-Sepharose bond with mild base or with neutral hydroxylamine.

Sica *et al.*[46] have recently shown that effective adsorbents for estradiol receptors could be prepared by coupling 17-β-estradiol hemisuccinate to agarose derivatives containing diaminodipropylamine, serum albumin, poly-L-lysine or poly-L-lysine-L-alanine copolymer. Between 60 and 80% of the estradiol-binding protein from a calf uterine supernatant was adsorbed by these columns. Furthermore, the adsorbents could be washed with large volumes of buffer and with 1M KCl without release of receptor. Elution was effected by incubating the adsorbent for 15 minutes at 30°C with 0·2 to 2 μg per ml of estradiol. With the best adsorbents the estradiol-binding protein was purified 10,000-fold with a recovery of about 50%.

Table IV.2 summarizes some properties of the vitamin- and hormone-binding proteins that have been purified by affinity chromatography.

The serum thyroxine-binding globulin has a dissociation constant of 10^{-10}M for thyroxine and is the major transport protein for the thyroid hormone *in vivo*. Despite its low concentration in serum (10–20 μg/ml) it has been purified on an L-thyroxine-agarose adsorbent[35] and an agarose substituted with *N*-(6-aminocaproyl)-thyroxine.[47]

Rosner and Bradlow[40] have developed a method for the purification of corticosteroid-binding globulin (CBG) by affinity chromatography. Cortisol

Table IV.2. Binding and carrier proteins

K_{diss}(M)[a]	Carrier protein	Ligand	Eluant	(1) Yield (2) Purification	Reference
$1–2 \times 10^{-4}$	L-Fucose-binding protein	*N*-(ε-Aminocaproyl)-β-L-fucosamine	40 mM L-Fucose at 25°C		41
5×10^{-7}	Corticosterone-binding globulin	Cortisol	200 μg/ml Cortisol 23°C		40
10^{-7}	Retinol-binding protein	Pre-albumin	Distilled deionized water adjusted to pH 8 with ammonia	(1) 31–42%	49
	Estradiol-binding protein	17-β-Estradiol hemisuccinate	0·2–2 μg/ml Estradiol 30°C	(1) 50% (2) 10,000-fold	46
10^{-8}	Testosterone-binding globulin	3-β-Aminoandrostan-17β-ol	1M Guanidine–HCl, pH 2·1	(1) 0·4% (2) 4-fold	36
	Folate-binding protein	Folic acid	8M Urea in 0·1M sodium acetate, pH 5·0	(1) 54% (2) 20-fold	39
	Granulocyte vitamin B_{12}-binding protein	Vitamin B_{12}	7·5M Guanidine–HCl, pH 7·5	(1) 90% (2) 9860-fold	38
10^{-1}	Thyroxine-binding globulin	L-Thyroxine	0·002M KOH, pH 9·3	(1) 18–37%	35
10^{-16}	Avidin	ε-*N*-Biotinyl-L-lysine	6M Guanidine–HCl, pH 1·5	(1) 90% (2) 4000-fold	37

[a] Dissociation constant for the ligand-binding protein complex.

hemisuccinate was coupled to 3,3′-diaminodipropylamino-Sepharose[43] with dicyclohexylcarbodiimide in dioxan. The cortisol-Sepharose was added batchwise to a 2-litre serum sample, allowed to equilibrate for 5 hours at 4°C and then thoroughly washed on a coarse Büchner funnel. Elution was effected with a sodium phosphate buffer containing cortisol (200 μg/ml) followed by a similar treatment at room temperature and stopping the flow through the column for 2 hours or more. Fig. IV.3 shows the elution of CBG from the cortisol-Sepharose.

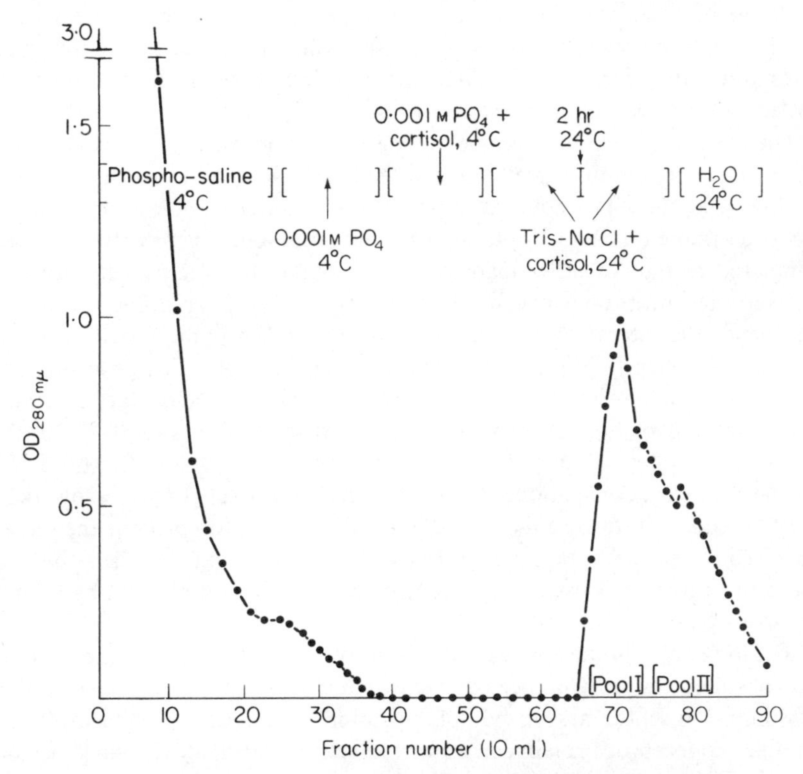

Figure IV.3. The elution of corticosteroid binding globulin (CBG) from cortisol-Sepharose. The conditions in brackets indicate what was added to the column at the stage indicated. Reproduced with permission from W. Rosner and H. L. Bradlow, *J. Clin. Endocrinol. Metab.*, **33**, 193 (1971)

Previous attempts to isolate granulocyte vitamin B_{12}-binding protein have been marred by the difficulty of collecting sufficient starting material and the relatively low yields obtained by conventional purification techniques.

However, Allen and Majerus[38] have isolated a human granulocyte vitamin B_{12}-binding protein using affinity chromatography as the sole purification technique. The protein was purified nearly 10,000-fold with a yield of over 90% and was homogeneous by several techniques. Furthermore, affinity chromatography on vitamin B_{12}-Sepharose resulted in a 24,000-fold purification of transcobalamin II from human plasma.[48]

3. Receptor Proteins

It has long been recognized that the primary action of some hormones is on the plasma membrane of target cells. The term 'receptor' is usually applied to designate all of the plasma membrane components which are involved in the action of a particular hormone. These specific membrane receptor structures must therefore be identified and purified in order to elucidate the molecular basis of action of hormones.

The electric organs of fishes belonging to the families of *Torpedinidae* and *Gymnotidae* are cholinergically innervated and have a high content of acetylcholine receptors. The concentration of receptor sites per gram fresh tissue has been quoted as 10^{14},[53] although the actual amount of receptors is small compared to that of other material present. Likewise, the concentration of glucagon-receptors on liver cell membranes is very low, 2·6 pmol/mg protein,[54] and hence the interaction of only a few sites with an immobilized hormone must be sufficiently effective to permit strong binding of large membrane fragments. The interaction of hormones with their complementary receptors is specific and of high affinity, with dissociation constants of 10^{-9}–10^{-10}M for glucagon,[54] 5×10^{-11}M for insulin[55] and 10^{-6}–10^{-7}M for norepinephrine.[56] Since a conventional isolation approach is likely to give a low yield, the low concentrations and high affinity of the receptor prompt the use of biospecific chromatography in selectively concentrating and isolating the receptor. Table IV.3 lists some receptors that have been purified by affinity chromatography.

The long delay in the application of affinity chromatography to the purification of such structures has been due to technical difficulties in the development of suitable receptor assay and solubilization techniques. For example, the cholinergic receptor protein can now be selectively labelled with snake venom α-toxins[50–52] and solubilized by mild detergents without impairing its ability to bind cholinergic ligands.[51] However, detergent-solubilized membrane proteins have similar physical properties and hence conventional procedures for protein fractionation are of limited value.[57]

Several methods are available for the affinity chromatography of acetylcholine receptors. Chromatography on cobrotoxin-agarose[58] yields a purified toxin–receptor complex rather than free receptor, although Karlsson *et al.*[59] demonstrate that venom–neurotoxin complexes can be partially dissociated by binding cholinergic substances. Thus, the nicotinic cholinergic receptor from a sólubilized extract of *Torpedo marmorata* was adsorbed to immobilized

Naja Naja siamensis (Thailand cobra) neurotoxin and subsequently desorbed with a linear gradient of carbamyl-choline at room temperature. A 1000-fold purification of two protein fractions with a high affinity for curare and neurotoxin was obtained with a yield of about 50%.

Schmidt and Raftery[58,60] have used a resin containing a quaternary ammonium function to purify α-bungarotoxin-binding membrane components from the electric organ of *Torpedo californica.* A 40-fold purification was achieved when the receptor was eluted with a linear salt gradient containing 0·1% Triton X-100.

Olsen *et al.*[52] describe an affinity adsorbent which gives an approximately 150-fold purification of the nicotinic receptor protein from *Electrophorus electricus* in a particularly high yield. Detergent extracted cholinergic receptor

Table IV.3. Some receptors that have been purified by affinity chromatography

	References
Acetylcholine	52, 58, 60
Nicotinic acetylcholine	59
Insulin	73–75, 81
Glucagon	54, 71
Estradiol	46, 64, 65
Penicillin	76
Morphine	77
Histamine	78
Concanavalin A	79
ACTH	72
Norepinephrine	56, 80

protein from the electric organ was applied to the affinity adsorbent (Fig. IV.4(b)) equilibrated with 0·1M tris-HCl buffer, pH 8·0, containing 0·1M NaCl and 1% Triton X-100. Non-adsorbed protein was washed off and the receptor protein eluted with a gradient of 0 to 10^{-3}M flaxedil (gallamine triethiodide, I), followed by 2M NaCl to desorb acetylcholinesterase (Fig. IV.4(a)). The purified receptor protein bound cholinergic ligands and a tritiated α-toxin from *Naja nigricollis*.

The cardiac β-adrenergic receptor has also been solubilized and extensively purified by affinity chromatography.[56,81] Effective solubilization of the norepinephrine binding activity was achieved with the non-ionic detergent Lubrol-PX or the ionic detergent sodium deoxycholate. Triton X-100 was less effective. The adsorbent was prepared by covalently binding norepinephrine to agarose via the 30 Å extension arm depicted in Fig. IV.5. Solubilized membrane fractions from ventricular microsomes were applied to the adsorbent and after washing through the unretarded proteins, the adsorbed receptors were eluted with 0·1M epinephrine at pH 3·8. The receptor-bound epinephrine was removed by exhaustive dialysis. The fraction eluted with epinephrine, which contained 110–120% of the original binding activity, was purified approximately 300-fold by a single passage through the affinity column.

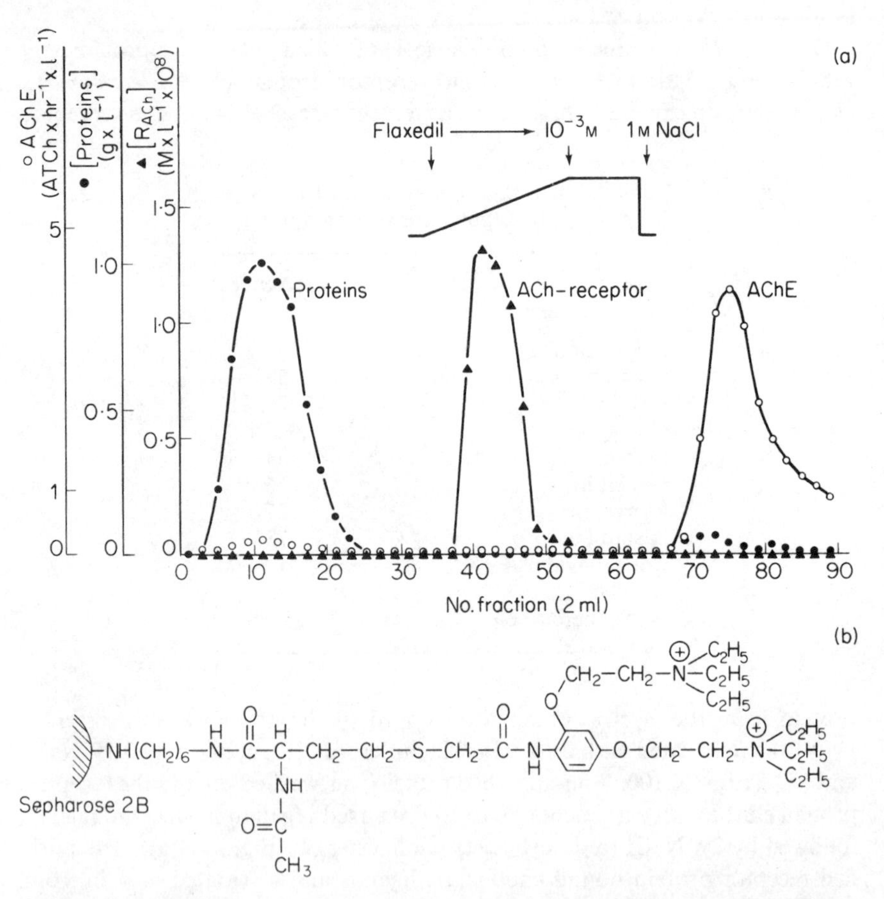

Figure IV.4. Purification of the nicotinic receptor protein from a Triton X-100 extract of membrane fragments from *Electrophorus electricus* electric organ. The extract (20 ml) was applied to the adsorbent (b) equilibrated with 0·1M tris-HCl, pH 8·0, containing 0·1M NaCl and 1% Triton X-100. The column was washed with 200 ml of the same buffer and the receptor protein eluted with a gradient of 0 to 10^{-3}M flaxedil. Finally, the column was washed with 2M NaCl to desorb acetylcholinesterase and other proteins. Reproduced with permission from R. W. Olsen, J. C. Meunier and J. P. Changeux, *FEBS Lett.*, **28**, 96 (1972)

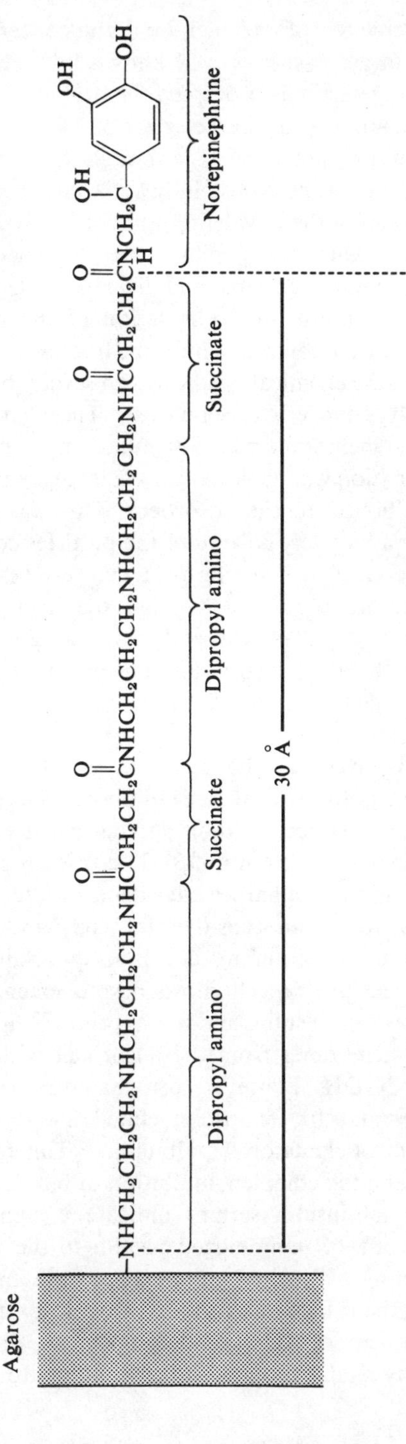

Figure IV.5. The structure of the norepinephrine-agarose adsorbent. Reproduced with permission from R. J. Lefkowitz, E. Haber and D. O'Hara, *Proc. Nat. Acad. Sci. U.S.A.*, **69**, 2828 (1972)

The presence of specific receptors which bind and concentrate estrogens in the uterus and other target tissues is well known.[61,62] The high affinity of these receptors for estrogens[63] has prompted the exploitation of immobilized estradiol for their isolation and characterization. A number of estradiol adsorbents have been prepared,[46,64,65] although their usefulness in the purification of the uterine receptor protein has not been realized.

Agarose beads containing the covalently attached polypeptide hormones, insulin,[66] ACTH[67] and glucagon,[68] retain their biological activity when tested on isolated mammary, adrenal and liver cells. Consequently, such adsorbents should prove useful in the isolation of the hormone receptor proteins from appropriate solubilized membrane fractions. Johnson *et al.*[69,70] have recently described the chemical synthesis and several biological activities of glucagon covalently attached to Sepharose. The agarose–glucagon adsorbent bound liver cell membrane particles containing glucagon receptors.[54] However, little fractionation was achieved since a large proportion of extraneous protein was also bound to the adsorbent. This was explained on the assumption that either a large proportion of the particles contained glucagon receptors[54] or that the glucagon-binding proteins were associated with high molecular weight membrane proteins. When the liver membranes were solubilized in 2% Lubrol-PX and subsequently chromatographed on glucagon–agarose equilibrated with 0·02% Lubrol-PX, a proportion of the protein was strongly adsorbed. The glucagon receptor proteins were eluted with 5M urea buffered at pH 6·0.[71]

ACTH-agarose derivatives may be useful in the purification of ACTH receptor proteins from both adrenal and adipose tissue cells. The ACTH binding proteins from an extract of rabbit adrenal tissue were quantitatively removed by passage through a column of β1-39-corticotropin-agarose.[72]

Similar studies with insulin-Sepharose have demonstrated its ability to bind firmly and selectively to insulin-sensitive fat cells and their membrane 'ghosts'.[73] Insulin-Sepharose columns consistently retained a significant proportion of insulin-sensitive fat cell ghosts despite extensive washing. The strong interaction between insulin and its receptor[55] is reflected in the inability to elute the membranes from the adsorbent with 1M NaCl, 0·05N acetic acid or 0·05N NaOH. However, 6M guanidine–HCl quantitatively eluted the membranes and 10^{-3}M insulin eluted about three times more membranes than a control eluate of 2% albumin.[74] This reversal of binding by the soluble insulin and the complete inhibition of binding by pretreatment of the columns with antiinsulin serum, indicate a significant degree of specificity for the binding of immobilized insulin to the fat cell receptors. Furthermore, the binding of isolated adipocytes to the immobilized insulin was abolished when their insulin-sensitivity was lost on treatment with trypsin. Likewise, treatment of the isolated adipocytes with trypsin abolished their insulin sensitivity and prevented their binding to the immobilized insulin. Incubation of the trypsinized cells with soybean trypsin inhibitor restored most of the binding capacity. The insulin receptor was further purified by solubilization of plasma membranes with Triton X-100, affinity chromatography on insulin-Sepharose, and subsequent elution with excess native insulin.[74]

Cuatrecasas[81] has recently reported a 250,000-fold purification of the insulin receptor from liver membranes. The receptor was solubilized with Triton X-100 and partially purified by ammonium sulphate fractionation and DEAE-cellulose chromatography prior to adsorption on immobilized insulin. The receptor macromolecule was eluted in high yield (50–80%) with 4·5M urea in a 0·05M sodium acetate buffer, pH 6·0, containing 0·1% (v/v) Triton X-100.

4. Cells and Viruses

Conventional cell-separation techniques exploit physical differences among cell types.[82] However, many tissues contain cells which represent a wide spectrum of differentiation and maturation. Such differentiating cell lines are difficult to separate by physical means because the physical characteristics of the various cells overlap.

It is known that cells have different antigenic properties[83] and that some surface antigens appear early in cell maturation.[84,85] Thus, such cells might be separated by immuno-specific procedures. The use of immunoadsorbents as a new approach to the specific fractionation of cells on immobilized antigens has been reported.[86,87] The major obstacle to developing matrices for receptor-specific cell separations is that these materials should be inert to the characteristic 'sticky' properties of cells. Cells are readily adsorbed onto glass and other charged or hydrophobic materials[86,87] and consequently the usefulness of such matrices is impaired by the non-specific adhesion of large numbers of cells. More recently, however, antigen-coated polyacrylamide[88,89] and agarose[90,91] have been used to separate cells and viruses. Nevertheless, activation of agarose or Sephadex with cyanogen bromide[92] fractures the polysaccharide beads and makes cell filtration through columns almost impossible.[88] On the other hand, polyacrylamide beads can be readily derivatized[93] without destruction of their integrity and consequently may be the preferred inert support for cell separations.

Truffa-Bachi and Wofsy[88] describe the specific resolution of antihapten antibody-producing cells from the spleens of mice immunized against an azo-coupled phenyl-β-lactoside (*lac*) antigen. The affinity adsorbent was prepared by coupling *p*-aminophenyl-β-lactoside to tightly cross-linked polyacrylamide P-6 beads and cell filtrations were performed in phosphate-saline buffer, pH 7·2. Of the number of cells applied to the columns (10^6–10^8), 99 ± 2% were recovered in the eluate whilst only 18 ± 7% of the anti-*lac* cells appeared in the eluate. In contrast, all the anti-*lac* cells were unbound when 1×10^{-5}M *p*-nitrophenyl-β-lactoside (PNP-*lac*) was included in the irrigating buffer. Consequently, anti-*lac* cells could be eluted from the *lac* immuno adsorbent by elution with the PNP-*lac* hapten.

Kenyon *et al.*[91] have applied the technique of immuno adsorption chromatography to the isolation of the Aleutian mink disease virus. Sepharose-antibody columns were charged with tissue extracts from mink infected with Aleutian disease. Dissociation of the adsorbent–virus complex with 0·75M NaCl and a gradient of glycine–HCl released infective particles resembling picornaviruses. Wood *et al.*[32] have reported the purification of Semliki Forest virus on an immuno adsorbent prepared from rabbit or chicken antisera.

5. Affinity Density Perturbation

The topological irregularities of plasma membranes make the isolation of cells by conventional physicochemical techniques considerably more difficult. Wallach *et al.*[94] have developed a general fractionation method for membrane fragments bearing specific receptors for drugs, toxins and hormones. The disrupted membrane fragments are separated ultracentrifugally according to a density increase occasioned by complexation of the receptor with its specific ligand to which a labelled particle of high density has been covalently coupled. Furthermore, electron microscopy of the complexes between the density-perturbing particles and the receptor-bearing membrane fragments allows the receptor topology to be mapped. In principle, this technique has as wide an application as affinity chromatography and is particularly useful for receptor localization. The approach is illustrated in Fig. IV.6.

As an example of the technique, concanavalin A (Con A) was converted into a density perturbant by glutaraldehyde coupling to purified coliphage K29, a stable icosahedron of diameter 450 Å. Membrane fragments were prepared from pig lymphocyte plasma membrane and contained a large number of Con A receptors.[95] Interaction of the receptor-bearing membrane fragments with the perturbant reversibly increased the buoyant density in a caesium chloride gradient from about 1·18 for untreated membranes to a broad layer with a maximal density at 1·30–1·40. This relatively broad density distribution of the membrane–ConA–K29 complex reflects the microheterogeneity in the distribution of receptor sites. Addition of excess α,α-trehalose dissociates the membrane–ConA–K29 complex and shifts the membrane distribution to lower densities.

In principle, therefore, density perturbants can be linked to hormones, transmitters, drugs, specific antigens or specific immunoglobulins and be used not only to isolate receptor domains but also to map membrane and cell topology.

Using a similar principle, Katzen and Soderman[96] have observed that lymphocytes, which normally float in physiological media, associate with sedimentable insulin-Sepharose beads to produce complexes that either floated or sank depending on the ratio of the concentrations of cells to beads. A mixture of unmodified Sepharose beads and viable fat cells rapidly separates into a top layer of cells, a clear infranatant and a sediment of beads. When insulin-Sepharose was used in place of Sepharose all of the beads floated with the cells, whereas when an excess of insulin-Sepharose was added,

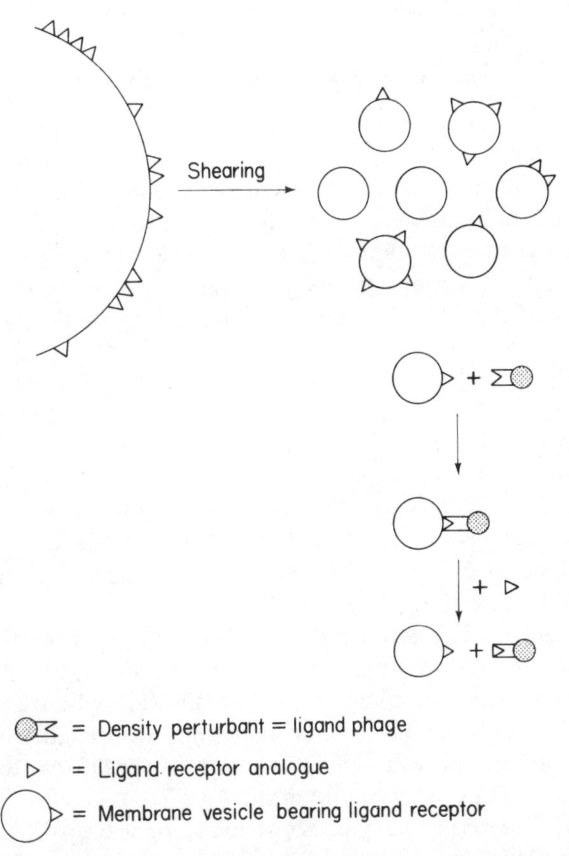

Figure IV.6. The principle of affinity density perturbation. A plasma membrane bearing multiple receptors (△) is sheared into membrane fragments carrying different numbers of receptors in varying distributions. These are reacted with the ligand (Σ) coupled to the density perturbant, i.e. K29 phage (●), producing a membrane-receptor–ligand–phage complex of higher density than the membrane itself and of lower density than the density perturbant. Addition of a low molecular weight dissociating agent (▲) returns the membrane and density perturbant to their original densities. Reproduced with permission from D. F. H. Wallach, B. Kranz, E. Ferber and H. Fischer, *FEBS Lett.*, **21**, 29 (1972)

cells sedimented with the beads. Clearly the number of cells bound per insulin-Sepharose bead determined the buoyancy of the resultant complex. The binding of whole lymphocytes to the beads was confirmed by interference microscopy.

Treatment of the cells with trypsin, antiinsulin serum or 10^{-5}M free insulin completely abolished the effect of the buoyant cells on the beads. These observations are consistent with the formation of a strong reversible bond between the insulin-Sepharose and specific insulin receptor sites on the cell membrane.

6. Concanavalin A and Glycoproteins

The ability of phytohaemagglutinins or lectins extracted from the seeds of certain plants,[97–99] *Helix pomatia*[100,101] or bacteria[102] to interact specifically with some carbohydrate groupings is well documented. These proteins, which are reminiscent of antibodies, react with components of cell membranes and agglutinate erythrocytes, tumour cells and embryonic cells.[103,104] Phyto-agglutinins with a defined sugar specificity may thus be a useful tool for the investigation of the surface structure of malignant or virally transformed cells.[105]

Tomita *et al.*[106] have reported the purification of galactose-binding phyto-agglutinins by affinity chromatography on the galactose-containing polymer, Sepharose. Extracts prepared from several plant seeds were applied to the Sepharose 4B column, and after washing off non-adsorbed protein, agglutinin activity was eluted with 0·1M galactose solution. Furthermore, a new galactose-binding haemagglutinin separated from cells of *Pseudomonas aeruginosa* has also been purified on a Sepharose 4B column by specific elution with 0·3M galactose[107] (Fig. IV.7).

Soybean agglutinin, a lectin isolated from soybean oil meal[108,109] which specifically binds *N*-acetyl-D-galactosamine and D-galactose,[110] has been purified on an adsorbent comprising *N*-ε-aminocaproyl-β-D-galactopyranosylamine linked to Sepharose.[111] The agglutinin was quantitatively ($>90\%$) eluted from the column with a 0·5% solution of D-galactose to yield 80 mg of protein from 50 g of soybean oil meal.

Lotan *et al.*[112] describe the synthesis of an analogous derivative of *N*-acetyl-D-glucosamine and its immobilized equivalent for the purification of the *N*-acetyl-D-glucosamine-specific wheat germ agglutinin[113] from commercial wheat germ. In contrast to other agglutinins, however, the adsorbed protein could not be eluted with its specific sugar, 0·1M *N*-acetyl-D-glucosamine, although it could be eluted with 0·1M acetic acid.

The purification of a glucose-binding phytoagglutinin, concanavalin A (Con A), on Sephadex columns has been reported.[114] This protein from the jack bean, *Canavalia ensiformis*, interacts specifically with α-D-glucopyranosyl,[115,116] α-D-mannopyranosyl,[115,117] β-D-fructofuranosyl,[118] α-D-glucosaminyl and other sterically related sugar residues. Furthermore, the resultant complexes are readily dissociated with methyl-α-D-mannopyranoside and methyl-α-D-glucopyranoside.[119]

Concanavalin A is easily isolated from jack bean meal in large amounts and its well-documented specificity makes it an ideal general ligand for use with a variety of polysaccharides and glycoproteins. The binding of polysaccharides and glycoproteins to concanavalin A, however, requires the presence of both Mn^{2+} and Ca^{2+} to maintain the stability to the subunit structure.[120]

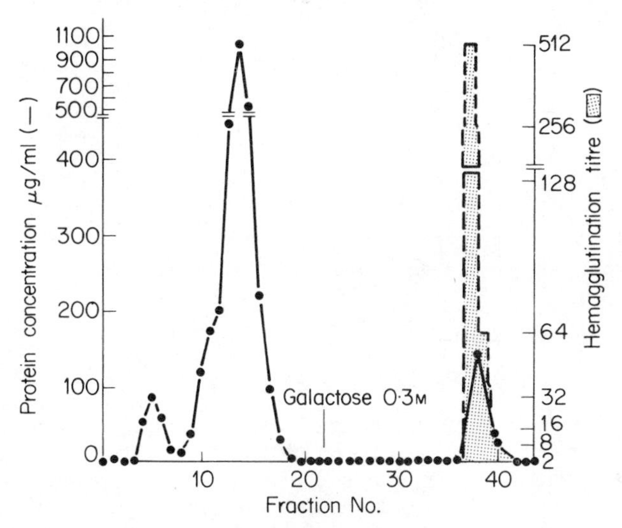

Figure IV.7. Chromatography of the haemagglutin of *Pseudomonas aeruginosa* on a Sepharose 4B column (2·5 cm × 40 cm). The arrow indicates the addition of 0·3M galactose. Reproduced with permission from N. Gilboa-Garber, L. Mizrahi and N. Garber, *FEBS Lett.*, **28**, 93 (1972)

Aspberg and Porath[121] have investigated the adsorption of serum glycoproteins to concanavalin A immobilized on Sepharose. Prealbumin, α- and β-globulins and IgM were adsorbed but not albumin or other immunoglobulins. The carbohydrate content of the adsorbed proteins was considerably higher than the non-adsorbed serum proteins. Furthermore, the α_1-antitrypsin has been isolated from human serum by affinity chromatography on Con A-Sepharose.[122] When a partially purified preparation from human serum was applied to a column of Con A-Sepharose, the protein that emerged in the void volume consisted largely of serum albumin (Fig. IV.8). Subsequent elution with 0·1M methyl-α-D-glucopyranoside displaced 75–80% of the antitrypsin activity together with a small amount of protein.

Hog blood group substance has been resolved into two components on a column of poly-L-leucyl-Con A[123] and Dufau *et al.*[124] have investigated the interaction of the glycoprotein hormones with Con A-Sepharose. Human chorionic gonadotropin, follicle-stimulating hormone and luteinizing hormone were strongly bound by Con A-Sepharose and could be eluted with 0·2M methyl-α-D-glucopyranoside. Chromatography on immobilized Con A was applied to the purification and isolation of human chorionic gonadotropin from crude extracts of urine and human pregnancy plasma. Unfortunately, only purifications of up to 3-fold were obtained and the most useful application of this technique may be the analysis of glycoprotein hormones in small samples of plasma or urine. For example, desialylated human chorionic

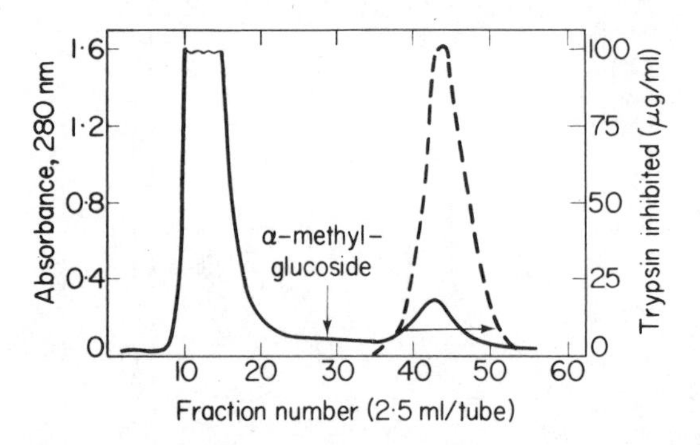

Figure IV.8. Affinity chromatography of human serum α_1-antitrypsin on immobilized-concanavalin A. A partially purified preparation of human serum was applied to a column (1·5 cm × 10 cm) of Con A-Sepharose equilibrated with 0·05M phosphate buffer, pH 7·6. Non-adsorbed protein was washed through and at the point denoted by the arrow 0·1M methyl-α-D-glucopyranoside in the same buffer was introduced. Fractions (2·5 ml) were collected at a flow rate of 15 ml/hour. Reproduced with permission from I. E. Liener, O. R. Garrison and Z. Pravda, *Biochem. Biophys. Res. Commun.*, **51**, 436 (1973)

gonadotropin was eluted more slowly than the intact hormone and comparison of plasma and urinary gonadotropin revealed an elution profile consistent with the partial desialylation of the urinary hormone.

The capacity of free concanavalin A to agglutinate certain cell lines[103–105] would suggest that immobilized Con A might be useful in specific cell separations. Thus, Edelman *et al.*[125] have shown that mouse erythrocytes, thymocytes and lymphocytes bind to insolubilized Con A. Furthermore, immobilized Con A was used to isolate the glycoprotein receptors specific for Con A from pig lymphocyte plasma membranes.[126] Lymphocyte plasma membranes were solubilized with 1% sodium deoxycholate and the solubilized membrane components applied to the column equilibrated with 1% sodium deoxycholate.

About 5% of the total protein was eluted with methyl-α-D-glucopyranoside and the glycoproteins were subsequently separated into five fractions by gel electrophoresis. Antibodies to these glycoproteins agglutinated whole lymphocytes.

In a related procedure, Hayman *et al.*[127] purified virus glycoproteins by affinity chromatography on *Lens culinaris* phytohaemagglutinin covalently attached to Sepharose. The value of the procedure for the identification and purification of the envelope glycoproteins from a variety of viruses was demonstrated. The same haemagglutinin was also used to isolate lymphocyte plasma membrane glycoproteins from deoxycholate-solubilized membranes.[128]

7. Applications to Molecular Biology

The isolation of genetic regulator proteins is frustrated by their presence in very low concentrations or by the lack of a practicable assay method. In principle, therefore, affinity chromatography offers considerable promise for the purification of such proteins.

The genes which determine the ability of *E. coli* to make and regulate the synthesis of the proteins concerned with lactose (glucose-4β-D-galactoside) metabolism are clustered in a small region of the *E. coli* chromosome termed the *lactose* (*lac*) *operon*. The operon codes for two proteins essential for the metabolism of lactose; β-galactosidase (*z* gene) and galactoside permease (*y* gene), and a third non-essential enzyme, thiogalactoside transacetylase (*a* gene). The transcription of the three genes is initiated at the lactose (*lac*) promoter (*p*) and, in the absence of an inducer of the *lac* operon, the transcription is prevented by the presence of a *lac* repressor protein. The repressor probably acts by binding to the operator (*o*) DNA which lies between the promoter and structural genes and hence prevents transcription (Fig. IV.9). The inducer allows transcription of the operon by interacting with the repressor and releasing it from the operator.

Tomino and Paigen[129] have covalently attached β-thiogalactoside ligands to insoluble protein polymers and isolated a series of proteins that recognize the β-thiogalactoside moiety. These proteins included the repressor proteins of the *lac* and *gal* operon, both of which bind β-thiogalactosides either as inducers or corepressors, and probably the *lac* permease. The adsorbents were prepared by coupling *p*-aminophenyl-thiogalactoside (PAPTG) and aminobutyl-thiogalactoside (ABTG) to polymerized bovine γ-globulin by a carbodiimide condensation. β-Galactoside activity was removed from a crude extract of *E. coli* by chromatography on the ABTG-adsorbent and could subsequently be eluted in an almost homogeneous state by 0·2M lactose. Extracts of a lac^+ inducible strain grown in the absence of inducer contained less galactosidase and almost no contaminating proteins when subjected to polyacrylamide electrophoresis. This suggests that β-galactosidase and the other proteins obtained when the *lac*$^+$ strain was grown in the presence of the inducer were coded by the *lac* operon.

The immobilized galactoside adsorbent is also capable of isolating the *lac* repressor protein when eluted with 0·1M isopropylthiogalactoside (IPTG). Extracts of the *lac* i^Q strain which contain high concentrations of the *lac* repressor show two marked differences from the uninduced wild-type cells when chromatographed on the PAPTG-adsorbent and the eluates subjected to electrophoresis. The *lac* repressor protein was partially purified by ammonium sulphate precipitation and quantitatively bound to the thiogalactoside-adsorbent. Elution with 0·2M KCl yielded 23% of the *lac* repressor as determined by its ability to bind IPTG, and only 0·7% of the original UV absorbing material. Polyacrylamide gel electrophoresis showed only one band.

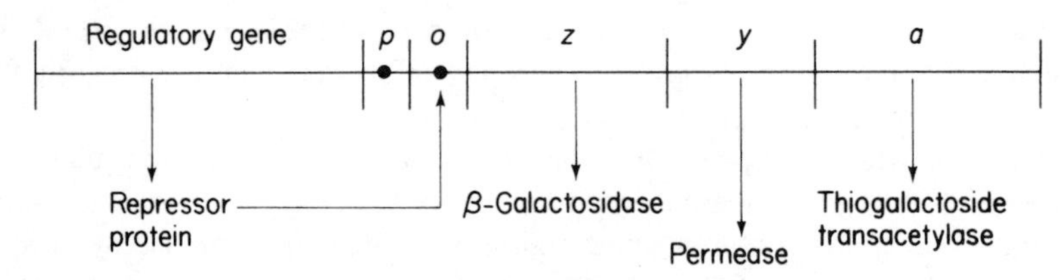

Figure IV.9. The lactose operon

Similar studies by Wilcox *et al.*[130] have shown that the *ara*C gene product, a regulatory protein required for the expression of the L-arabinose operon, can be purified by affinity chromatography. 4-Amino-phenyl-β-D-6-deoxy-galactopyranoside was found to be a very powerful anti-inducer of the *ara* operon and was attached to Sepharose via a 4-aminophenyl-butanamido side-chain. A protein, about 20% pure, that had the characteristics of the *ara*C gene protein was isolated from extracts of an F^1 homogenote containing at least two copies of the *ara*C gene. Elution was effected with 0·1M borate, pH 10, and the eluted protein binds not only to the anti-inducer, 4-amino-phenyl-β-D-6-deoxygalactopyranoside, but also specifically to *ara*DNA. These results strongly suggest that the purified protein is the product of the L-arabinose C gene.

Many *z* gene mutant strains of *E. coli* produce defective β-galactosidases in that only a portion of the polypeptide chain is produced which, although enzymically inactive, will cross-react with an antibody produced to the wild-type enzyme.[131] Villarejo and Zabin[132] have used affinity chromatography to establish whether the incomplete polypeptide chains from a variety of *z* gene mutants are folded in the conformation necessary for recognition of the substrate. Fig. IV.10(a) shows the behaviour of wild-type β-galactosidase on an adsorbent comprising *p*-aminophenyl-β-D-thiogalactoside covalently attached to Sepharose.[133]

Mutant β-galactosidase polypeptides shown in Fig. IV.10(b) are indistinguishable in their behaviour from the wild-type enzyme and include all of the middle part and at least some of the NH_2-terminal sequences. The mutant

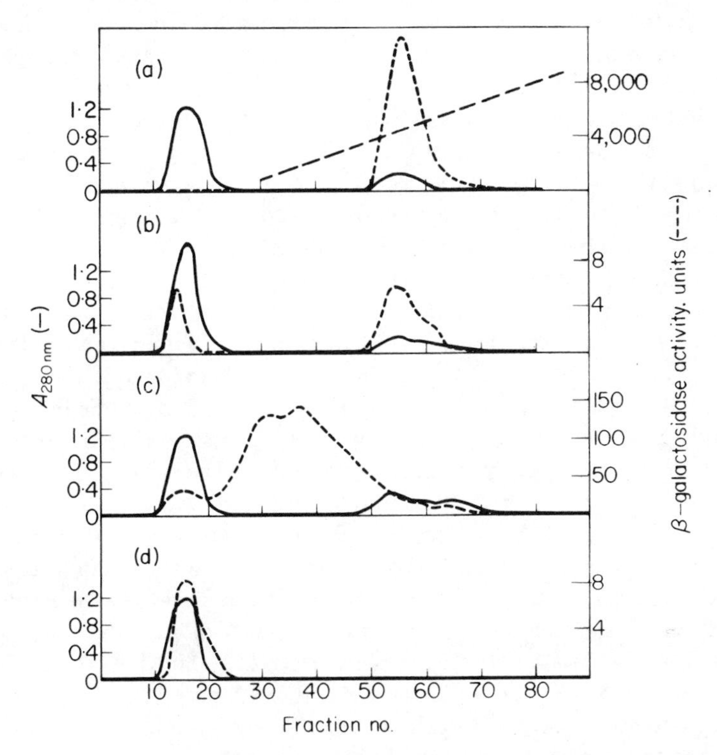

Figure IV.10. Affinity chromatography of *E. coli* wild-type β-galactosidase and mutant polypeptide chains on an adsorbent comprising *p*-aminophenyl-β-D-thiogalactoside covalently attached to Sepharose. (a) Wild-type enzyme; (b) mutant β-galactosidase polypeptides containing all of the middle part and at least some of the NH_2-terminal sequence; (c) mutant polypeptides containing at least 40% of the molecule beginning at the NH_2-terminus; (d) mutant polypeptides containing the COOH-terminal region of the β-galactosidase sequence. Reproduced with permission from M. R. Villarejo and I. Zabin, *Nature New Biol.*, **242,** 50 (1973)

peptide chains that are retarded by the adsorbent (Fig. IV.10(c)) contain at least 40% of the molecule beginning at the NH_2-terminus. A third group of mutant peptides which contain the COOH-terminal region of the β-galactosidase sequence, but in which most of the middle region is absent, are neither bound nor retarded by the adsorbent (Fig. IV.10(d)). These results suggest that the complete polypeptide chain of this protein is not necessary for the achievement of the native or nearly native conformation.

Miller *et al.*[134] have utilized a similar principle for partially purifying tyrosine aminotransferase-synthesizing ribosomes from Hepatoma tissue culture (HTC) cells. Affinity columns of immobilized pyridoxamine phosphate will selectively bind tyrosine aminotransferase (TAT) from HTC cells with some 100-fold purification on elution.[135] Since TAT contains four identical subunits each binding one molecule of pyridoxal-phosphate,[136] the synthesis of one TAT subunit on a polysome may be sufficiently complete to bind the immobilized pyridoxamine phosphate. If the entire protein synthesis system for TAT remained intact and could be eluted undamaged, then synthesis of TAT might continue to completion in a cell-free system. The TAT-synthesizing ribosomes were adsorbed onto the support and eluted with 10 mM pyridoxal phosphate. The enriched fractions were enzymically active and contained immunoprecipitable material, suggesting that TAT-synthesizing capacity was higher in the fractionated than in the unfractionated ribosomes. A related approach by Faust *et al.*[137] may be more generally applicable to the isolation of a specific type of polysome from a heterologous mixture. The adsorbent, prepared by covalently coupling the antibody to Sepharose (the antibody was raised to the desired polysomal-synthesized protein), binds only to the immunochemically competent peptides. The specifically adsorbed polysomes can be recovered by treatment with moderate salt and puromycin.

Stimulation of RNA transcription from pea and corn DNA has been observed by a protein retained on an affinity adsorbent comprising the synthetic plant hormone, 2,4-dichlorophenoxy acetic acid (2,4-D) attached to Sepharose.[138] Chromatography of crude extracts of pea and corn on the immobilized hormone and elution with 1M NaCl, 2 mM KOH or 50 mM 2,4-D yielded a protein factor that stimulated DNA-dependent RNA synthesis in the presence of added *E. coli* polymerase by 40–200%. Some evidence suggested that the protein may be influencing the initiation of RNA chains.

Gyenge *et al.*[139] have shown that Sepharose-bound ribosomal proteins will form specific complexes with ribosomal RNA.

B. ANALYTICAL APPLICATIONS

Whereas affinity chromatography has, so far, been most fruitfully exploited in preparative methods, the study and exploration of complex interacting systems can often be greatly facilitated by immobilizing one of the components on an inert solid support. The method should, in principle, be applicable to any binding system and permit quantitative evaluation of the nature and mechanism of the binding process. The approach could thus be evoked for the resolution of chemically modified, affinity labelled or inactive enzymes from their native counterparts, for the study of enzyme subunit or kinetic interactions, or for the separation and resolution of nucleic acids.

1. Resolution of Chemically Modified Enzymes

The general principles of affinity chromatography make it ideally suited to the resolution of native and chemically or otherwise modified proteins. Thus, fully functional xanthine oxidase has been separated from non-functional enzyme despite the fact that both forms of the enzyme possess a full complement of the redox components, molybdenum, FAD, iron and acid-labile sulphur.[163] Non-functional xanthine oxidase does not possess the cyanolysable persulphide group that is required for catalysis, but can be partially reactivated by incubation with sodium sulphide.

Pyrazolo (3,4-*d*) pyrimidines are powerful inhibitors of xanthine oxidase[164] and form inactive complexes with the enzyme when the molybdenum component is in the reduced state (Mo^{IV}). The fact that only the functional enzyme formed tight complexes with these inhibitors was used as a basis for the chromatographic separation. 6-Amino-1-propyl-3(1*H*-pyrazolo(3,4-*d*)pyrimidin-4-ylamino) hexanoate was synthesized, coupled to Sepharose and converted to the 6-hydroxy derivative shown in Fig. IV.11 by prolonged aerobic incubation with xanthine oxidase.

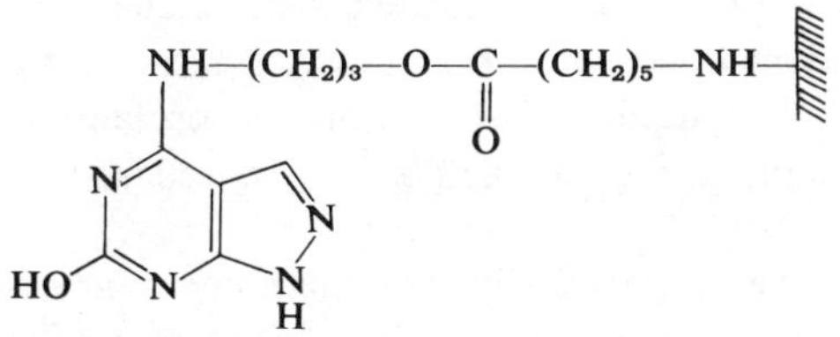

$NH-(CH_2)_3-O-C(=O)-(CH_2)_5-NH-$

Figure IV.11. The structure of the affinity adsorbent used for the separation of functional and non-functional xanthine oxidase

Milk xanthine oxidase containing approximately 70% functional enzyme was applied to the adsorbent, equilibrated with a buffer containing $Na_2S_2O_4$ under an atmosphere of nitrogen and allowed to equilibrate overnight at room temperature. A fraction with very little activity was eluted with the $Na_2S_2O_4$-containing buffer whilst the remaining enzyme was tightly bound to the adsorbent. The column was then equilibrated aerobically for several days at 4°C to allow a slow oxidation of Mo^{IV} to Mo^{VI}, with the concomitant dissociation of the enzyme from its complex with the pyrazolo (3,4-*d*) pyrimidine adsorbent.[165] Subsequent elution with the aerobic buffer showed that this eluate contained $\geqslant 95\%$ functional enzyme.[164]

Likewise, immobilized derivatives of *p*-aminophenyl mercuric acetate have been used to resolve active papain from enzymically inactive enzyme.[166] A similar problem can arise by chemical modification of purified proteins, which rarely leads to complete loss of enzymic activity. Affinity chromatography can often be used to determine if the residual activity reflects unmodified native protein or a quantitatively altered protein with diminished catalytic activity. For example, 83% of the enzyme activity of a Staphylococcal nuclease preparation was lost on stoichiometric reaction of an active-site tyrosyl residue with the diazonium derivative of deoxythymidine-3′-*p*-aminophenyl-phosphate 5′-phosphate.[167] Affinity chromatography of the affinity-labelled nuclease on a nuclease-specific column[92] revealed two equally yellow protein fractions, the major one appearing in the void volume and devoid of activity, the other strongly adsorbed and enzymically active. The residual activity was thus caused by a 20% contamination of non-active site-labelled protein. Similar resolution of active and non-active Staphylococcal nuclease could be achieved following chemical modification with bromoacetamidophenyl active-site reagents.[168] Furthermore, affinity chromatography was employed to separate monocarboxamidomethyl carbonic anhydrase from unreacted enzyme on Sepharose-bound *p*-aminomethylbenzene sulphonamide.[169]

2. Purification of Affinity Labelled Active-site Peptides

Active-site-directed irreversible inhibitors are used extensively in protein chemistry.[170] Analogues of the substrate possess a chemically reactive group that forms a covalent bond with an amino acid residue at or near the active site of the protein. The affinity labelled protein is proteolytically digested and the peptides, including the labelled one, are separated by conventional techniques.

Givol *et al.*[171] have described a general method for the isolation of labelled peptides by affinity chromatography that utilizes the affinity of the native protein for the ligand used for the labelling of the same protein. The labelled peptide is the only one in the digest that displays affinity for the immobilized protein and can be eluted under conditions that dissociate the protein–ligand bonds.

A uridine diphosphate-labelled peptide was isolated from tryptic digests of bovine pancreatic ribonuclease (RNase) on a RNase-Sepharose column[171] and a dinitrophenyl- (DNP)-labelled peptide from anti-DNP antibody by an anti-DNP Sepharose column. Elution of the labelled peptides was effected with 0·8M NH_4OH and 20% formic acid respectively. Furthermore, Wilchek[172] has used this method to isolate affinity labelled peptides from Staphylococcal nuclease reacted with the bromoacetyl derivative of deoxy-thymidine-3′-*p*-aminophenyl-phosphate-5′-phosphate and with bromoacetyl-*p*-aminophenyl-phosphate. Fig. IV.12 shows the affinity chromatography of the affinity-labelled peptides on a nuclease column.

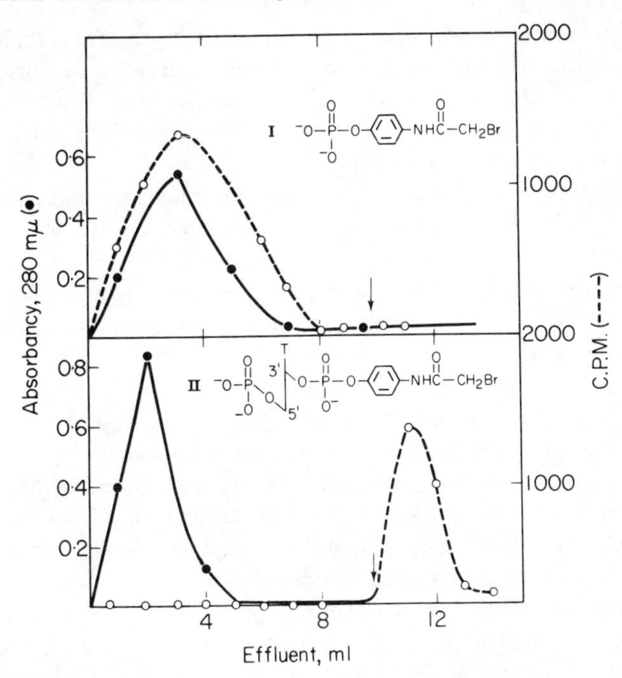

Figure IV.12. Affinity chromatography on nuclease-Sepharose column (0·5 cm × 2 cm) of affinity labelled peptides with reagents I and II. The columns were equilibrated with 0·05M borate buffer, pH 8·0, containing 10 mM $CaCl_2$. Tryptic digests of modified nuclease (1·7 mg) were applied in 0·5 ml of the same buffer. After 10 ml of buffer had passed through, the bound peptides were eluted with NH_4OH, pH 11·0 (arrow). Reproduced with permission from M. Wilchek, *FEBS Lett.*, **7**, 161 (1970)

3. Purification of Complementary and Synthetic Peptides and Proteins

Affinity chromatography can readily be exploited in cases where both partners of the interacting system are either proteins or large peptides. However, for the immobilization of proteins it is important that the tertiary structure is not disrupted and some considerations with regard to this are given elsewhere in the book.

Bovine pancreatic ribonuclease A consists of a peptide chain comprising 124 amino acid residues which can be cleaved by proteases into residues 1–20, the *S*-peptide, and residues 21–124, the *S*-protein.[140] The two components are not active individually but combine at neutral pH to yield an active enzyme.

This interaction has been used by Hofmann *et al.*[141] to purify ribonuclease *S*-peptide synthesized by conventional techniques of peptide chemistry.

Furthermore, immobilized ribonuclease *S*-protein bound the complementary *S*-peptide, selectively and with a high capacity, from crude preparations arising from solid phase synthesis.[143] Gawronski and Wold[142,144] have investigated the kinetics and thermodynamics of the interaction of ribonuclease *S*-protein and agarose-*S*-peptide.

Tryptic digestion of the major extracellular nuclease of *Staphylococcus aureus* also yields an enzyme derivative consisting of two non-covalently bound polypeptide chains, termed Nuclease-T.[145,146] The solid-phase synthesis of the smaller 42-residue peptide, termed Fragment P_2, has been reported.[147] The binding of synthetic analogues of P_2 to its complementary native fragment, P_3, containing residues 49–149, has been investigated by affinity chromatography on an adsorbent comprising P_3 covalently attached to agarose. The native fragment P_2 bound tightly to the P_3-agarose adsorbent and could be eluted under conditions known to dissociate P_2 and P_3, 0·1M acetic acid, pH 3. Likewise, synthetic analogues of P_2 containing residues 6–47 and 9–47 were bound whilst those containing only residues 18–47 or 33–47 were not adsorbed. The synthetic 6–47 polypeptide isolated in this way exhibited 30–50% of the enzymic activity of the native material.

The advantage accrued in the exploitation of the interaction of complementary peptides and proteins has been clearly recognized in the functional purification of specific protein inhibitors. For example, Gribnau *et al.*[148] describe how a highly purified ribonuclease (RNase) inhibitor from rat liver can be obtained by chromatography on immobilized-RNase. A partially purified extract from rat liver supernatant was applied to a column of RNase linked to CM-cellulose and, when practically all the protein had emerged in the void volume, elution with 0·9M NaCl eluted the inhibitor with a very high specific activity. In a related procedure, the inhibitor of pancreatic deoxyribonuclease I (DNase I) has been purified from crude extracts of calf thymus by affinity chromatography on immobilized DNase.[159] The solid-phase adsorbent removed one major protein from crude extracts which could be eluted with 3M guanidine–HCl in 1M sodium acetate and 30% glycerol. However, the yield of active DNase inhibitor was only 8%, probably due to its instability in the solvent system which is necessary to effect its elution from the adsorbent. Furthermore, the high affinity of the natural proteinase inhibitors for their complementary enzymes, and the relative ease of dissociation of the resultant complexes at low pH values, have been exploited in affinity chromatography. These attempts are summarized in Table IV.4 and discussed in detail in Chapter III of this book.

Pradelles *et al.*[160] have purified the antidiuretic hormone vasopressin by affinity chromatography on Sepharose-bound neurophysins. The nonapeptide hormone vasopressin and the related hormone oxytocin are stored in neurosecretory granules of the bovine pituitary posterior lobes in association with a number of relatively small proteins.[161] These proteins, termed neurophysins,

Table IV.4. The purification of proteases and natural protease inhibitors by affinity chromatography

Adsorbent	Complementary protein	References
Chicken ovomucoid	Trypsin (several sources)	149, 150
Soybean trypsin inhibitor	Proteases from activated pancreatic juice	151, 152
Potato inhibitor	Elastin	153
Kallikrein inhibitor	Kallikrein, plasmin	154
Soybean inhibitor	Kallikrein	155
α-Chymotrypsin	Chicken ovo inhibitor	156
Trypsin	Peanut and soybean trypsin inhibitors	157
Trypsin	Bovine ovary trypsin inhibitor	158

form non-covalent stoicheiometric 1:1 complexes with the hormones.[162] Isotopically labelled 8-lysine-vasopressin was purified on an adsorbent comprising bovine neurophysins immobilized on agarose and eluted with 0·1N formic acid, pH 2·5.

4. Exploration of Enzyme Mechanisms

The effectiveness of affinity chromatography as a tool for enzyme purification can be improved by taking advantage of kinetic characteristics. The technique can also be used to study the mechanism and nature of the interaction since it is capable of yielding unequivocal information about multi-component reaction mechanisms. For example, the elegant work of O'Carra and Barry[173] has clearly confirmed the compulsory-ordered mechanism of lactate dehydrogenase, in which the pyridine nucleotide binds first (Fig. IV.13). Affinity chromatography of the enzyme on ligands kinetically analogous to pyruvate or lactate would not be expected to be fruitful since it is the binary enzyme–nucleotide complex that interacts with them. An immobilized oxamate derivative, structurally analogous to pyruvate, was prepared by coupling oxalate to 6-aminohexyl-Sepharose with a water-soluble carbodiimide. In the presence of 0·5M NaCl, lactate dehydrogenase had no affinity for the pyruvate analogue and was eluted within one column volume. However, when NADH was included in the irrigant at concentrations as low as 10 μm, the enzyme was strongly retained. When NADH was removed from the irrigant, lactate dehydrogenase was eluted approximately one column later (Fig. IV.14). These data imply that NADH induces a binding site for pyruvate in lactate dehydrogenase.

In contrast to the pronounced effect of NADH, lactate dehydrogenase is retarded by the immobilized oxamate in the presence of saturating NAD^+

concentrations. The comparatively weak binding in the presence of NAD^+ is consistent with kinetic studies with free pyruvate and oxamate.[174] When free pyruvate is included with the NAD^+ in the equilibrating buffer the retardation is virtually abolished and chromatography at several NAD^+ and pyruvate concentrations suggests a direct competition between pyruvate and the immobilized oxamate. These results are consistent with the formation of an 'abortive' ternary complex of enzyme–NAD^+–pyruvate.

The results presented in Fig. IV.14 indicate that there is no interaction between the immobilized oxamate and lactate dehydrogenase in the presence of near saturating concentrations of 5′-AMP or 5′-ADP. It seems reasonable to conclude that ligand binding to lactate dehydrogenase follows the sequence: the AMP half of the nucleotide followed by the nicotinamide end, followed by oxamate or pyruvate, where each ligand induces the binding site for the next.

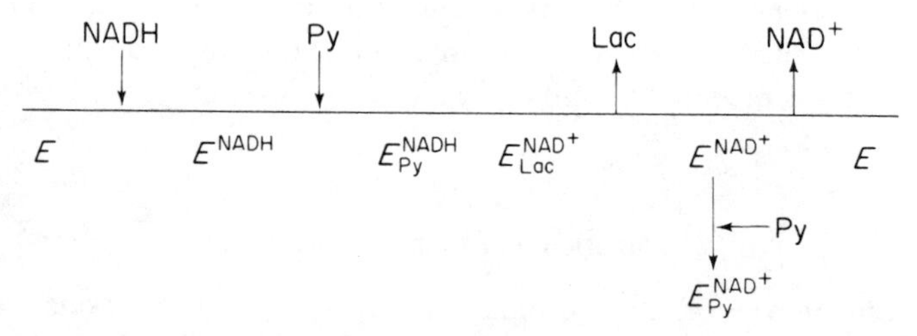

Figure IV.13. The ordered reaction sequence of lactate dehydrogenase. E = enzyme, Py = pyruvate, Lac = lactate

This conclusion has been supported by the work of Lowe and Dean[175] with the interaction of lactate dehydrogenase with immobilized fragments of NAD^+. Table IV.5 shows that the binding of rabbit muscle lactate dehydrogenase was relatively strong to immobilized adenine-containing fragments of the coenzyme, until the 5′-phosphate was removed. No binding to the immobilized adenosine or adenine was observed and the binding to immobilized ADP-ribose was stronger than that to NAD^+.

These data[175] and those of O'Carra and Barry[173] are entirely consistent with the free kinetics of adenosine, AMP, ADP, ADP-ribose and NAD^+ for dogfish muscle lactate dehydrogenase.[176] The K_i values are in the order: adenosine $\gg$ ADP $\simeq$ AMP $>$ ADP-ribose, suggesting that the 5′-phosphate produced a significant increase in binding affinity. All three phosphate-containing fragments produced the conformation change in crystalline lactate dehydrogenase that was characteristic of NAD^+ and NADH. It was suggested that the 5′-phosphate group anchored by the adenosine moiety interacted with charged groups on the protein to induce the conformation change.[177]

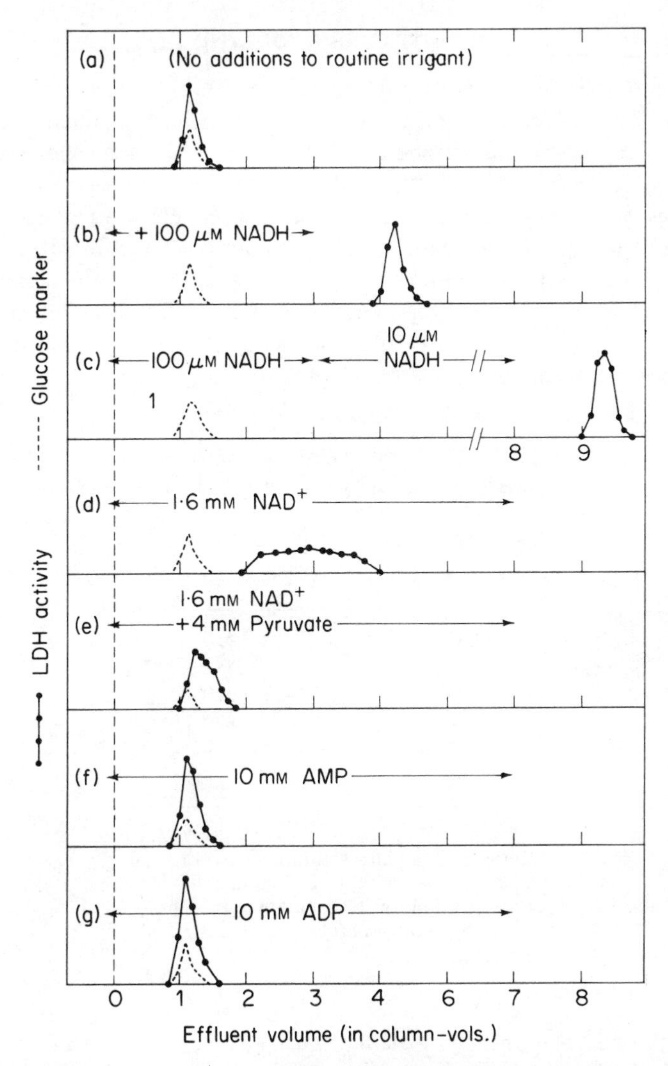

Figure IV.14. Affinity chromatography of lactate dehydrogenase (H_4) from pig heart on the Sepharose-linked oxamate derivative. Irrigant throughout was 0·5M NaCl in 0·02M sodium phosphate buffer, pH 6·8, with the nucleotide additions indicated on the individual elution diagrams. The points of commencement and termination of such additions are indicated by the points of the horizontal arrows which are correlated with volume of effluent collected when such changes were made. Effluent volume was measured from time of application of sample, this zero-point in the effluent being marked with the vertical dashed line in the figure. As indicated, additions to the irrigant were commenced a little ahead of sample application. The same additions were made to the applied samples, which were made up in the routine irrigant. A little glucose was added to all applied samples and was analysed in the effluent fractions to provide a reliable measure of 'straight-through' (i.e. unretarded) volume. (AMP = adenosine 5′-monophosphate.) Reproduced with permission from P. O'Carra and S. Barry, *FEBS Lett.*, **21**, 281 (1972)

Similar studies on the galactosyltransferase (EC 2.4.1.22) of bovine skim milk have detected enzyme-reactant complexes by affinity chromatography.[178] Kinetic studies on the enzyme indicate the ordered reaction mechanism shown in Fig. IV.15.[179,180] Furthermore, UDP-glucose and L-arabinose are inhibitory analogues of UDP-galactose and glucose respectively, and *N*-acetylglucosamine and glucose are subject to substrate inhibition. The substrate inhibition increased as the α-lactalbumin concentration increased and the formation of the dead-end or abortive complexes, *E*-glucose (or *E*-*N*-acetylglucosamine) and *E*-MnUDP-glucose (or *E*-MnUDP-*N*-acetylglucosamine) was evoked to explain the phenomena.

Table IV.5. The binding of rabbit muscle lactate dehydrogenase to immobilized fragments of NAD^+

Immobilized fragment	Binding (mM)[a]
NAD^+	410
ADP-Ribose	540
ADP	290
AMP	270
Adenosine	0
Adenine	0

[a] The binding represents a measure of the strength of the interaction between the enzyme and the immobilized ligand and is the concentration of KCl (mM) required to elute the peak of activity on a linear KCl gradient.

Since the galactosyltransferase can be purified on columns of α-lactalbumin-Sepharose in the presence of glucose or *N*-acetylglucosamine[181,182] the effects of various reactants in retarding the passage of the enzyme through a similar column were tested. In the presence of glucose and Mn^{2+} the galactosyltransferase was retarded but was eluted when glucose was removed from the irrigant. The enzyme was less readily eluted in the presence of Mn^{2+} than with glucose alone, suggesting the formation of a stronger complex when Mn^{2+} was present. Similar results were obtained with *N*-acetylglucosamine as substrate. Furthermore, no retardation was observed when lactose and Mn^{2+}, sucrose or arabinose were present in the irrigant.

Although this technique provides evidence for the occurrence of enzyme-reactant complexes it does not shed light on whether the complex is important in the catalytic sequence, is an abortive complex or is formed under the conditions for studying the overall reaction. At first sight it would appear that glucose and *N*-acetylglucosamine combine firstly with free enzyme or *E*–Mn^{2+} complex and then react with the immobilized α-lactalbumin as part of the normal catalytic mechanism. However, these complexes are responsible for the substrate inhibition by glucose and *N*-acetylglucosamine.[180] The enzyme is more strongly retarded when Mn^{2+}, UDP and glucose are present in the buffers than it is in the presence of glucose alone. This finding confirms the formation of an abortive *E*–MnUDP–glucose complex which interacts with α-lactalbumin and contributes to the substrate inhibition by glucose. The specificity of the complexes was shown by the lack of effect of lactose or sucrose.

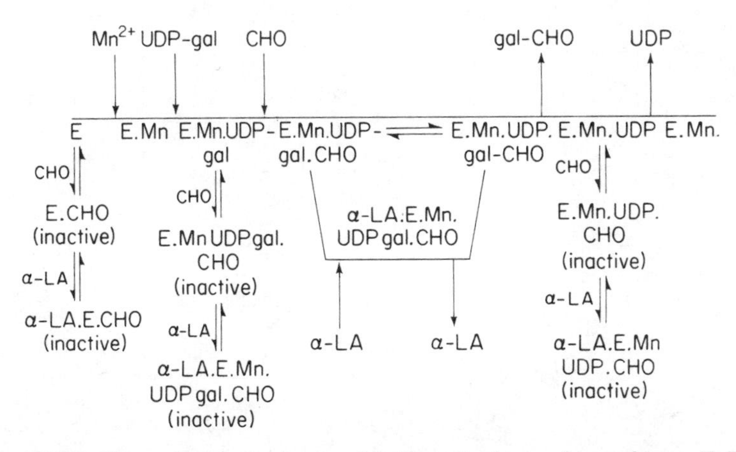

Figure IV.15. The ordered reaction mechanism of galactosyl transferase. E, UDP, UDP-gal, α-LA and CHO represent enzyme, uridine 5′-diphosphate, UDP-galactose, α-lactalbumin and carbohydrate respectively. Reproduced with permission from R. Mawal, J. F. Morrison and K. E. Ebner, *J. Biol. Chem.*, **246**, 7106 (1971)

These results support the conclusions obtained with previous kinetic studies[179,180] and detect complexes that do not appear to be catalytically significant. Clearly, some caution must be exercised in the interpretation of kinetic mechanisms on the basis of affinity chromatography, although the method can give unequivocal evidence for some interactions. For example, Akanuma *et al.*[183] have studied the binding of basic and aromatic amino acid analogues to bovine carboxypeptidase B. The method revealed that these ligands occupy different sites on the enzyme and that the affinity for the aromatic residues was profoundly influenced by ε-aminocaproic acid. Furthermore, Doellgast and Kohlhaw[184] have provided some evidence for a conformational change in α-isopropylmalate synthase. The affinity of the *Salmonella typhimurium* enzyme for an immobilized feedback-inhibitor, L-leucine-Sepharose, increased as the concentration of potassium phosphate increased from 0·2 to 1·0M. The order of binding to several immobilized amino acids, L-leucine > L-isoleucine ≫ D-leucine > L-valine, was identical to that observed for inhibition of enzyme activity. It has been assumed that the high salt concentrations induce a conformational change affecting the leucine binding site. Similar results were obtained with the biosynthetic threonine deaminase from *E. coli.*[185]

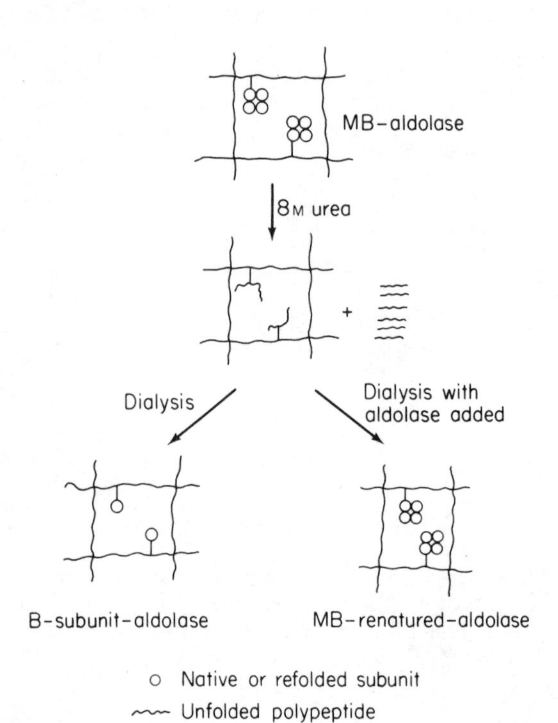

Figure IV.16. Scheme of matrix-bound aldolase derivatives. Reproduced with permission from W. W. C. Chan, *Biochem. Biophys. Res. Commun.*, **41**, 1198 (1970)

Insoluble derivatives of proteins can also be used to study enzyme mechanisms and the effects of subunit aggregation and polymerization. Thus, Wider de Xifra *et al.*[186] have studied the reaction of immobilized soybean callus succinyl-CoA synthetase with ATP and furnished evidence for a stable phosphorylated form of the enzyme. Furthermore, immobilized ATP is an enzymically active coenzyme for the enzyme and generates the same phosphoenzyme intermediate. Chan[187] has covalently attached protein subunits to a rigid matrix and thus prevented them from reassociation. Native rabbit muscle aldolase can be dissociated by 8M urea[188] into four identical subunits which reassociate when the urea is subsequently removed by dialysis. Aldolase was attached covalently to agarose under conditions such that only one link between the enzyme and the matrix backbone was formed (Fig. IV.16). Exhaustive washing with 8M urea led to dissociation and removed all subunits not covalently attached to the matrix.

Furthermore, immobilized aldolase subunits could be renatured by adding back excess aldolase subunits in 8M urea prior to dialysis. The immobilized subunits displayed about one-third the specific activity of the original immobilized tetramer. In contrast, Feldman *et al.*[189] have shown that complete dissociation of Sepharose-bound rabbit muscle phosphorylase to monomers leads to a loss of almost all of the activity of the original matrix-bound dimer. Stable hybrids between phospho- and dephospho-subunits and between phosphorylase subunits containing the active cofactor, pyridoxal phosphate, and inactive analogues can be prepared. The hybrid enzyme, composed of an active subunit containing pyridoxal phosphate and an intrinsically inactive subunit binding the analogue, had half the specific activity of the original dimer.

Ikeda and Fukui[190] have described a novel way of immobilizing tryptophanase based on its interaction with pyridoxal 5′-phosphate. Sepharose-bound pyridoxal 5′-phosphate was prepared by coupling pyridoxal 5′-phosphate to diazotized *p*-aminobenzamidohexyl-Sepharose. Tetramic apotryptophanase was immobilized by incubation with the Sepharose-bound cofactor for 20 minutes at 37°C and the resulting complex reduced with sodium borohydride. The immobilized tryptophanase retained about 60% of the enzymic activity of the free tryptophanase used. This technique was superior to immobilizing tryptophanase with CNBr-activated agarose or with diazotized *p*-aminobenzamidohexyl-Sepharose.[190]

5. Applications to Nucleic Acid Biochemistry

The binding of mononucleotides, oligonucleotides and nucleic acids to inert matrices provides a useful approach both to the study of the physical and chemical properties of nucleic acids and to the enzymes involved in their synthesis and degradation. Insoluble oligonucleotides can be used to separate, fractionate and determine the structure of various nucleic acids.

Early methods for immobilizing nucleic acids involved drying DNA onto cellulose[191] by ultraviolet irradiation. The method suffered the disadvantage that the adsorbed DNA slowly leached out and it was not possible to insolubilize low molecular weight DNA. Nevertheless, single-stranded DNA bound to nitrocellulose filters will bind homologous RNA,[192] and the technique has been developed into a general procedure for the isolation of gene-specific *m*RNA.[193,194] For these studies the insolubilized polynucleotide is ideally attached by specific covalent linkages such that the positions and orientation of the polynucleotide chains are known. Thus, nucleic acids can be covalently attached to acetylated phospho-cellulose by their hydroxyl groups with dicyclohexyl-carbodiimide in pyridine[195] or methanol.[196] Poonian *et al.*[197] coupled nucleic acids to CNBr-activated agarose although the method only readily linked single-stranded deoxy- and ribonucleic acids. However, they also described a method whereby single-stranded ends could be introduced into double-stranded molecules to effect covalent attachment of the latter to agarose.

The terminal nucleotides of polynucleotides are different from those present in the internal parts of the polymer chain, and hence it would be expedient to develop methods for the covalent binding of these polymers at one or the other terminus. Such insoluble polymers can be utilized in new fractionation methods in which the resolution is based on the stabilities of base-paired complexes formed between the free and insolubilized nucleotides.[198,199] There are two general synthetic approaches to the terminal linkage of polynucleotides. First, mononucleotides are polymerized chemically in anhydrous solution, and on exposure to cellulose powder the activated terminal phosphate groups condense with the cellulose hydroxyl groups.[198] Homopolymer chains containing up to about 10 nucleotides can be coupled by phosphodiester linkages to the 3′- or 5′-terminus depending on whether the 3′ or 5′-nucleotides were polymerized. Secondly, the terminal phosphate group is activated in aqueous solution under mild conditions.[200] Thus, the terminal monosubstituted phosphate or polyphosphate of the nucleotide or polynucleotide can be specifically activated in aqueous solution at pH 6 by a water-soluble carbodiimide.

Immobilized polynucleotides have been used as primers and templates for nucleotide-polymerizing enzymes.[201] Thymidine polynucleotide cellulose serves as a primer and template for DNA polymerase, a template for RNA polymerase and as an initiator for terminal deoxynucleotidyl transferase. Furthermore, Cozzarelli *et al.*[202] have used the same insolubilized polynucleotide, together with its complementary polynucleotide, to assay for enzymic joining of polynucleotide strands.

Robberson and Davidson[203] describe a procedure for coupling RNA to a modified agarose resin with a final objective of isolating high molecular weight single strands of DNA containing *r*DNA genes. The method involves coupling periodate oxidized RNA to a hydrazide derivative of agarose. Poly(U) on the resin surface will hybridize with poly(A) in solution and the poly(A) can subsequently be eluted with a denaturing solvent, dimethylsulphoxide. Furthermore, a 20-fold enrichment of *r*DNA from sheared denatured *E. coli* DNA of single-stranded molecular weight 7×10^4 can be achieved on hybridization with immobilized *E. coli* 16S *r*RNA and elution with sodium hydroxide.

The unsubstituted 2′,3′-diol groups located at the 3′-terminus of polynucleotides and RNA molecules are capable of complex formation with the borate anion. The changes in physical characteristics exhibited by the diol in the presence of borate have been exploited for the chromatographic and electrophoretic separation of these components.[204] Weith *et al.*[205,206] have

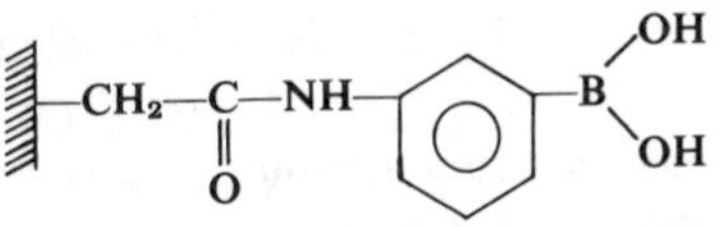

Figure IV.17. The structure of the dihydroxyboryl derivative

synthesized cellulose derivatives containing the dihydroxyboryl group (Fig. IV.17) and studied their capacity to form specific complexes with sugars, polyols and nucleic acid components. Fig. IV.18 illustrates the elution pattern of a complex mixture of nucleotides when chromatographed on a column of *N*-(*m*-dihydroxyborylphenyl)-carbamylmethyl-cellulose.

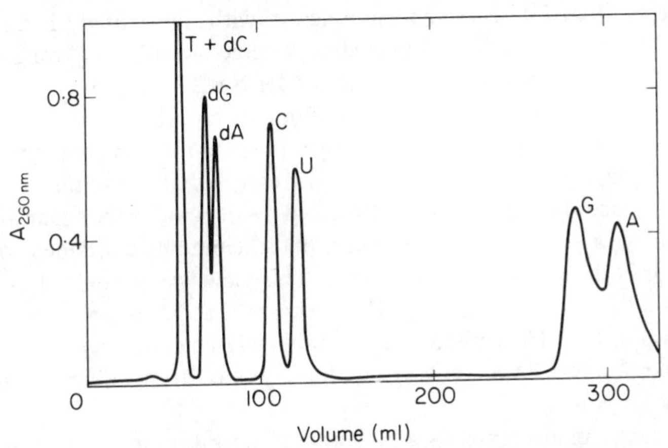

Figure IV.18. Elution pattern obtained from the chromatography of a mixture of thymidine (T), deoxycytidine (dC), deoxyguanosine (dG), cytidine (C), uridine (U), guanosine (G) and adenosine (A) on a column of *N*-(*m*-dihydroxyborylphenyl)-carbamylmethylcellulose. The column had dimensions 55 cm × 1 cm and the elution was effected with 0·1M sodium phosphate buffer, pH 7·5, at 20°C and at a flow rate of about 10 ml/hour. Reproduced with permission from H. L. Weith, J. L. Wiebers and P. T. Gilham, *Biochemistry*, **9**, 4396 (1970)

The retention volume for a particular polyol was found to depend on (i) the presence of a suitable glycol on the compound in the appropriate configuration and conformation, (ii) the pH of the eluant, (iii) the ionic strength and nature of the eluant and (iv) with nucleotides, the nature of the base attached to the glycol moiety.

C. SPECIAL TECHNIQUES

1. Hydrophobic Affinity Chromatography

Hydrophobic bonding significantly contributes to the strength of non-covalent interactions in aqueous media and may function in maintaining the native tertiary structure of proteins. The occurrence of accessible hydrophobic binding sites on many enzymes,[207,208] and on a disparate array of proteins such as immunoglobulins,[209] serum albumin,[210] β-lactoglobulin and ovalbumin[211] has been recognized only recently.

Proteins with binding sites for hydrophobic moieties could in principle be purified on adsorbents containing only hydrocarbon groups. In aqueous media such an adsorbent would have some of the characteristics of an oil–water interface, including the capacity to denature proteins. Nevertheless, hydrophobic ligands have been attached to Sepharose and utilized for affinity chromatography.[212,213]

Hofstee[212] has investigated the binding of several purified proteins to 4-phenylbutylamine (PBA) covalently attached to Sepharose and deduced some factors that influence their interaction with the strongly hydrophobic phenyl-$(CH_2)_4$-residue. The 4-phenylbutylamine adsorbent strongly binds α-chymotrypsin[214] even in the presence of 1M NaCl,[212] although the enzyme is readily released by 1M tetraethylammonium chloride. Furthermore, the binding of α-chymotrypsin is reversed by a polarity-reducing compound such as 50% ethylene glycol, an effect which is greatly enhanced in the presence of 1M NaCl. Since the salt would be expected to increase rather than decrease hydrophobic interactions, it would seem that the binding contained contributions from both apolar and polar forces. This view was supported by studies with several other proteins. Fig. IV.19 shows that 7*S* γ-globulin was very tightly bound to PBA-Sepharose and could only be eluted by 50% ethylene glycol and 1M NaCl together. In contrast, some serum albumin could be eluted by 1M NaCl.

Serum albumin also exhibits a high affinity for bilirubin[215] and long-chain fatty acids.[210] Peters *et al.*[216] describe the synthesis of several affinity adsorbents by coupling oleic and palmitic acids to aminoalkyl-Sepharose with a water-soluble carbodiimide reaction. The immobilized fatty acid adsorbents bound about 10 mg albumin per ml of agarose and retained some other proteins to a small extent and with a lower affinity. When a serum sample was applied to a column of oleyl-aminoethylamino-agarose, all of the albumin and a small proportion of the α- and β-globulin fractions were removed. Albumin could be eluted virtually free of contaminants in a yield of 80% with ethanol/1·4M acetic acid (1:1) or with solutions of oleate in 10% ethanol. Acidic solutions were ineffective in the absence of ethanol. Increasing the length of the alkyl arm from 2 to 10 carbon atoms had little effect on the capacity for albumin although it did increase the binding of other serum proteins.

A homologous series of hydrocarbon-Sepharoses of general type, Sepharose-NH$(CH_2)_n$-H, where $n = 1$–8, have been synthesized. Lysozyme, bovine serum albumin and bovine γ-globulin were not bound by Sepharose containing covalently attached 4-aminobutane (Seph-C4), although rabbit muscle glycogen phosphorylase *b* was retained and could be eluted with deforming buffers. Phosphorylase *b* was not retained by methyl-Sepharose ($n = 1$), was retarded by propyl-Sepharose ($n = 3$), adsorbed on butyl-Sepharose ($n = 4$)

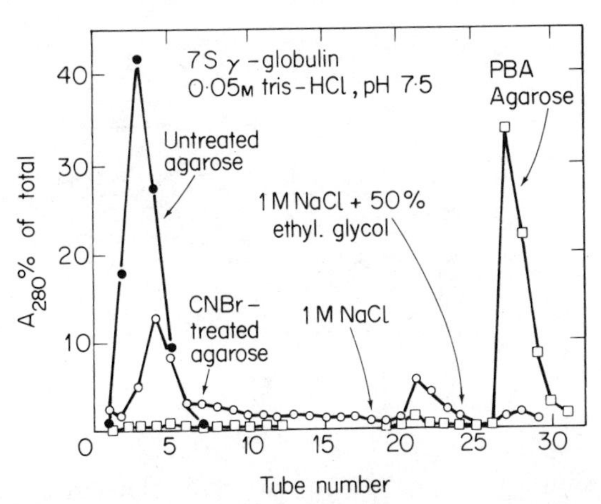

Figure IV.19. Hydrophobic affinity chromatography: behaviour of 7S γ-globulin (1–2 mg) on 5 ml columns of untreated, CNBr-activated and 4-phenylbutylamine-Sepharose 4B, equilibrated at room temperature with 0·05M tris-HCl buffer, pH 7·5. The loaded columns were washed 12–18 times with 2 ml portions of buffer and then successively with several 2 ml portions of buffer containing either 1M NaCl or 1M NaCl plus 50% ethylene glycol. Reproduced with permission from B. H. J. Hofstee, *Anal. Biochem.*, **52**, 430 (1973)

and so tightly adsorbed to hexyl-Sepharose ($n = 6$) that it could only be eluted with 0·2N acetic acid in a denatured form. The enzyme could be purified 100-fold by chromatography of a crude muscle extract on a small column of butyl-Sepharose.

The effects of pH and ionic strength on the interaction of several proteins with adsorbents containing apolar ligands indicated the involvement of both polar and apolar forces.[212] This can be turned to advantage for chromatography of lipophilic proteins on adsorbents containing mixed hydrophobic and ionic species.[217] Careful selection of ionic groups and ambient pH should introduce electrostatic repulsion to counteract the hydrophobic bonding between the protein and the adsorbent. Mixed polar–apolar adsorbents should be able to discriminate between lipophilic proteins to a greater extent than apolar adsorbents. Fig. IV.20 illustrates the chromatography of bovine serum albumin on *N*-(3-carboxypropionyl)-aminodecyl-agarose (CPAD-agarose) at several pH values. Elution was effected with 0·5M NaCl followed by 1·5% sodium dodecyl sulphate or 10% (v/v) butan-1-ol. Near the isoelectric point of serum albumin (pH 4·7–4·9) nearly all the protein was bound hydrophobically, whilst at higher pH values the contribution of charge-repulsion increases and albumin appears in the void volume of the column.

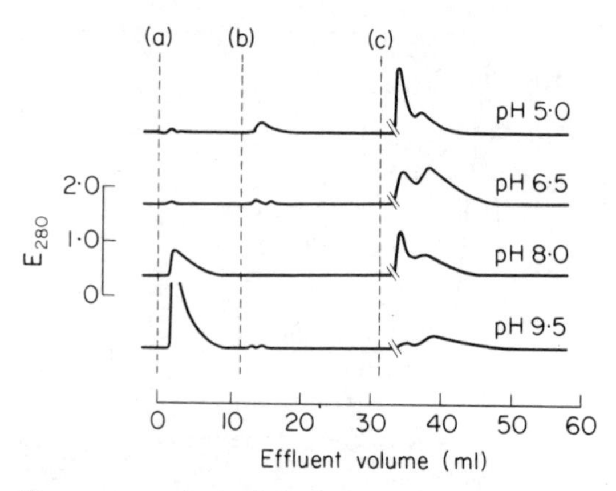

Figure IV.20. Chromatography of bovine serum albumin on a column of *N*-(3-carboxypropionyl)aminodecyl-agarose (CPAD-agarose). Small columns (4 cm × 0·8 cm) of CPAD-agarose were equilibrated with buffer. A 0·1 ml portion of a 10% (w/v) solution of bovine serum albumin was introduced at (a). From (a) to (b) the column was developed with buffer only. Between (b) and (c) the buffer contained 0·5M NaCl. From (c) to the end of the run the buffer contained 1·5% (w/v) sodium dodecyl sulphate. Chromatography was carried out at room temperature (22°C) and buffer compositions were: pH 5·0, 50 mM sodium citrate; pH 6·5 and pH 8·0, 50 mM sodium phosphate; pH 9·5, 50 mM sodium glycinate. Reproduced with permission from R. T. Yon, *Biochem. J.*, **126**, 765 (1965)

Such adsorbents may be useful in resolving mixtures of lipophilic proteins, especially in combination with pH or detergent gradients. A preliminary extraction of lipid may be necessary in the chromatography of crude biological extracts. Thus a crude preparation of aspartate transcarbamoylase from defatted wheat germ was applied to a column of CPAD-agarose. Washing the column with equilibration buffer and 0·2M NaCl removed most of the inert protein whilst 0·2% (w/v) sodium deoxycholate in the absence of salt released 5% of the protein containing 40% of the enzyme activity. Enzyme activity could also be eluted by raising the pH to 10 or by making the buffer 2% (v/v) in acetone, to achieve an overall 8-fold purification.

Weiss and Bücher[218] have described the influence of various immobilized aliphatic chains on the chromatographic separation of membrane proteins. Aliphatic amines of differing chain length (butyl, capryl, lauryl and palmityl) and degree of unsaturation (stearyl, oleyl and linoleyl) were immobilized on polyacrylic acid by amide bonds. *Neurospora crassa* mitochondrial membrane proteins, solubilized with cholate or deoxycholate, were resolved most efficiently on capryl or linoleyl resins and least effectively on free polyacrylic acid and the stearyl resin. It is suggested that the membrane proteins require a mobile aliphatic chain for optimal interaction.

The observations suggest that hydrophobic affinity chromatography may represent a new approach to the purification of proteins endowed with accessible hydrophobic binding sites.

2. Covalent Affinity Chromatography

The selective purification of biologically active macromolecules by affinity chromatography is based on the reversible interaction between an immobilized ligand and the free macromolecule. In principle, affinity chromatography could be performed on adsorbents containing an 'irreversible' inhibitor such that the complementary macromolecule was covalently trapped by the column. Release could subsequently be effected by suitable chemical treatment.

Acetylcholinesterases and other serine esterases are readily inhibited 'irreversibly' by organophosphate triesters that contain a good leaving group to yield a covalent phosphoryl-enzyme derivative. Nevertheless, specific nucleophiles such as 2-(hydroximinomethyl)-1-methyl pyridinium iodide (2PAM),[219,220] 1,1′-trimethylene-*bis*-(4-hydroximinomethyl pyridinium) dibromide (TMB4)[221,222] or non-specific nucleophiles such as fluoride ion[223] or hydroxylamine[224] will reactivate inhibited acetylcholinesterase.

Ashani and Wilson[225] have coupled a potential inhibitor of acetylcholinesterase, 2-aminoethyl-*p*-nitro-phenyl-methylphosphonate, to Sepharose and generated a covalent affinity support competent for acetylcholinesterase (Scheme IV.1). Almost quantitative binding of acetylcholinesterase from electric eel was achieved and the covalently bound enzyme could not be removed by extensive washing with buffers containing high salt concentrations. The enzyme was removed from the gel by contact with 10^{-2}M TMB4 or 2PAM for 16 hours at room temperature prior to elution. However, significant recoveries of enzyme were only obtained after equilibration for up to 112 hours.

The adsorbent is potentially capable of binding all serine esterases and will bind α-chymotrypsin although it displays no affinity for human serum albumin. Ashani and Wilson[225] suggest that specificity could be achieved by including reversible inhibitors in the equilibrating buffers to allow some esterases to pass through unretarded, or by changing the leaving group of the ligand.

Blumberg and Strominger[226] have isolated the penicillin-binding components from solubilized membranes of *Bacillus subtilis* by covalent affinity chromatography. It has been proposed that penicillin exerts its effect on microorganisms by irreversibly inactivating the enzyme responsible for the

$$\text{Sepharose}\sim\text{arm}-\underbrace{NH(CH_2)_2O\overset{\displaystyle O}{\overset{\|}{\underset{\underset{\displaystyle CH_3}{|}}{P}}}O\phi NO_2\text{-}p}_{\text{ligand}} + E \quad \text{I}$$

$$\downarrow$$

$$\sim\text{—}NH(CH_2)_2O\overset{\displaystyle O}{\overset{\|}{\underset{\underset{\displaystyle CH_3}{|}}{P}}}\text{-}E + HO\phi NO_2\text{-}p$$

$$\downarrow \text{2PAM}$$

$$\sim\text{—}NH(CH_2)_2O\overset{\displaystyle O}{\overset{\|}{\underset{\underset{\displaystyle CH_3}{|}}{P}}}\text{-2PAM} + E$$

$$\text{arm} = -\left[NH(CH_2)_5NH\overset{\displaystyle O}{\overset{\|}{C}}(CH_2)_2\overset{\displaystyle O}{\overset{\|}{C}}\right]_2-$$

last stage of cell wall biosynthesis.[227,228] The inactivation is believed to be due to the penicilloylation[229] of the enzyme and is reversed by neutral hydroxylamine.

B. subtilis membranes were solubilized with 2% Nonidet P-40 and the extract added batchwise to the 6-amino-penicillanic acid-Sepharose. The slurry was poured into a squat column at room temperature, washed thoroughly and one column volume of elution buffer, containing 0·8M neutral hydroxylamine, applied, the flow stopped for 30 minutes and the process repeated two or three times. The eluted proteins were finally washed off. This method could be conveniently scaled up to give 50–100 mg of enzyme by batchwise purification with a 50% overall recovery of enzymatic activity.

REFERENCES

1. Obermeyer, F., and Pick, E. P. (1904): *Wien Klin. Wochschr.*, **17**, 265.
2. Landsteiner, K., and Lampl, H. (1917): *Z. Immunitaetsforsch.*, **26**, 258.
3. Eisen, H. N., Simms, E. S., Little, J. R., and Steiner, L. A. (1964): *Fed. Proc.*, **23**, 559.
4. Campbell, D. H., Luescher, E., and Lerman, L. S. (1951): *Proc. Nat. Acad. Sci. U.S.A.*, **37**, 575.
5. Isliker, H. C. (1953): *Ann. N.Y. Acad. Sci.*, **57**, 225.
6. Yagi, Y., Engel, K., and Pressman, D. (1960): *J. Immunol.*, **85**, 375.
7. Boegman, R. J., and Crumpton, M. J. (1970): *Biochem. J.*, **120**, 373.
8. Cuatrecasas, P. (1969): *Biochem. Biophys. Res. Commun.*, **35**, 531.
9. Axen, R., Porath, J., and Ernbäck, S. (1967): *Nature (London)*, **214**, 1302.
10. Singer, S. J., Fothergill, J. E., and Shainoff, J. R. (1960): *J. Amer. Chem. Soc.*, **82**, 565.
11. Singer, S. J. (1964): in *Methods of Medical Research*, Vol. 10 (Ed. H. N. Eisen), Year Book Medical Publishers, Chicago, p. 87.
12. Akanuma, Y., Kuzuya, T., Hayashi, M., Ide, T., and Kuzuya, N. (1970): *Biochem. Biophys. Res. Commun.*, **38**, 947.
13. Givol, D., Weinstein, Y., Gorecki, M., and Wilchek, M. (1970): *Biochem. Biophys. Res. Commun.*, **38**, 825.
14. Tanigaki, N., Kitagawa, M., Yagi, Y., and Pressman, D. (1967): *Cancer Res.*, **27**, 747.
15. Kleinschmidt, W. J., and Boyer, P. D. (1952): *J. Immunol.*, **69**, 247.
16. Kleinschmidt, W. J., and Boyer, P. D. (1952): *J. Immunol.*, **69**, 257.
17. Dandliker, W. B., Alonso, R., de Saussure, U. A., Kierszenbaum, F., Levison S. A., and Schapiro, H. C. (1967): *Biochemistry*, **6**, 1460.
18. Avrameas, S., and Ternynck, T. (1967): *Biochem. J.*, **102**, 37C.
19. Avrameas, S., and Ternynck, T. (1969): *Immunochem.*, **6**, 53.
20. Ternynck, T., and Avrameas, S. (1971): *Biochem. J.*, **125**, 297.
21. Melchers, F., and Messer, W. (1970): *Eur. J. Biochem.*, **17**, 267.
22. Weintraub, B. D. (1970): *Biochem. Biophys. Res. Commun.*, **39**, 83.
23. Beaumont, J. L., and Delphanque, B. (1969): *Immunochem.*, **6**, 489.
24. Karush, F., and Mark, R. (1957): *J. Immunol.*, **78**, 296.
25. Porter, R. R., and Press, E. M. (1962): *Ann. Rev. Biochem.*, **31**, 625.
26. Silman, I. H., and Katchalski, E. (1966): *Ann. Rev. Biochem.*, **35**, 873.
27. Wofsy, L., and Burr, B. (1969): *J. Immunol.*, **103**, 380.
28. Goetzl, E. J., and Metzger, H. (1970): *Biochemistry*, **9**, 1267.
29. Omenn, G., Ontjes, D. A., and Anfinsen, C. B. (1970): *Nature (London)*, **225**, 189.
30. Bing, D. H. (1971): *J. Immunol.*, **107**, 1243.
31. Spitzer, R. H., Kaplan, M. A., and Leija, J. G. (1968): *Int. Arch. Allergy*, **34**, 488.
32. Wood, K. R., Stephen, J., and Smith, H. (1968): *J. Gen. Virol.*, **2**, 313.
33. Moudgal, N. R., and Porter, R. R. (1963): *Biochim. Biophys. Acta*, **71**, 185.
34. Allen, R. H., and Majerus, P. W. (1972): *J. Biol. Chem.*, **247**, 7695.
35. Pensky, J., and Marshall, J. S. (1969): *Arch. Biochem. Biophys.*, **135**, 304.
36. Burstein, S. H. (1969): *Steroids*, **14**, 263.
37. Cuatrecasas, P., and Wilchek, M. (1968): *Biochem. Biophys. Res. Commun.*, **33**, 235.
38. Allen, R. H., and Majerus, P. W. (1972): *J. Biol. Chem.*, **247**, 7702.

39. Salter, D. N., Ford, J. E., Scott, K. J., and Andrews, P. (1972): *FEBS Lett.*, **20**, 302.
40. Rosner, W., and Bradlow, H. L. (1971): *J. Clin. Endocrinol. Metab.*, **33**, 193.
41. Blumberg, S., Hildesheim, J., Yariv, J., and Wilson, K. J. (1972): *Biochim. Biophys. Acta*, **264**, 171.
42. Ludens, J. H., DeVries, J. R., and Fanestil, D. D. (1972): *J. Biol. Chem.*, **247**, 7533.
43. Cuatrecasas, P. (1970): *J. Biol. Chem.*, **245**, 3059.
44. Vonderhaar, B., and Mueller, G. C. (1969): *Biochim. Biophys. Acta*, **176**, 626.
45. Cuatrecasas, P. (1972): *Adv. Enzymol.*, **36**, 29.
46. Sica, V., Nola, E., Puca, G. A., Parikh, I., and Cuatrecasas, P. (1973): *Fed. Proc.*, **32**, 1297.
47. Cuatrecasas, P. (1970): cited in *Adv. Enzymol.*, **36**, 29.
48. Allen, R. H., and Majerus, P. W. (1972): *J. Biol. Chem.*, **247**, 7709.
49. Vahlquist, A., Nilsson, S. F., and Peterson, P. A. (1971): *Eur. J. Biochem.*, **20**, 160.
50. Lee, C. Y. (1970): *Clin. Toxicol.*, **3**, 457.
51. Changeux, J. P., Kasai, M., and Lee, C. Y. (1970): *Proc. Nat. Acad. Sci. U.S.A.*, **67**, 1241.
52. Olsen, R. W., Meunier, J. C., and Changeux, J. P. (1972): *FEBS Lett.*, **28**, 96.
53. La Torre, J. L., Lunt, G. S., and de Robertis, E. (1970): *Proc. Nat. Acad. Sci. U.S.A.*, **65**, 716.
54. Krug, F., Desbuquois, B., and Cuatrecasas, P. (1971): *Nature New Biology*, **234**, 268.
55. Cuatrecasas, P. (1971): *J. Biol. Chem.*, **246**, 7265.
56. Lefkowitz, R. J. (1972): in *The Role of Membranes in Metabolic Regulation* (Eds. M. A. Mehlman and R. W. Hanson), Academic Press, New York, p. 264.
57. Raftery, M. A., Schmidt, J., Clark, D. G., and Wolcott, R. G. (1971): *Biochem. Biophys. Res. Commun.*, **45**, 1622.
58. Schmidt, J., and Raftery, M. A. (1973): *Biochemistry*, **12**, 852.
59. Karlsson, E., Heilbronn, E., and Widlung, L. (1972): *FEBS Lett.*, **28**, 107.
60. Schmidt, J., and Raftery, M. A. (1972): *Biochem. Biophys. Res. Commun.*, **49**, 572.
61. Talwar, G. P., Segal, S. J., Evans, A., and Davidson, D. W. (1964): *Proc. Nat. Acad. Sci. U.S.A.*, **52**, 1059.
62. Noteboom, W. D., and Gorski, J. (1965): *Arch. Biochem. Biophys.*, **111**, 559.
63. Toft, D., Shaymala, G., and Gorski, J. (1967): *Proc. Nat. Acad. Sci. U.S.A.*, **57**, 1740.
64. Vonderhaar, B., and Mueller, G. C. (1969): *Biochim. Biophys. Acta*, **176**, 626.
65. Cuatrecasas, P. (1972): *Methods Enzymol.*, **36**, 29.
66. Turkington, R. W. (1970): *Biochem. Biophys. Res. Commun.*, **41**, 1362.
67. Selinger, R. C. L., and Civen, M. (1971): *Biochem. Biophys. Res. Commun.*, **43**, 793.
68. Pohl, S. L., Birnbaumer, L., and Rodbell, M. (1971): *J. Biol. Chem.*, **246**, 1849.
69. Johnson, C. B., Blecher, M., and Giorgio, N. A. (1972): *Fed. Proc.*, **31**, 439.
70. Johnson, C. B., Blecher, M., and Giorgio, N. A. (1972): *Biochem. Biophys. Res. Commun.*, **46**, 1035.
71. Blecher, M., Giorgio, N. A., and Johnson, C. B. (1972): in *The Role of Membranes in Metabolic Regulation* (Eds. M. A. Mehlman and R. W. Hanson), Academic Press, New York, p. 367.

72. Civen, M. (1972): in *The role of Membranes in Metabolic Regulations* (Eds. M. A. Mehlman and R. W. Hanson), Academic Press, New York, p. 313.
73. Soderman, D. D., Germershausen, J., and Katsen, H. M. (1972): *Fed. Proc.*, **31**, 486.
74. Katsen, H. M., and Soderman, D. D. (1972): in *The Role of Membranes in Metabolic Regulations* (Ed. M. A. Mehlman and R. W. Hanson), Academic Press, New York, p. 195.
75. Cuatrecasas, P. (1972): *Proc. Nat. Acad. Sci. U.S.A.*, **69**, 318.
76. Blumberg, P. M., and Strominger, J. L. (1972): *Proc. Nat. Acad. Sci. U.S.A.*, **69**, 3751.
77. Simon, E. J., Dole, W. P., and Hiller, J. M. (1972): *Proc. Nat. Acad. Sci. U.S.A.*, **69**, 1835.
78. Melmon, K. L., Bourne, H. R., Weinstein, J., *et al.* (1972): *Science*, **177**, 707.
79. Allan, D., Auger, J., and Crumpton, M. J. (1972): *Nature New Biology*, **236**, 23.
80. Lefkowitz, R. J., Haber, E., and O'Hara, D. (1972): *Proc. Nat. Acad. Sci. U.S.A.*, **69**, 2828.
81. Cuatrecasas, P. (1972): *Proc. Nat. Acad. Sci. U.S.A.*, **69**, 1277.
82. Rabinowitz, Y. (1964): *Blood*, **23**, 811.
83. Thorsby, E. (1967): *Vox Sang.*, **13**, 194.
84. Yunis, J. J., and Yunis, E. J. (1964): *Blood*, **24**, 522.
85. Yunis, E. J., and Yunis, J. J. (1964): *Blood*, **24**, 531.
86. Wigzell, H., and Anderson, B, (1969): *J. Exp. Med.*, **129**, 23.
87. Evans, W. H., Mage, M. G., and Peterson, E. A. (1969): *J. Immunol.*, **102**, 899.
88. Truffa-Bachi, P., and Wofsy, L. (1970): *Proc. Nat. Acad. Sci. U.S.A.*, **66**, 685.
89. Henry, C., Kimura, J., and Wofsy, L. (1972): *Proc. Nat. Acad. Sci. U.S.A.*, **69**, 34.
90. Davie, J. M., and Paul, W. E. (1970): *Cell. Immunol.*, **1**, 404.
91. Kenyon, A. J., Gander, J. E., Lopez, C., and Good, R. A. (1973): *Science*, **179**, 187.
92. Cuatrecasas, P., Wilchek, M., and Anfinsen, C. B. (1968): *Proc. Nat. Acad. Sci. U.S.A.*, **61**, 636.
93. Inman, J. K., and Dintzis, H. M. (1969): *Biochemistry*, **8**, 4074,
94. Wallach, D. F. H., Kranz, B., Ferber, E., and Fischer, H. (1972): *FEBS Lett.*, **21**, 29.
95. Edelman, G., and Milette, C. F. (1971): *Proc. Nat. Acad. Sci. U.S.A.*, **68**, 2436.
96. Katzen, H. M., and Soderman, D. D. (1972): in *The Role of Membranes in Metabolic Regulation* (Eds. M. A. Mehlman and R. W. Hanson), Academic Press, New York, p. 205.
97. Sumner, J. B., and Howell, S. F. (1936): *J. Bacteriol.*, **32**, 227.
98. Goldstein, I. J., Hollerman, C. E., and Smith, E. F. (1965): *Biochemistry*, **4**, 876.
99. Yariv, J., Kalb, A. J., and Levitzki, A. (1968): *Biochim. Biophys. Acta*, **165**, 303.
100. Kühnemund, O., and Köhler, W. (1969): *Experientia*, **25**, 1137.
101. Ishiyama, I., and Uhlenbruck, G. (1972): *Z. Immun.-Forsch.*, **143**, 147.
102. Gilboa-Garber, N. (1972): *FEBS Lett.*, **20**, 242.
103. Inbar, M., and Sachis, L. (1967): *Proc. Nat. Acad. Sci. U.S.A.*, **57**, 359.
104. Moscona, A. A. (1971): *Science*, **171**, 905.
105. Burger, M. M. (1970): *Nature* (*London*), **228**, 512.

106. Tomita, M., Kurokawa, T., Onozaki, K., Ichiki, N., Osawa, T., and Ukita, T. (1971): *Experientia*, **28**, 84.
107. Gilboa-Garber, N., Mizrahi, L., and Garber, N. (1972): *FEBS Lett.*, **28**, 93.
108. Liener, I. E., and Pallansch, M. J. (1952): *J. Biol. Chem.*, **197**, 29.
109. Lis, H., Sharon, N., and Katchalski, E. (1966): *J. Biol. Chem.*, **241**, 684.
110. Lis, H., Sela, B. A., Sachs, L., and Sharon, N. (1970): *Biochim. Biophys. Acta*, **211**, 582.
111. Gordon, J. A., Blumberg, S., Lis, H., and Sharon, N. (1973): *FEBS Lett.*, **24**, 193.
112. Lotan, R., Gussin, A. E. S., Lis, H., and Sharon, N. (1973): *Biochem. Biophys. Res. Commun.*, **52**, 656.
113. Allen, A. K., Neuberger, A., and Sharon, N. (1973): *Biochem. J.*, **131**, 151.
114. So, L. L., and Goldstein, I. J. (1968): *Biochim. Biophys. Acta*, **165**, 398.
115. Goldstein, I. J., Hollerman, C. E., and Merrick, J. M. (1965): *Biochim. Biophys. Acta*, **97**, 68.
116. Sumner, J. B., and Howell, S. F. (1936): *J. Biol. Chem.*, **115**, 583.
117. Sumner, J. B., and O'Kane, D. J. (1948): *Enzymologia*, **12**, 251.
118. Goldstein, I. J., and So, L. L. (1965): *Arch. Biochem. Biophys.*, **111**, 407.
119. So, L. L., and Goldstein, I. J. (1967): *J. Immunol.*, **99**, 158.
120. Yariv, J., Kalb, A. J., and Levitzki, A. (1968): *Biochim. Biophys. Acta*, **165**, 303.
121. Aspberg, K., and Porath, J. (1970): *Acta Chem. Scand.*, **24**, 1839.
122. Liener, I. E., Garrison, O. R., and Pravda, Z. (1973): *Biochem. Biophys. Res. Commun.*, **51**, 436.
123. Lloyd, K. O. (1970): *Arch. Biochem. Biophys.*, **137**, 460.
124. Dufau, M. L., Tsuruhara, T., and Catt, K. J. (1972): *Biochim. Biophys. Acta*, **278**, 281.
125. Edelman, G. M., Rutishauser, U., and Milette, C. F. (1971): *Proc. Nat. Acad. Sci. U.S.A.*, **68**, 2153.
126. Allan, D., Auger, J., and Crumpton, M. J. (1972): *Nature New Biology*, **236**, 23.
127. Hayman, M. J., Skehel, J. J., and Crumpton, M. J. (1973): *FEBS Lett.*, **29**, 185.
128. Hayman, M. J., and Crumpton, M. J. (1972): *Biochem. Biophys. Res. Commun.*, **47**, 923.
129. Tomino, S., and Paigen, K. (1970): in *The Lac Operon* (Eds. D. Zipser and J. Beckwith), Cold Spring Harbour Laboratory for Quantitative Biology, p. 223.
130. Wilcox, G., Clemetson, K. J., Santi, D. V., and Englesberg, E. (1971): *Proc. Nat. Acad. Sci. U.S.A.*, **68**, 2145.
131. Fowler, A. V., and Zabin, I. (1968): *J. Mol. Biol.*, **33**, 35.
132. Villarejo, M. R., and Zabin, I. (1973): *Nature New Biology*, **242**, 50.
133. Steers, E., Cuatrecasas, P., and Pollard, H. B. (1971): *J. Biol. Chem.*, **246**, 196.
134. Miller, J. V., Cuatrecasas, P., and Thompson, E. B. (1971): *Proc. Nat. Acad. Sci. U.S.A.*, **68**, 1014.
135. Miller, J. V., Cuatrecasas, P., and Thompson, E. B. (1972): *Biochim. Biophys. Acta*, **276**, 407.
136. Valeriote, F. A., Auricchio, F., Tomkins, G. M., and Riley, D. (1969): *J. Biol. Chem.*, **244**, 3618.
137. Faust, C. H., Vassalli, P., and Mach, B. (1972): *Experientia*, **28**, 745.

138. Venis, M. A. (1971): *Proc. Nat. Acad. Sci. U.S.A.*, **68**, 1824.
139. Gyenge, L., Spiridonova, V. A., and Bogdanov, A. A. (1972): *FEBS Lett.*, **20**, 209.
140. Richards, F. M., and Vithayathil, P. J. (1959): *J. Biol. Chem.*, **234**, 1459.
141. Hofmann, K., Smithers, M. J., and Finn, F. M. (1966): *J. Amer. Chem. Soc.*, **88**, 4107.
142. Gawronski, T. H., and Wold, F. (1972): *Biochemistry*, **11**, 442.
143. Cuatrecasas, P. (1972): *Adv. Enzymol.*, **36**, 29.
144. Gawronski, T. H., and Wold, F. (1972): *Biochemistry*, **11**, 449.
145. Taniuchi, H., Anfinsen, C. B., and Sodja, A. (1967): *Proc. Nat. Acad. Sci. U.S.A.*, **58**, 1235.
146. Taniuchi, H., and Anfinsen, C. B. (1968): *J. Biol. Chem.*, **243**, 4778.
147. Ontjes, D. A., and Anfinsen, C. B. (1969): *J. Biol. Chem.*, **244**, 6316.
148. Gribnau, A. A. M., Schoenmakers, J. G. G., Van Kraaikamp, M., and Bloemendal, H. (1970): *Biochem. Biophys. Res. Commun.*, **38**, 1064.
149. Feinstein, G. (1970): *FEBS Lett.*, **7**, 353.
150. Robinson, N. C., Tye, R. W., Neurath, H., and Walsh, K. A. (1971): *Biochemistry*, **10**, 2743.
151. Reeck, G. R., Walsh, K. A., and Neurath, H. (1971): *Biochemistry*, **10**, 4690.
152. Porath, J., and Sundberg, L. (1971): in *Protides of the Biological Fluids* (Ed. H. Peeters), Pergamon Press, New York, p. 401.
153. Sundberg, L., and Christiansen, T. (1972): *Biotechnol. Bioeng. Symp.*, **3**, 165.
154. Fritz, H., Brey, B., Schmal, S., and Werle, E. (1969): *Hoppe-Seyler's Z. Physiol. Chem.*, **350**, 617.
155. Fritz, H., Wunderer, G., and Dittmann, B. (1972): *Hoppe-Seyler's Z. Physiol. Chem.*, **353**, 893.
156. Feinstein, G. (1971): *Biochim. Biophys. Acta*, **236**, 73.
157. Stewart, K. K., and Doherty, R. F. (1971): *FEBS Lett.*, **16**, 226.
158. Chauvet, J., and Acher, R. (1972): *FEBS Lett.*, **23**, 317.
159. Lindberg, U., and Erickson, S. (1971): *Eur. J. Biochem.*, **18**, 474.
160. Pradelles, P., Morgat, J. L., Fromageot, P., Camier, M., Bonne, D., Cohen, P., Bockaert, J., and Jard, S. (1972): *FEBS Lett.*, **26**, 189.
161. Rauch, R., Hollenberg, M. D., and Hope, D. B. (1969): *Biochem. J.*, **115**, 473.
162. Breslow, E., and Abrash, L. (1966): *Proc. Nat. Acad. Sci. U.S.A.*, **56**, 640.
163. Edmondson, D., Massey, V., Palmer, G., Beacham, L. M., and Elion, G. B. (1972): *J. Biol. Chem.*, **247**, 1597.
164. Massey, V., Komai, H., Palmer, G., and Elion, G. B. (1970): *J. Biol. Chem.*, **245**, 2837.
165. Massey, V., Edmondson, D., Palmer, G., Beacham, L. M., and Elion, G. B. (1972): *Biochem. J.*, **127**, 10P.
166. Sluyterman, L. A. E., and Wijdenes, J. (1970): *Biochim. Biophys. Acta*, **200**, 593.
167. Cuatrecasas, P. (1970): *J. Biol. Chem.*, **245**, 574.
168. Cuatrecasas, P., Wilchek, M., and Anfinsen, C. B. (1969): *J. Biol. Chem.*, **244**, 4316.
169. Whitney, P. L. (1971): *Fed. Proc.*, **30**, 1291.
170. Singer, S. J. (1967): *Adv. Prot. Chem.*, **22**, 1.
171. Givol, D., Weinstein, Y., Gorecki, M., and Wilchek, M. (1970): *Biochem. Biophys. Res. Commun.*, **38**, 825.

172. Wilchek, M. (1970): *FEBS Lett.*, **7**, 161.
173. O'Carra, P., and Barry, S. (1972): *FEBS Lett.*, **21**, 281.
174. Schwert, G. W. (1970): in *Pyridine Nucleotide-Dependent Dehydrogenases* (Ed. H. Sund), Springer-Verlag, Heidelberg, p. 133.
175. Lowe, C. R., and Dean, P. D. G. (1973): *Biochem. J.*, **133**, 515.
176. McPherson, A. (1970): *J. Mol. Biol.*, **51**, 39.
177. Adams, M. J., McPherson, A., Rossmann, M. G., Shevitz, R. W., Smiley, I.E., and Wonacott, A. J. (1970): in *Pyridine Nucleotide-Dependent Dehydrogenases* (Ed. H. Sund), Springer-Verlag, Heidelberg, p. 157.
178. Mawal, R., Morrison, J. F., and Ebner, K. E. (1971): *J. Biol. Chem.*, **246**, 7106.
179. Ebner, K. E., Morrison, J. F., and Mawal, R. (1971): *Fed. Proc.*, **30**, 1265.
180. Morrison, J. F., and Ebner, K. E. (1971): *J. Biol. Chem.*, **246**, 3977.
181. Trayer, I. P., Mattock, P., and Hill, R. L. (1970): *Fed. Proc.*, **29**, 597.
182. Andrews, P. (1970): *FEBS Lett.*, **9**, 297.
183. Akanuma, H., Kasuga, A., Akanuma, T., and Yamasaki, M. (1971): *Biochem. Biophys. Res. Commun.*, **45**, 27.
184. Doellgast, G., and Kohlhaw, G. (1972): *Fed. Proc.*, **31**, 424.
185. Rahimi-Laridjani, I., Grimminger, H., and Lingens, F. (1973): *FEBS Lett.*, **30**, 185.
186. Wider de Xifra, E. A., Mendiara, S., and Batlle, A. M. del C. (1972): *FEBS Lett.*, **27**, 275.
187. Chan, W. W. C. (1970): *Biochem. Biophys. Res. Commun.*, **41**, 1198.
188. Stellwagen, E., and Schachman, H. K. (1962): *Biochemistry*, **1**, 1056.
189. Feldman, K., Zeisel, H., and Helmreich, E. (1972): *Proc. Nat. Acad. Sci. U.S.A.*, **69**, 2278.
190. Ikeda, S., and Fukui, S. (1973): *Biochem. Biophys. Res. Commun.*, **52**, 482.
191. Alberts, B. M., Amodio, F. J., Jenkins, M., Gutmann, E. D., and Ferris, F. L. (1968): *Cold Spring Harbour Quant. Biol.*, **33**, 289.
192. Goldhaber, P. (1965): *Science*, **147**, 407.
193. Bautz, E. K. F., and Reilly, E. (1966): *Science*, **151**, 328.
194. Nyggard, A. P., and Hall, B. D. (1963): *Biochem. Biophys. Res. Commun.*, **12**, 98.
195. Alder, A. J., and Rich, A. (1962): *J. Amer. Chem. Soc.*, **84**, 3977.
196. Bautz, E. L. F., and Holt, B. D. (1962): *Proc. Nat. Acad. Sci. U.S.A.*, **48**, 400.
197. Poonian, M. S., Schlaback, A. J., and Weissbach, A. (1971): *Biochemistry*, **10**, 424.
198. Gilham, P. T. (1964): *J. Amer. Chem. Soc.*, **86**, 4982.
199. Gilham, P. T., and Robinson, W. E. (1964): *J. Amer. Chem. Soc.*, **86**, 4985.
200. Gilham, P. T. (1968): *Biochemistry*, **7**, 2809.
201. Jovin, T. M., and Kornberg, A. (1968): *J. Biol. Chem.*, **243**, 250.
202. Cozzarelli, N. R., Melechen, N. E., Jovin, T. M., and Kornberg, A. (1967): *Biochem. Biophys. Res. Commun.*, **28**, 578.
203. Robberson, D. L., and Davidson, N. (1972): *Biochemistry*, **11**, 533.
204. Khym, J. X. (1967): *Methods Enzymol.*, **12**, 93.
205. Weith, H. L., Wiebers, J. L., and Gilham, P. T. (1970): *Biochemistry*, **9**, 4396.
206. Rosenberg, M., Wiebers, J. L., and Gilham, P. T. (1972): *Biochemistry*, **11**, 3623.
207. Dixon, M., and Webb, E. C. (1964): in *Enzymes*, 2nd Ed., Academic Press, New York, p. 255.
208. Baker, B. R. (1967): in *Design of Active-Site-Directed Irreversible Enzyme Inhibitors*, John Wiley, New York.

209. Karush, F. (1962): *Adv. Immunol.*, **2**, 1.
210. Spector, A. A., John, K., and Fletcher, J. E. (1969): *J. Lipid Res.*, **10**, 56.
211. McClure, W. O., and Edelman, G. M. (1966): *Biochemistry*, **5**, 1908.
212. Hofstee, B. H. J. (1973): *Anal. Biochem.*, **52**, 430.
213. Er-El, Z., Zaidenzaig, Y., and Shaltiel, S. (1972): *Biochem. Biophys. Res. Commun.*, **49**, 383.
214. Stevenson, K. J., and Landman, A. (1971): *Can. J. Biochem.*, **49**, 119.
215. Jacobsen, J. (1969): *FEBS Lett.*, **5**, 112.
216. Peters, T., Taniuchi, H., and Anfinsen, C. B. (1973): *J. Biol. Chem.*, **248**, 2447.
217. Yon, R. J. (1972): *Biochem. J.*, **126**, 765.
218. Weiss, H., and Bücher, T. (1970): *Eur. J. Biochem.*, **17**, 561.
219. Wilson, I. B., and Ginsburg, S. (1955): *Biochim. Biophys. Acta*, **18**, 168.
220. Davies, D. R., and Green, A. L. (1955): *Disc. Faraday Soc.*, **20**, 269.
221. Hobbiger, F., O'Sullivan, D. G., and Saddler, P. W. (1958): *Nature* (*London*), **182**, 1498.
222. Poziomek, E. J., Hackley, B. E., and Steinberg, G. M. (1958): *J. Org. Chem.*, **23**, 714.
223. Walsh, K. A., and Wilcox, P. E. (1970): *Methods Enzymol.*, **19**, 37.
224. Wilson, I. B. (1951): *J. Biol. Chem.*, **190**, 111.
225. Ashani, Y., and Wilson, I. B. (1972): *Biochim. Biophys. Acta*, **276**, 317.
226. Blumberg, P. M., and Strominger, J. L. (1972): *Proc. Nat. Acad. Sci. U.S.A.*, **69**, 3751.
227. Izaki, K., Matsuhashi, M., and Strominger, J. L. (1968): *J. Biol. Chem.*, **243**, 3180.
228. Tipper, D. J., and Strominger, J. L. (1968): *J. Biol. Chem.*, **243**, 3169.
229. Tipper, D. J., and Strominger, J. L. (1965): *Proc. Nat. Acad. Sci. U.S.A.* **54**, 1133.

Chapter V

The Chemistry of Affinity Chromatography

This chapter describes some chemical procedures commonly employed in the preparation of affinity adsorbents. They are described in the order that one would expect to prepare the adsorbent, i.e., activation of the matrix and attachment of the spacer arm followed by coupling the ligand to the spacer-arm matrix conjugate. It is, however, important to realize that the chemistry employed in the activation and functionalization of the inert support matrices is largely dictated by their nature and stability. For example, glass beads can be activated by refluxing in chloroform with the appropriate silane agent under conditions that would totally destroy agarose. Unfortunately, the relatively mild conditions necessary for the 'ideal' matrix, agarose, seriously limit the number of chemical procedures that can be used for its derivatization. Some consideration of the chemistry of inert matrices is thus relevant to their subsequent activation and functionalization. Spacer arms of various lengths and polarities can be attached to the activated matrix and the specific ligand bearing common functional groups such as amino, carboxyl, phenolic, aldehyde, sulphydryl, etc., can be coupled using a variety of mild procedures. In some cases, it may be necessary to modify the ligand prior to attachment to the support in order to facilitate coupling. The reader is advised to consult textbooks of organic chemistry for this purpose.

Some idea of the potential capacity and chromatographic behaviour of the affinity adsorbent can be gained from a knowledge of the extent of ligand substitution. The method employed for determining the ligand concentration depends on the nature of the ligand itself. For example, if the ligand absorbs light in the UV or visible regions, direct spectroscopy on the gel or solubilized gel can be used, whereas elemental analysis can be used for ligands containing sulphur, phosphorus or other characteristic atoms.

The last section of this chapter deals with some practical aspects on the handling of agarose, its activation with cyanogen bromide and the preparation of a common spacer-arm agarose. Details of the water-soluble carbodiimide coupling and the preparation of the derivatives for diazonium coupling are also given. For more detailed accounts of these procedures and others the reader is referred to the appropriate references cited in this chapter.

A. SUPPORT MATRICES

1. Cellulose

Cellulose is the main constituent of plant cell walls and consists primarily of a linear polymer of β-1,4-linked D-glucose units with an occasional 1,6-linkage (Fig. V.1). The microcrystalline cellulose available commercially is generally cross-linked with a bifunctional reagent such as epichlorohydrin and is remarkably stable to chemical attack. However, the glycosidic linkage

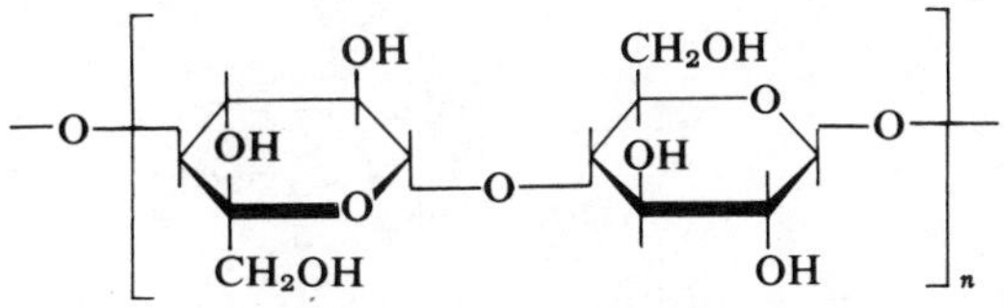

Figure V.1. The structure of cellulose

is susceptible to acid hydrolysis which under extreme conditions can yield almost quantitative amounts of pure crystalline D-glucose. Oxidizing agents such as sodium periodate can generate aldehyde groups and thence carboxyl groups. Furthermore, the polymer is susceptible to enzymic attack by plant or bacterial cellulases.

2. Dextran Gels

Dextran is a branched-chain glucose polysaccharide produced by fermentation of sucrose by microorganisms such as *Leuconostoc mesenteroides*.[1] The raw dextran is extensively purified, partially hydrolysed and then fractionated by ethanol precipitation to yield a product suitable for gel chromatography. This soluble dextran contains more than 90% α-1,6-glucosidic linkages and is branched by 1,2-, 1,3- and 1,4-glucosidic linkages.[2] The soluble polymer chains are cross-linked by glycerin ether bonds on reaction of dextran with epichlorohydrin in alkaline solution to yield a three-dimensional gel which solidifies exothermically.[3] A partial structure of a cross-linked gel is shown in Fig. V.2.

Dextran gels, or Sephadex gels as they are known commercially, are very stable to chemical attack. The gels are not affected by treatment with 0·25M NaOH for 2 months at 60°C,[4] and hence contaminants such as denatured proteins or lipids that are irreversibly adsorbed to the matrix can be removed by treatment with strong alkali. The glucosidic linkages are, however, susceptible to hydrolysis at low pH values, although the material will withstand 0·02M HCl for 6 months,[5] 0·1M HCl for 1–2 hours or 88% formic acid.[6]

Prolonged exposure of the gels to oxidizing agents can generate aldehyde and carboxyl groups and impair the chromatographic behaviour of the matrix. The problem is readily circumvented by maintaining the ionic strength of the eluant above 0·01.

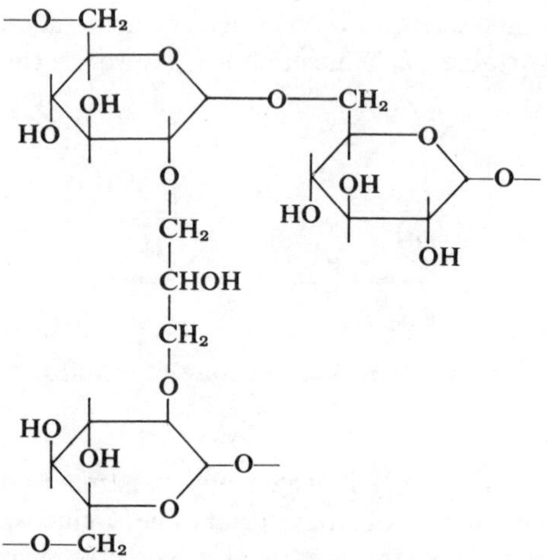

Figure V.2. Partial structure of a cross-linked dextran gel illustrating the glycerin ether bond

Dextran gels in solution can be heated to 110°C in an autoclave for 40 minutes or to 120°C in the dry state without impairment of properties. Furthermore, the drying and swelling are reversible. The gels also swell to some extent in ethanol, ethylene glycol, formamide, *N,N'*-dimethylformamide and dimethylsulphoxide.[7]

3. Agarose

Agar agar has been isolated from various species of red sea weed and consists of two principal polysaccharide components: (i) Agaropectin, which contains 6% charged sulphate groups, pyruvic acid and glucuronic acid, and (ii) Agarose, which is a neutral linear poly-galactose consisting of alternating residues of D-galactose and 3,6-anhydro-L-galactose[8] (Fig. V.3). These polysaccharides have been resolved by fractional precipitation with polyethylene glycol[9] or by precipitation of the agaropectin with cetylpyridium chloride.[10]

Agarose is freely soluble in boiling water and forms an insoluble gel on cooling below 40°C, even at concentrations as low as 0·4%. In contrast to dextran and polyacrylamide gels, the polymer chains are held together by hydrogen bonds and not by covalent linkages. This means that agarose gels are less stable chemically than the covalently cross-linked dextran or polyacrylamide gels. Thus, agarose gels are less stable to extremes of pH, and exposure to media with pH values outside the range 4–9 should be avoided. Reagents which disrupt hydrogen bonds can decrease the mechanical stability of the gels[11,12] if the agarose content is low (<2%) although, in general, moderate concentrations of urea or guanidine–HCl are well tolerated. Furthermore, the gels are not disrupted by organic solvents such as ethanol or acetone,[14] whereas dextran and polyacrylamide gels shrink completely

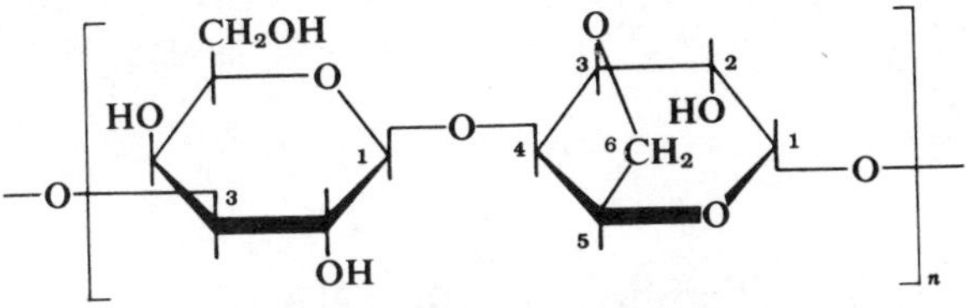

CH_2OH HO OH O D-Galactose 3,6-Anhydro-L-galactose HO $^6CH_2$

Figure V.3. The structure of agarose

when treated with acetone. However, it should be noted that agarose is a thermally reversible gel and on heating the gels lose stability and eventually melt. Thus, these gels must be kept at low ambient temperatures, although not frozen, since freezing also results in irreversible structural disruption. In practical terms, heat sterilization is impossible and the gels must be stored in the wet state since they cannot be dried and reswollen. Table V.1 summarizes

Table V.1. Solvents and media suitable for use with agarose

	Conditions	References
0·1M NaOH	2–3 hours at room temperature	14
1M HCl		14
6M Guanidine–HCl		11, 12
7M Urea		11, 12
Ethanol		13
Methanol		15
Butanol		15
Acetone		13
Aqueous pyridine	80% v/v	15
Aqueous dimethylformamide	50% v/v	14
Aqueous ethylene glycol	50% v/v	14
Anhydrous dioxan	6 months	16

some of the solvents and media that have been used with native agarose and that leave the gel undamaged. However, derivatized agaroses are often more stable to heat or solvents than the parent gel. Nevertheless, the chemical lability of agarose is an important factor in the choice and execution of a derivatization procedure. All coupling procedures used must be performed under conditions innocuous to the structure of the gel.

4. Polyacrylamide Gels

The neutral hydrophilic polyacrylamides are entirely synthetic gels formed by copolymerization of acrylamide ($H_2C{=}CH.CONH_2$) with the bifunctional cross-linking agent, *N,N'*-methylene-*bis*-acrylamide ($H_2C{=}CH.CO.$

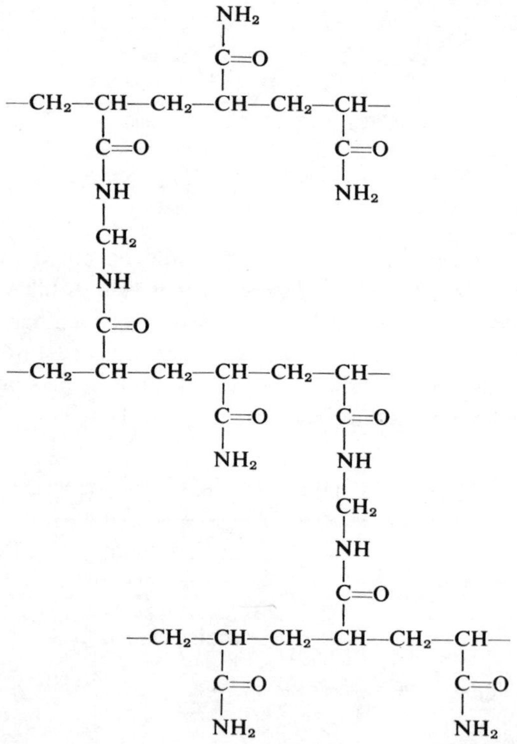

```
                       NH2
                        |
                       C=O
                        |
 —CH2—CH—CH2—CH—CH2—CH—
       |                    |
      C=O                  C=O
       |                    |
      NH                   NH2
       |
      CH2
       |
      NH
       |
      C=O
       |
 —CH2—CH—CH2—CH—CH2—CH—
                |           |
               C=O         C=O
                |           |
               NH2         NH
                            |
                           CH2
                            |
                           NH
                            |
                           C=O
                            |
           —CH2—CH—CH2—CH—CH2—CH—
                 |                   |
                C=O                 C=O
                 |                   |
                NH2                 NH2
```

Figure V.4. Partial structure of a polyacrylamide matrix

$NH.CH_2.NH.CO.CH{=}CH_2$).[17] The ratio of the concentration of acrylamide in the reaction mixture to that of the cross-linking agent can be varied to give an infinite series of insoluble, covalently bonded gel products which differ in their average pore size. Fig. V.4 represents part of a polyacrylamide matrix.

Like dextran gels, polyacrylamide is a xerogel in that the dry polymers swell immediately on addition of water to form a porous gel suitable for chromatography. The polyacrylamide gels are chemically stable to a wide range of eluants commonly employed by biochemists. They are chemically inert to dilute solutions of salts, detergents, guanidine–HCl, urea and most organic solvents, although under extreme conditions some change in porosity might be experienced. Furthermore, the gels are biologically inert and are not generally subject to enzymic digestion or to metabolism by microorganisms.

The amide groups on the polyacrylamide matrix are subject to hydrolysis at extremes of pH or by nitrous acid to yield carboxyl functions. Prolonged exposures to eluants with pH values outside the range 2–11 should be avoided. Likewise, the use of strong oxidizing agents, such as hypochlorite or hydrogen peroxide, should be avoided.

5. Glass

Haller[18] has described the preparation of a glass powder of regular pore size by heat treatment of sintered alkali borosilicate glass. The resulting glass beads comprise a system of anastomosing channels of uniform diameter. Such materials can now be obtained commercially from the Corning Biological Products Group, Medfield, Mass. 02052.

Glass beads may be particularly useful in cases where the ligand to be immobilized is highly insoluble in water. Organic solvents can be used without altering the pore size, and the material is not changed or degraded by exotic eluants (guanidine–HCl, sodium dodecyl sulphate), corrosive solvents (perchloroethylene) or concentrated acids (nitric or sulphuric acids) that may be required for its derivatization or utilization as an affinity adsorbent.

6. Other Supports

Levin *et al.*[19] first used the anionic carrier, poly-(ethylene/maleic anhydride) cross-linked with hexamethylenediamine to immobilize enzymes. Several workers have reported the use of hydrophilic carriers based on acrylamide and called Enzacryls.[20–22] Furthermore, several vinyl copolymers,[23,24] nylon and polystyrene matrices[25] and amino acid polymers[26,27] have also been used.

B. ACTIVATION AND FUNCTIONALIZATION OF SUPPORT MATRICES

1. Polysaccharide Matrices

The hydrophilic polysaccharide polymers have been widely used as chromatographic supports for the separation and isolation of sensitive biological compounds. Affinity chromatography requires that the ligand to be immobilized is covalently attached to the matrix backbone under conditions innocuous to the structure of the matrix. It is important that the procedure for the activation of the polymer is mild such that the subsequent chromatographic properties of the gel are not adversely altered.

a. Cyanogen Halides

The reaction of cyanogen bromide with hydrophilic polysaccharide matrices was first mentioned by Patty in 1949[28] as a method for the technological processing of cellulose products. Axen *et al.*[29] introduced the method to the field of affinity chromatography in 1967 and since then the method has been widely exploited. The chemical activation of the polysaccharides by cyanogen halides in alkaline media is a simple, convenient, easily reproducible one-step method[29,30] that does not significantly alter the physical nature of the polymer. Furthermore, the extent of activation and subsequent reaction with small ligands is a direct function of the amount of cyanogen halide utilized and the activation pH.

The precise chemistry of the activation process and the binding reaction has been only partially elucidated[29] in the case of Sephadex, although the same structures are believed to arise when other hydroxylic polymers are used. Activation of Sephadex with cyanogen bromide (CNBr) leads to the formation of cyclic imidocarbonates and carbamic acid esters of the polysaccharide,[31] as shown in Fig. V.5. The cyclic imidocarbonates are responsible for most of the coupling capacity to primary amino groups, whilst the carbamic acid esters are stable, inert and electrically neutral.

Parallel studies on the mechanism of activation of cellulose by CNBr[32] have confirmed this general reaction by IR analysis of the activated matrices and by the observation that ammonia was liberated in the coupling reaction.[29] IR analysis of the activated matrix reveals a strong band at 1730 cm^{-1} which is not present in unactivated cellulose and which is destroyed by prolonged reaction in base. In contrast, brief hydrolysis of the activated matrix under acid conditions produced an additional strong band at 1810 cm^{-1} which is destroyed by prolonged acid hydrolysis. The band at 1730 cm^{-1} remains unchanged under these conditions. These observations are consistent with the IR studies on Sephadex G-200 by Axen *et al.*[29] and support the mechanism depicted in Fig. V.5. Thus, the reactive intermediate in the activation of cellulose by CNBr is a cyclic imidocarbonate (I) which is readily hydrolysed in acid media to a cyclic carbonate (II). Treatment of I with aqueous base or prolonged hydrolysis of II regenerates the original cellulose vicinal-diol.

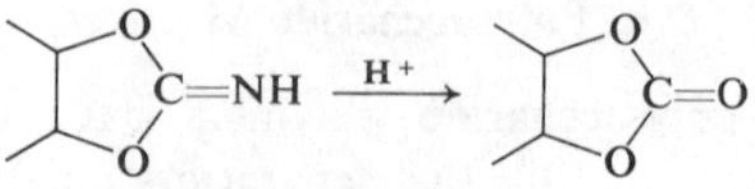

O, C=NH, O $\xrightarrow{H^+}$ O, C=O, O

I II

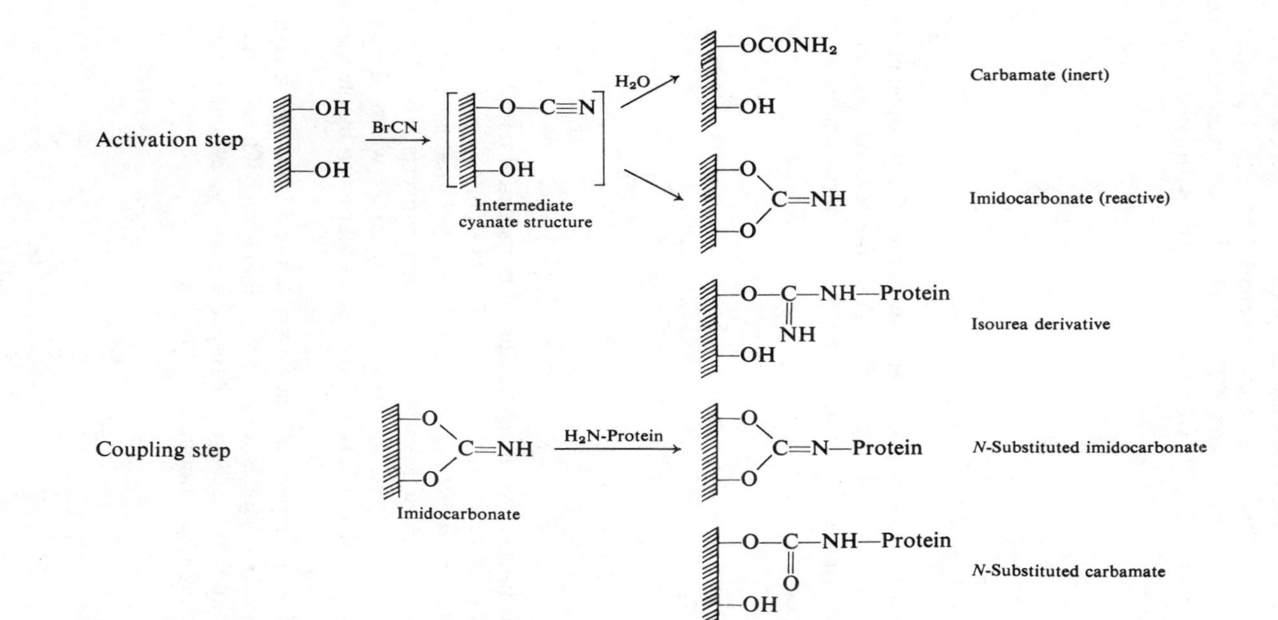

Figure V.5. Chemical activation of polysaccharides by means of cyanogen halides and chemical coupling of proteins to cyanogen-halide-activated polysaccharides. Reproduced with permission from R. Axen and S. Ernbäck, *Eur. J. Biochem.*, **18**, 351 (1971)

Studies with model compounds have led to a more rigorous identification of the intermediates in the activation of hydroxylic polymers by CNBr. The *vic*-diol function of cellulose has been represented by *trans*-1,2-cyclohexane-diol.[32] Subsequent reaction with CNBr yielded a product with spectral properties consistent with III[33] which was smoothly hydrolysed to the cyclic carbonate (IV) by dilute HCl.

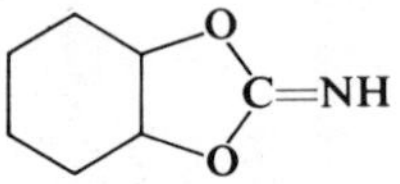

C=NH   C=O

III 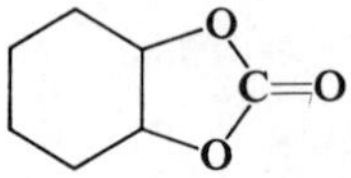   IV

When cyclohexylmethanol (V) is reacted with CNBr in aqueous solution small amounts (0·1%) of the carbamate (VII) are formed, presumably by hydrolysis of the cyanate.[34] The appearance of the carbamate in this model

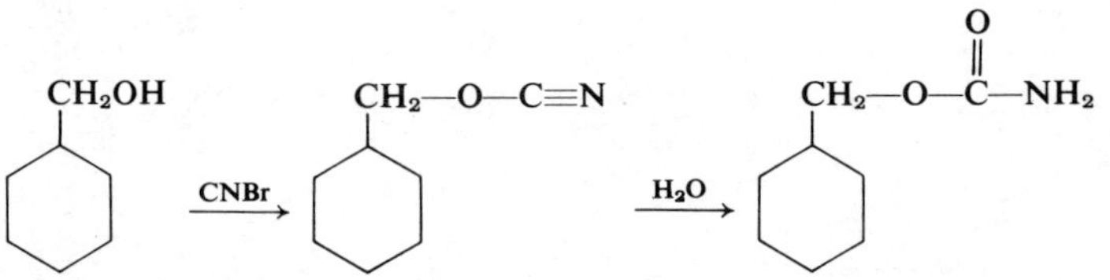

CH_2OH —CNBr→ $CH_2—O—C≡N$ —H_2O→ $CH_2—O—C(=O)—NH_2$

V   VI   VII

reaction is important in the interpretation of the spectral data for cellulose. The adsorption near 1730 cm^{-1} in the activated cellulose could be a composite of the imidocarbonate function (C=N) and the carbamate C=O stretching frequency. Brief hydrolysis with acid converts I to II with the concomitant appearance of the band at 1810 cm^{-1}, whilst the residual slightly decreased band at 1730 cm^{-1} represents that given by the carbamate stretching frequency.

Ahrgen *et al.*[37] have pursued the investigation further by studying the reaction of cyanogen halides with methyl-4,6-*O*-benzylidene-α-D-glucoside (VIII) and prepared the 2,3-cyclic carbonate and the 2- and 3-carbamate derivatives as reference compounds.

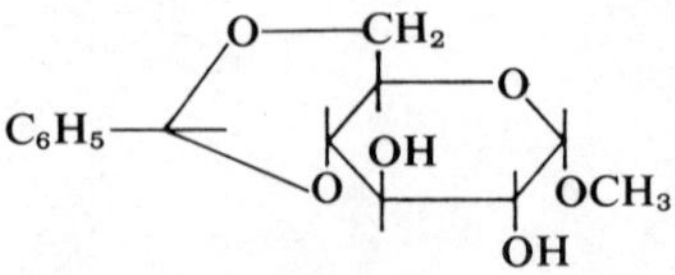

O—CH_2 O C_6H_5 O OH OCH_3 OH

VIII

IR and NMR spectroscopy of the main products of the reaction are consistent with the formation of the cyclic imidocarbonate (I). Likewise, activation of dextran or Sephadex with CNBr probably generates 2,3- or 3,4-linked cyclic imidocarbonates, although 4,6-linkages at the terminal D-glucose residues and interchain imidocarbonates may also be formed. The latter conclusion is supported by the observation that the molecular weight of soluble dextran preparations increases on treatment with cyanogen bromide[35] or cyanates.[36] Furthermore, the reaction between Sephadex G-200 and CNBr is accompanied by a progressive decrease in water regain as the substitution increases.[29] This cross-linking is consistent with the formation of iminoesters. Cyanogen bromide activation of agarose,[30,36] which contains no vicinal hydroxyl groups, is assumed to proceed via 4,6-cyclic and interchain imidocarbonates.

There is also some evidence for the formation of *iso*-urea derivatives (Fig V.5).[37] Svensson[38] has shown that isoelectric focusing of a water-soluble enzyme, prepared by attachment of subtilisin to CNBr-activated amylodextrin, is consistent with the formation of *N*-substituted isoureas (IX).

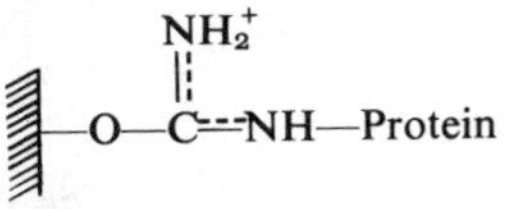

NH_2^+

—O—C—NH—Protein

IX

Axen and Vretblad[39] have confirmed the presence of reactive imidocarbonate groups in CNBr-activated Sephadex. However, they found that after prolonged acid hydrolysis there was a small residual coupling capacity which could not be attributed to imido-carbonate structures. Cyanogen bromide will trimerize under alkaline conditions to trihalo-triazines (X), and triazines will react with polysaccharides under the same experimental conditions as for CNBr.[40,41] The low but significant bromine content of the activated gel may reflect the presence of a low concentration of tribromotriazine in the cyanogen bromide, which should also participate in the coupling reactions.

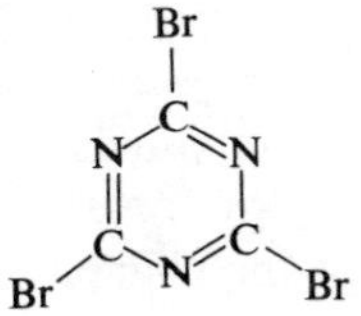

Br

N C N

C N C

Br Br

X

Polysaccharides can also be activated with cyanogen iodide (CNI) although the activation process proceeds more slowly, as does the hydrolysis of cyanogen iodide. The ligand–polysaccharide conjugates obtained after CNI activation have the same properties as those generated by CNBr activation.[29,31] Cyanogen chloride does not appear to have any practical advantages[29] although its derivatives, the triazines, are useful.[41]

The activation of the polysaccharide is performed in alkaline solution and the coupling capacity evaluated by means of a small model peptide such as glycyl-leucine[29] or alanine.[14] The higher the pH employed during the activation process the greater the coupling capacity. Similar results were obtained

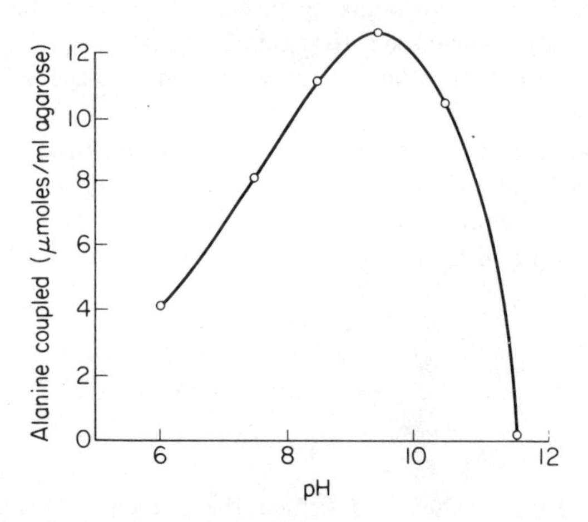

Figure V.6. The effect of pH on the coupling of ^{14}C-alanine to activated Sepharose. Packed Sepharose 4B (60 ml) was mixed with 60 ml water and treated with 15 g CNBr. To the cold, washed activated Sepharose were added 2·2 mmoles of ^{14}C-alanine (0·1 μC_i per μM) in 60 ml cold distilled water, and 20 ml samples of the mixed suspension were added rapidly to beakers containing 5 ml of cold 0·5M buffer. The final concentration of alanine was 15 mM. After 24 hours the suspensions were thoroughly washed, aliquots were hydrolysed by heating at 110°C in 6N HCl for 24 hours, and the amount of ^{14}C-alanine released was determined. Buffers: 0·1M sodium citrate (pH 6·0), phosphate (pH 7·5), borate (pH 8·5, 9·5), and carbonate (pH 10·5, 11·5). Reproduced with permission from P. Cuatrecasas, *J. Biol. Chem.*, **245**, 3059 (1970)

for the relationship of glycyl-leucine coupling yield and the activation pH for Sephadex G25.[29] In the case of compounds carrying an α-amino group such as alanine, an optimum coupling pH of 9·5 to 10·0 is observed (Fig. V.6). This is a reflection of two competing phenomena, the charge on the α-amino group and the stability of the activated complex. The pH at which the coupling stage is performed determines the extent of coupling since it is the unprotonated form of the amino group which is reactive. Compounds with primary aliphatic amino alkyl groups, such as the α-amino group of lysine or the 6-amino group of 6-aminohexanoic acid, are best coupled at pH values around 10. The relatively low pK of the amino group of aromatic amines produces a very facile coupling with most high efficiency at pH values between 8 and 9.[42]

The decreased coupling efficiency at pH values above 9·5–10·0 (Fig. V.6) reflects a sharp decline in stability of the activated complex. The activated product is stable for at least 4 hours at +4°C in slightly alkaline solution with only a small decrease in coupling potential. An activated product stable on storage for several months can be prepared by gradually replacing the water around the gel with acetone until finally the gel is suspended in pure acetone. Following evaporation of the acetone *in vacuo*, the dried gel can be stored in a sealed vessel in the refrigerator.[31] A freeze-dried powder containing a lactose and dextran for stability is available commercially. The powder is swollen and then extensively washed to remove the contaminants prior to use.

The amount of ligand substitution can also be controlled to a certain extent by the amount of ligand added to the activated gel. For highly substituted gels a large excess (20–30 times) of the ligand should be added. The amount of cyanogen bromide per ml of packed Sepharose should also be increased to achieve high ligand densities.

The activated Sepharose is unstable at elevated temperatures and the coupling reaction is generally performed at +4°C. The reaction is usually complete in 2–3 hours, although it is advisable to allow the reaction to stand overnight to ensure complete loss of reactive groups on the matrix. Alternatively, subsequent to the reaction of the desired ligand, the residual active groups can be neutralized by reaction with 1-amino-propan-2,3-diol or D-glucosamine.[43]

Cyanogen halide-activated polysaccharides have a remarkably high capacity to fix proteins covalently under slightly alkaline conditions. Thus, chymotrypsin and papain can be almost quantitatively (90%) attached to Sephadex G-200, cellulose or agarose to yield conjugates containing as much as 30% protein.[31] Trypsin, under the same conditions, pH 8·3 and room temperature, couples relatively inefficiently, a difference attributed to the availability of amino groups. Use of a higher pH for coupling trypsin leads to a higher fixation whilst the high specific activity is still retained. The trypsin is rapidly taken up by the activated gel and thenceforth the enzyme-matrix conjugate becomes quite stable. However, exposure of the enzyme to a high-density activated polymer under conditions where most of the reactive amino groups of the protein are unprotonated can lead to multiple attachment and a decreased activity for the fixed enzyme. This problem can be circumvented by coupling at a less favourable pH, preferably such that a single residue is left unprotonated.

Some practical aspects on the use of cyanogen halides and the activation procedures are given in Section H.

b. Triazines

Polyhydroxylic support matrices can also be activated by cyanuric chloride and dichloro-*sym*-triazines.[41] The latter are conveniently prepared as stable compounds with solubilizing groups such as carboxy-methoxy (XI) and carboxymethylamino (XII). Like the cyanogen halide reaction, the activation and coupling procedure is effectively a two-step reaction: (*i*) rapid reaction of one of the chlorines in a few minutes at pH 9–11 and 20°C with the matrix to give a monochloro-*s*-triazinyl-polysaccharide complex and (*ii*) a slower reaction (16–20 hours) with strong nucleophiles such as the primary amino groups of amino acids or proteins under mild conditions (0–20°C, pH 7–9). The reaction can be represented by the following formula:

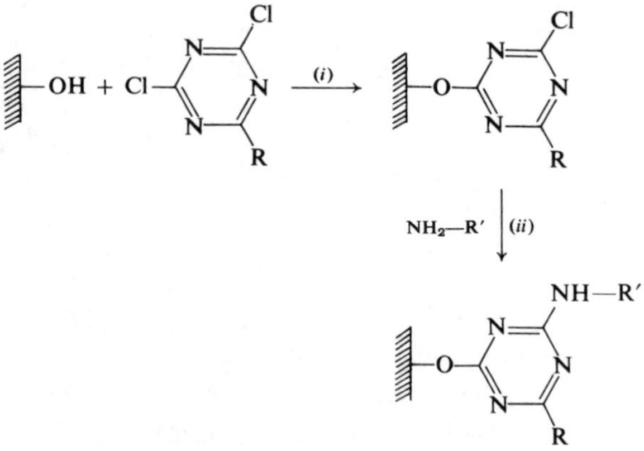

R = —O—CH_2—COOH (XI)
R = —NH—CH_2—COOH (XII)
R = —Cl (XIII)

Furthermore, methods have been developed to react cyanuric chloride (XIII) with cellulose and still retain two of the chlorine atoms.[41] The dichloro-*s*-triazine so formed is exceedingly reactive towards proteins and will react with chymotrypsin in a few minutes at pH 7.

c. Periodate Oxidation

Sanderson and Wilson[44] describe a simple method for the activation of polysaccharides by oxidation with sodium periodate to produce reactive aldehyde groups for the attachment of proteins. The periodate is soluble in aqueous media, easily standardized and readily removed by washing after reaction. Furthermore, the 'activated' polysaccharide can be stored at +4°C for at least a month without loss of coupling potential.

The β-D-glucopyranoside residue of the polysaccharide matrix (XIV) generates aldehydes (XV) on oxidation with 0·1M sodium periodate for

24 hours. Nucleophilic attack by ε-amino groups of lysine under mild conditions would give rise to carbinolamines (XVI) or the corresponding Schiff bases, and subsequent reduction with sodium borohydride would produce the more stable alkylamines (XVII). Such treatment would also block further coupling by reducing residual aldehyde functions.

The substitution potential for periodate-oxidized Sephadex G-25 and G-200 were comparable to those obtained by CNBr activation.[29] However, with

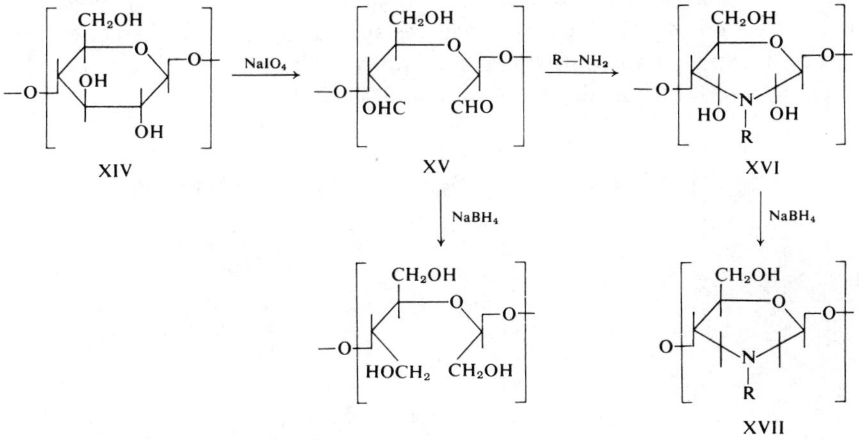

$$\text{XIV} \xrightarrow{NaIO_4} \text{XV (OHC, CHO)} \xrightarrow{R-NH_2} \text{XVI (HO, N–R, OH)}$$

$$\text{XV} \xrightarrow{NaBH_4} \text{(HOCH}_2\text{, CH}_2\text{OH)} \qquad \text{XVI} \xrightarrow{NaBH_4} \text{XVII (N–R)}$$

agarose, periodate oxidation leads to considerably less bound protein than does the equivalent CNBr-activated gel. Sodium periodate oxidation of cellulose leads to the conversion of anhydro-D-glucose units to dialdehyde units without the simultaneous occurrence of side-reactions.[45]

d. Epoxides

Hydroxylic polymers such as agarose, Sephadex and cellulose can be converted to oxirane derivatives by reaction with epichlorohydrin[46] in hot concentrated alkaline solution:

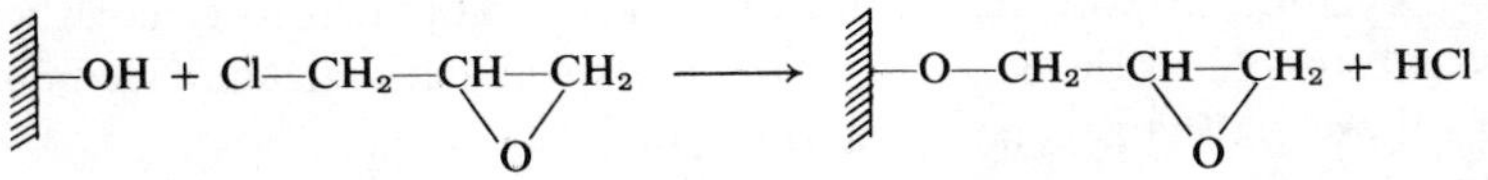

$$\text{Matrix}-OH + Cl-CH_2-\underset{\backslash O/}{CH-CH_2} \longrightarrow \text{Matrix}-O-CH_2-\underset{\backslash O/}{CH-CH_2} + HCl$$

When excess reagent has been removed by washing, coupling to a primary amino group of an amino acid or protein can be effected in alkaline solution:

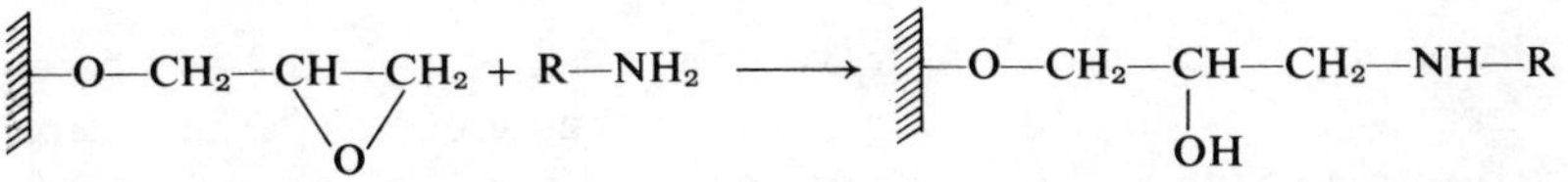

$$\text{Matrix}-O-CH_2-\underset{\backslash O/}{CH-CH_2} + R-NH_2 \longrightarrow \text{Matrix}-O-CH_2-\underset{OH}{CH}-CH_2-NH-R$$

A similar reaction occurs when *bis*-epoxides (*bis*-oxiranes) are utilized. However, one major disadvantage of the epoxy method is that the agarose is simultaneously cross-linked by this procedure. This is advantageous to the extent that cross-linked agarose is more stable in hot concentrated alkali but may impose severe restrictions as an affinity adsorbent since the permeability is compromised.

e. Bifunctional Reagents

Several bifunctional reagents have been utilized in the activation of polyhydroxylic polymers. For example, divinylsulphone has many advantages for the activation of polysaccharide matrices such as agarose.[52] The activation proceeds rapidly under fairly mild conditions and can be exploited to couple amino groups, carbohydrates, phenols and alcohols.

$$\text{—OH} + CH_2{=}CH\text{—}SO_2\text{—}CH{=}CH_2 \quad \text{Divinylsulphone}$$

$$\downarrow$$

$$\text{—O—}CH_2\text{—}CH_2\text{—}SO_2\text{—}CH{=}CH_2$$

$$\downarrow \text{R—NH}_2$$

$$\text{—O—}CH_2\text{—}CH_2\text{—}SO_2\text{—}CH_2\text{—}CH_2\text{—NH—R}$$

Divinylsulphone polymerizes, however, to some extent under the conditions used to activate agarose and hence either the activated polymer or the final product should be extensively washed with water or dimethylsulphoxide before use. Furthermore, like most bifunctional activating agents, cross-linking of the matrix can occur. Indeed, the rigidity of the agarose beads increases considerably[52] after activation. *Bis*-epoxides give similar effects.[53]

f. Other Methods

The alkylation of polysaccharides with chloroacetic acid to give carboxymethyl substituents is the basis for many subsequent derivatization techniques such as those via the acyl azide intermediate.[47]

$$\text{—OH} + \text{Cl—}CH_2\text{—COOH} \longrightarrow \text{—O—}CH_2\text{—COOH}$$

Alternately, polysaccharide hydroxyl groups can be acylated by haloacetyl halides followed by alkylation of a suitable amino group.[48,49] The matrix is stirred with a solution of bromoacetic acid in dioxan for 16 hours at 30°C, after which bromoacetyl bromide is added and the stirring continued for a further 7 hours. The product requires thorough washing. Chloroacetyl-

$$\text{Matrix}{-}OH + Br{-}CO{-}CH_2Br \longrightarrow \text{Matrix}{-}O{-}CO{-}CH_2Br \xrightarrow{R{-}NH_2} \text{Matrix}{-}O{-}CO{-}CH_2{-}NH{-}R$$

derivatives are readily prepared by substituting chloroacetic acid and chloroacetyl chloride, and the iodoacetyl-derivative by the exchange reaction between chloro- or bromoacetyl-matrix and iodine. Substitutions as high as 2·3–2·5 mM per gram polymer can be reached.[49] The lability of the ester linkage produced in the first reaction even at neutral pH can be a disadvantage.

Campbell *et al.*[50] have attached an aromatic nitro compound to cellulose which can subsequently be reduced to the corresponding aromatic amine and then diazotized. Cellulose can be acylated by *p*-nitrobenzyl chloride in 10% sodium hydroxide.

$$\text{Matrix}{-}OH + Cl{-}CH_2{-}C_6H_4{-}NO_2 \xrightarrow{10\%\ NaOH} \text{Matrix}{-}O{-}CH_2{-}C_6H_4{-}NO_2 \xrightarrow{\text{Sodium dithionite}} \text{Matrix}{-}O{-}CH_2{-}C_6H_4{-}NH_2$$

Axen and Porath[51] prepared *p*-nitrophenoxy-hydroxypropyl ethers of Sephadex G-200, reduced the nitro groups to amino groups with sodium dithionite, and converted the amino groups to isothiocyanato groups with thiophosgene. The resultant isothiocyanato-Sephadex was then coupled to several peptides and proteins.

2. Polyacrylamide

The carboxamide side-groups attached to the polyacrylamide hydrocarbon skeleton are chemically stable and resistant to hydrolysis over a wide pH range. The amide nitrogen is, however, readily displaced by certain nitrogen compounds with the liberation of free ammonia.[63] These reactions permit the synthesis of a series of derivatives which can serve as parent compounds for a further range of substituents. Furthermore, the abundance of modifiable carboxamide groups permits a high degree of functionalization.

Polyacrylamide can be aminoethylated by stirring dry beads with anhydrous ethylenediamine at 90°C (Fig. V.7: Reaction I). The number of millimoles of aminoethyl groups introduced per gram of polymer was linearly related to the time of heating at 90°C over a 7-hour time period. Substitution densities as high as 2 mM/g polymer could be achieved. Higher reaction rates were observed by dilution of the ethylenediamine, although a greater proportion of hydroxyl groups were formed by alkaline hydrolysis.

The hydrazide derivative of polyacrylamide was prepared by stirring swollen beads and a calculated amount of hydrazine hydrate or a dilute aqueous solution in a water bath for several hours (Fig. V.7: Reaction II). Nearly 4 mM hydrazide per gram of polymer could be introduced by heating the beads with 6M hydrazine hydrate for 8 hours at 47°C. The proportion of

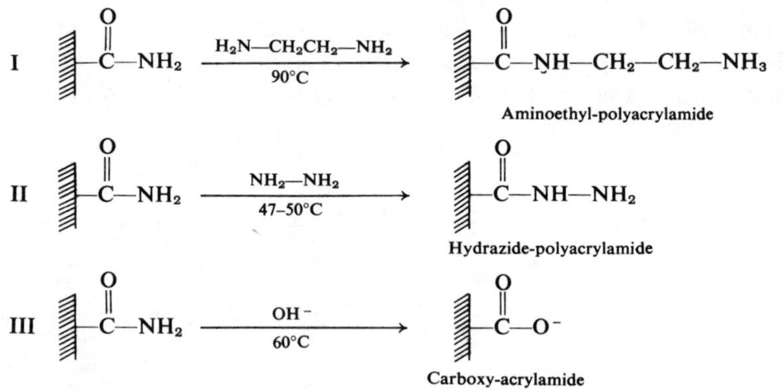

I $$-C(=O)-NH_2 \xrightarrow[90°C]{H_2N-CH_2CH_2-NH_2} -C(=O)-NH-CH_2-CH_2-NH_3$$

Aminoethyl-polyacrylamide

II $$-C(=O)-NH_2 \xrightarrow[47\text{–}50°C]{NH_2-NH_2} -C(=O)-NH-NH_2$$

Hydrazide-polyacrylamide

III $$-C(=O)-NH_2 \xrightarrow[60°C]{OH^-} -C(=O)-O^-$$

Carboxy-acrylamide

Figure V.7. The primary derivatization of polyacrylamide

carboxyl groups generated during hydrazinolysis (2–3%) was smaller than that during aminoethylation (8%). Polyacrylamide-hydrazide can thus be formed at any predetermined level of substitution and converted into a wide range of other derivatives by reaction with suitable aliphatic amines. For example, succinylhydrazide, diethylaminoethyl, phosphoethyl, sulphoethyl and 3-picolyl substituents can readily be introduced.[63]

The formation of cross-links between the diamine and the polyamides is statistically unfavourable in view of the large excess of diamine used in the reaction. Furthermore, no significant cross-linking was apparent since shrinkage of the beads and smaller void volumes was not observed.

Partial alkaline hydrolysis of polyacrylamide leads to deamidation and the appearance of carboxyl functions (Fig. V.7: Reaction III). Thus, when polyacrylamide beads are heated for 3 hours at 60°C in $NaHCO_3/Na_2CO_3$ buffer pH 10·5, almost 2 mM of titratable carboxyl groups are introduced per gram of polymer.

Weston and Avrameas[64] have shown that glutaraldehyde in excess reacts by one of its two aldehyde groups with the amide side-chains of polyacrylamide. The remaining free aldehyde group is then available for Schiff's base formation with a primary amino group, although the reaction is probably more complex. Polyacrylamide beads were incubated with glutaraldehyde at 37°C for 17 hours at several pH values. Activation at pH 7·7 or above either aggregated or destroyed the beads and only at pH 6·9 or less was the physical nature of the beads unaltered. The glutaraldehyde linkages have been shown to be relatively stable.[54]

3. Porous Glass

The physicochemical action of several silane coupling agents on porous silica glass has been investigated.[55–57] The silanization process was found to be a reaction between the glass surface hydroxyls and the amino-functional silane coupling agent:

$$\underset{\text{Glass}}{-O-\overset{\overset{|}{O}}{\underset{\underset{|}{O}}{Si}}-OH} + \underset{\gamma\text{-Aminopropyltriethoxy-silane } (\gamma\text{-ABTES})}{CH_2-CH_3-O-\overset{O-CH_2-CH_3}{\underset{O-CH_2-CH_3}{Si}}-O-(CH_2)_3-NH_2} \longrightarrow$$

$$\underset{\text{Alkylamino-glass } (NH_2\text{-glass})}{-O-\overset{\overset{|}{O}}{\underset{\underset{|}{O}}{Si}}-O-\overset{O-CH_2-CH_3}{\underset{O-CH_2-CH_3}{Si}}-(CH_2)_3-NH_2}$$

The reaction proceeds in methanol and water at 25°C and the resulting alkylamino-glass can subsequently be used to immobilize ligands and proteins by azo, thiourea or diimide coupling procedures. Weibel *et al.*[58] used 10% γ-ABTES in toluene to silanize both solid and porous glass. The glass beads were refluxed for 24 hours although the time required for complete silanization is unknown.

4. Other Support Matrices

The hydrophobic polystyrenes can be nitrated, reduced to form polyaminostyrene and then diazotized. However, the strongly hydrophobic nature of the polystyrene matrices limits their usefulness for affinity chromatography, since many proteins denature at apolar surfaces.

In contrast, nylon is more hydrophilic. It is first treated to remove amorphous regions[59,60] and then partially hydrolysed to liberate free carboxyl and amino groups. The free amino groups can be destroyed[61] and the carboxyl groups converted via the acid hydrazide to the azide and hence reacted with the free amino group of the ligand or protein. Alternatively, the free amino groups of the nylon could be linked to the amino group of the ligand through the bifunctional reagent, glutaraldehyde.[62]

C. SPACER ARMS

It is now well established that the affinity ligand must be placed sufficiently distant from the matrix skeleton to minimize steric interference with the binding process. This is particularly critical for small ligands, for interacting systems of low affinity or for those involving multi-subunit or high molecular weight proteins.

This necessity for a spacer group has since been corroborated by numerous reports in the literature describing successful purifications using adsorbents in which some spacer group has been interposed between the ligand and the matrix backbone. The length and nature of the spacer arm have been discussed elsewhere in this book. Ideally, the ligand should be attached to the spacer arm prior to immobilization on the activated matrix. This method often requires considerable expertise in organic synthesis and up to this time the preferred method has been to couple the ligand to a spacer-arm–matrix conjugate.

The most general and extensively used technique for the introduction of spacer arms is to couple ω-aminoalkyl compounds of the general type $NH_2(CH_2)_nR$ to the activated gel, where R may be a carboxyl or amino function and $n = 2$–12. Thus, spacer arms containing two,[65] four,[66] six[67–69] and twelve[70] carbon atoms interposed between the matrix and the ligand have been reported. For example, spacer arms containing six carbon atoms with amino or carboxyl termini may be synthesized by covalent coupling of 1,6-diaminohexane and 6-aminohexanoic acid, respectively, to cyanogen bromide-activated Sepharose (Fig. V.8(a) and (b)). Statistically, one might expect cross-linking to occur with bifunctional amines although in practice this is drastically reduced by using a large excess of diamine during the coupling stage.

For longer ligand extension arms, 3,3′-diaminodipropylamine has been attached to the gel matrix with cyanogen bromide[71,72] (Fig. V.8(c)). The terminal amino group can subsequently be coupled directly to a carboxyl function on the ligand with a carbodiimide condensation or succinylated by treatment with succinic anhydride in aqueous solution to generate an ω-carboxyl group, potentially some 21 Å distant from the matrix skeleton.[71] Furthermore, retreatment of the succinylated 3,3′-diaminodipropylamine derivative of Sepharose with the amine and succinic anhydride produces a terminal carboxyl group at least 30 Å from the matrix backbone (Fig. V.8(e)). This exceptionally long extension arm has been attached to norepinephrine by a carbodiimide condensation and utilized to purify the cardiac β-adrenergic receptor from solubilized ventricular microsomal particles.[73]

One limitation on the use of flexible hydrophobic polymethylene extension arms is their tendency to fold back on themselves in aqueous media. This

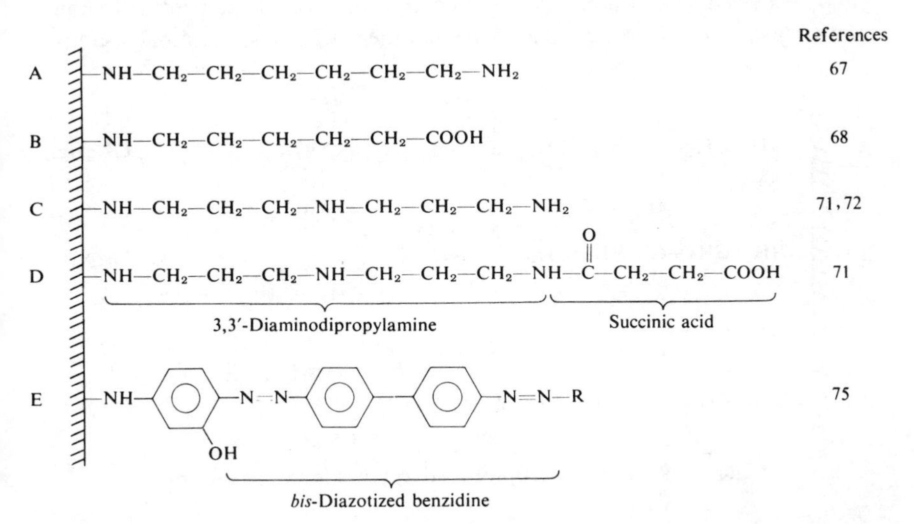

Figure V.8. Some common extension arms used in affinity chromatography

phenomenon can have serious consequences on the interaction with complementary macromolecules[74] and for this reason the use of long hydrocarbon arms of twelve or so carbon atoms is not recommended.[70] Not only is there no increase in the strength of the interaction[70,74] but the use of such long hydrocarbon arms has attendant problems of solubility and the possible denaturation of proteins at hydrophobic interfaces.

The problem can be circumvented in part by using more rigid or more hydrophilic extension arms. Thus, either polyamide, benzenoid or polyene spacer arms are capable of imparting a rigidity to the linkage between the polymer and the ligand which is not easily achieved with other types of arm. *bis*-Diazotized benzidine can form a rigid bridge between a phenol immobilized on the matrix and a suitable group on the ligand and push the ligand some 15 Å out into the free solvent[75] as depicted in Fig. V.8(f). A similar approach has been utilized by other investigators for the purification of enzymes utilizing *p*-aminophenyl or *p*-aminobenzyl esters of nucleotides or sugars.[71,76–78] Furthermore, a spacer arm comprising a 4-aminophenyl residue has been used covalently to attach a ligand for the purification of a phosphodiesterase from *Bothrops atrox* venom.[79]

The recent emphasis on hydrophobic affinity chromatography, particularly with residues containing benzenoid[80] or aliphatic polymethylene[81] hydrocarbons, prompts the use of more hydrophilic extension arms. Lowe *et al.*[74] have interposed polyglycine extension arms between the matrix and the ligand. These oligopeptides, containing up to four glycine units, serve the dual purpose of being both more hydrophilic and, because of the peptide linkage, more rigid than their polymethylene counterparts. The terminal α-amino

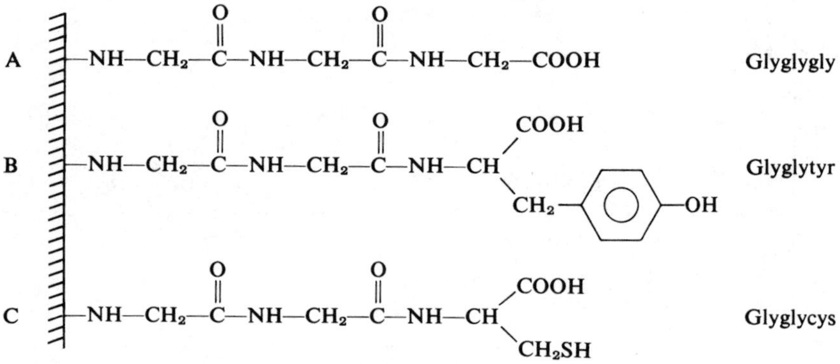

A $$\text{—NH—CH}_2\text{—}\overset{\text{O}}{\overset{\|}{\text{C}}}\text{—NH—CH}_2\text{—}\overset{\text{O}}{\overset{\|}{\text{C}}}\text{—NH—CH}_2\text{—COOH}$$ Glyglygly

B $$\text{—NH—CH}_2\text{—}\overset{\text{O}}{\overset{\|}{\text{C}}}\text{—NH—CH}_2\text{—}\overset{\text{O}}{\overset{\|}{\text{C}}}\text{—NH—CH(COOH)—CH}_2\text{—C}_6\text{H}_4\text{—OH}$$ Glyglytyr

C $$\text{—NH—CH}_2\text{—}\overset{\text{O}}{\overset{\|}{\text{C}}}\text{—NH—CH}_2\text{—}\overset{\text{O}}{\overset{\|}{\text{C}}}\text{—NH—CH(COOH)—CH}_2\text{SH}$$ Glyglycys

Figure V.9. Some hydrophilic oligo-peptide extension arms

group of the peptide is coupled to CNBr-activated agarose and the free carboxyl group subsequently utilized for linkage to the ligand. Furthermore, oligopeptide sequences ending in tyrosine or phenylalanine such as glyglytyr enable hydrophilic spacers with phenolic or benzenoid termini to be prepared.[14] Likewise cysteine, serine or glutamic acid as terminal residues suggest a facile method for obtaining matrix-bound sulphydryl, hydroxyl or carboxyl functions for the linkage of a variety of ligands. One serious criticism of these methods, however, is that a terminal carboxyl group is also introduced along with the desired function and could impair the interaction with the complementary macromolecule (Fig. V.9).

Extension of this logic suggests that polypeptides or even proteins themselves may be useful hydrophilic spacer groups.[82] For example, hapten–protein conjugates used for eliciting an antibody response have been coupled to Sepharose and the resulting matrix–hapten–protein complex has been competent for the purification of the antihapten antibody.

D. THE PREPARATION OF HIGH-CAPACITY ADSORBENTS

Interacting systems of low affinity ($K_L \geqslant 10^{-4}$M) are often not amenable to affinity chromatography unless the ligand concentration can be made sufficiently high to retard the downward migration of the enzyme through the column bed. Porath and Sundberg[52] have introduced a second generation of chemisorbents whose significant feature is a greatly increased capacity for ligand substitution. The number of hydroxyl groups native to the polysaccharide or other matrix is increased by the introduction of polyhydric phenols and alcohols. This increases the number of hydroxyl groups available for subsequent reaction with bifunctional reagents to form affinity adsorbents. For example, the polyhydric phenol, phloroglucinol, can be coupled to Sepharose with epichlorohydrin as shown below.

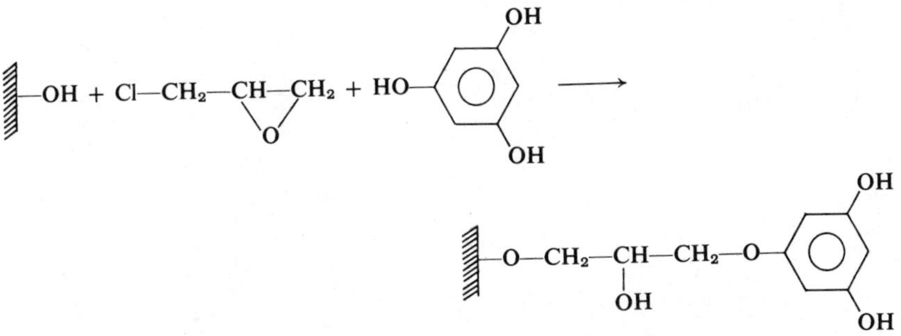

$$\text{—OH} + \text{Cl—CH}_2\text{—}\underset{\diagdown\,\text{O}\,\diagup}{\text{CH—CH}_2} + \text{HO—C}_6\text{H}_3(\text{OH})_2 \longrightarrow$$

$$\text{—O—CH}_2\text{—}\underset{\text{OH}}{\underset{|}{\text{CH}}}\text{—CH}_2\text{—O—C}_6\text{H}_3(\text{OH})_2$$

The resulting phloroglucinol-substituted polymer has three distinct advantages over the original parent gel:

(i) the total number of hydroxyl groups is increased,

(ii) the phenolic hydroxyl groups are more reactive, and

(iii) the hydroxyl groups are less sterically hindered since they are remote from the matrix backbone.

The hyper-hydroxylated polymer can now be activated with cyanogen halides,[29] *bis*-epoxides or epihalohydrins,[53] triazines[41] or divinylsulphone[52] in the usual way, prior to attachment of a ligand. The higher reactivity of the phenolic groups ensures their preferential activation. Preliminary studies have shown that the capacity of these new adsorbents is considerably greater than that of those prepared in the usual fashion.[52]

E. REACTIONS FOR COUPLING LIGANDS TO SPACER ARMS

The primary derivatives described in Sections C and D of this chapter can be used to attach a wide variety of ligands to the support. The procedures involved are as prolific and diverse as the ligands themselves, although in essence they consist only of a few basic, well-documented reactions.

Ligands containing primary aliphatic or aromatic amino groups can be coupled directly to agarose by the cyanogen bromide activation procedure.[30]

1. Carbodiimide Condensations

Ligands carrying primary aliphatic or aromatic amines, and carboxylic acid functions, can be coupled to ω-carboxyl-alkyl-Sepharose derivatives and ω-aminoalkyl-Sepharose derivatives respectively by a carbodiimide-promoted condensation reaction.[83,84]

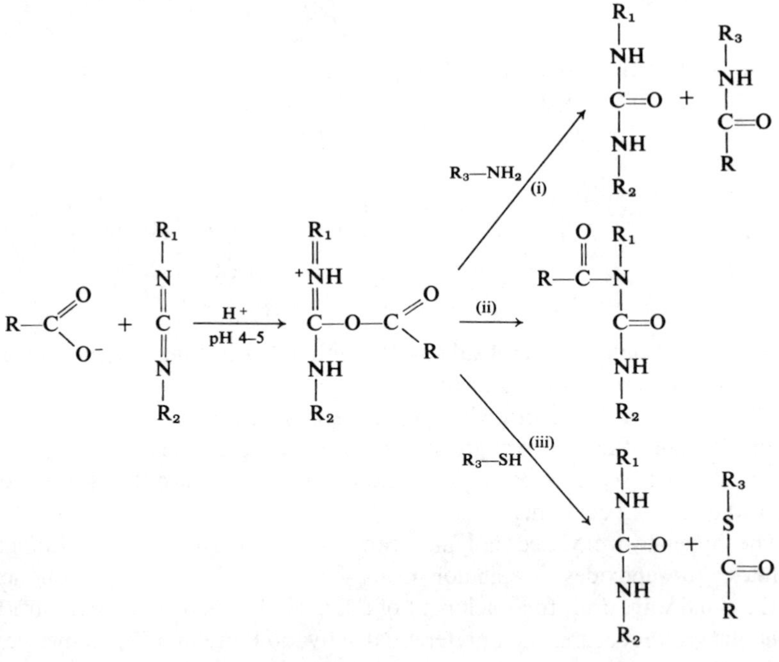

$$R{-}C(=O){-}O^- + R_1{-}N{=}C{=}N{-}R_2 \xrightarrow[\text{pH 4–5}]{H^+} R_1{=}{}^+NH{=}C(NH{-}R_2){-}O{-}C(=O){-}R$$

$$\text{(i)}\quad \xrightarrow{R_3{-}NH_2} R_1{-}NH{-}C(=O){-}NH{-}R_2 + R_3{-}NH{-}C(=O){-}R$$

$$\text{(ii)}\quad \longrightarrow R{-}C(=O){-}N(R_1){-}C(=O){-}NH{-}R_2$$

$$\text{(iii)}\quad \xrightarrow{R_3{-}SH} R_1{-}NH{-}C(=O){-}NH{-}R_2 + R_3{-}S{-}C(=O){-}R$$

Figure V.10. The carbodiimide-promoted reaction: (i) and (ii) the mechanism, (iii) the formation of thiol ester bonds

The mechanism of the carbodiimide-assisted reaction is illustrated in Fig. V.10. The reaction is initiated by the addition of the carboxyl across one of the diimide double bonds to generate an *O*-acylisourea.[83] The activated carboxyl of this intermediate can then react in one of two ways.

(i) Nucleophilic attack can yield an acyl-nucleophile product and the urea corresponding to the carbodiimide.

(ii) The *O*-acylisourea intermediate can undergo an intramolecular acyl transfer to yield an *N*-acylurea.

In the special case where the nucleophile is water, the carboxyl is regenerated and the carbodiimide is stoicheiometrically converted to its corresponding urea. Hoare and Koshland[83] have shown that if the concentration of nucleophile is sufficiently high the intramolecular rearrangement can be reduced almost to zero. Thus, the coupling reaction between carboxyl and nucleophile can be almost quantitative in the presence of excess carbodiimide and nucleophilic reagent.

Several carbodiimides have been used for the synthesis of adsorbents for affinity chromatography. Dicyclohexyl-carbodiimide (XVIII) is insoluble in water but has been used very effectively in 80% (v/v) aqueous pyridine for the synthesis of immobilized NAD^+[15,74] and for the preparation of an adsorbent for mushroom tyrosinase.[85] One disadvantage in the use of water-insoluble

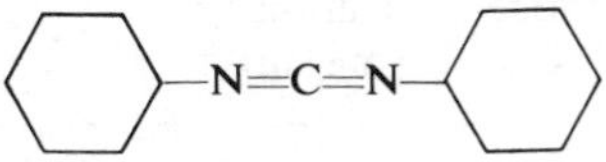

$$C_6H_{11}{-}N{=}C{=}N{-}C_6H_{11}$$

XVIII

carbodiimides is that the derived urea is also insoluble in water and consequently must be removed from the gel by extensive washing with organic solvents such as ethanol and butanol.[15] However, the problem can be circumvented by using one of two soluble carbodiimides, 1-ethyl-3-(3-dimethylaminopropyl) carbodiimide hydrochloride (EDC, XIX) and 1-cyclohexyl-3-(2-morpholinoethyl) carbodiimide metho *p*-toluene sulphonate (CMC, XX).

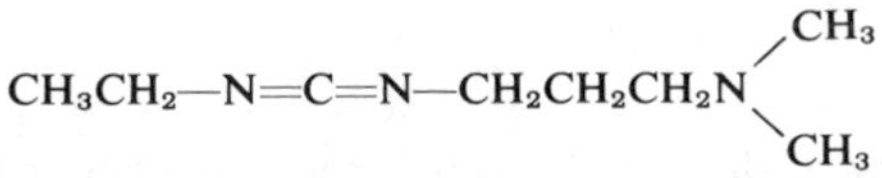

$$CH_3CH_2{-}N{=}C{=}N{-}CH_2CH_2CH_2N(CH_3)_2$$

XIX

$$C_6H_{11}{-}N{=}C{=}N{-}CH_2CH_2{-}{}^{+}N(CH_3)(CH_2CH_2)_2O$$

$$^{-}SO_3{-}C_6H_4{-}CH_3$$

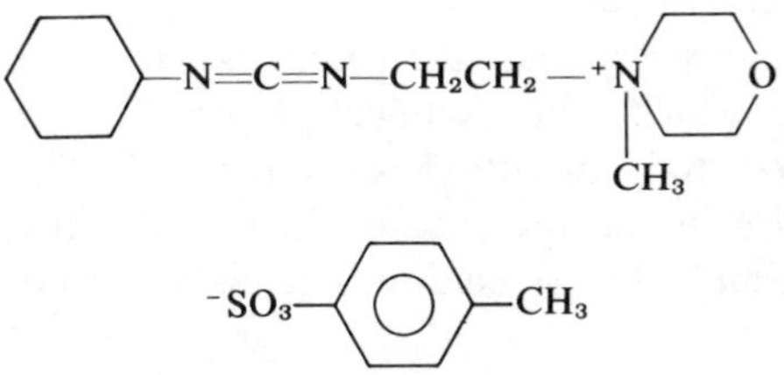

XX

These are convenient to use since their corresponding urea derivatives are soluble in water and are hence easily removed from the gel product by washing with water. These two carbodiimides have been used under a variety of experimental conditions (Table V.2), although it is important that the pH, diimide concentration and reaction time be optimized for each individual case. In some cases EDC gives better coupling yields than CMC and, furthermore, CMC can introduce non-specific ion-exchange groups into matrices. This effect presumably arises by immobilization of the charged morpholino moiety.[86]

In addition to carboxyl functions, carbodiimides react with water, alcohols, amines, phenols, sulphydryls and many other compounds with an active hydrogen. For example, thiol ester bonds can be formed by coupling sulphydryl ligands to ω-carboxyl-alkyl-agarose with EDC at pH 4·7.[14] The resultant thiol ester bond is stable at neutral pH but is readily cleaved by exposure to

Table V.2. Carbodiimide condensations in affinity chromatography

Coupled ligand	Carbodiimide	Concentration (mg/ml)	pH	Reaction time (hours)	Reference
Methotrexate	EDC	18	6·4	1·5	69
Amethopterin	EDC	20	6·5	72	89
Folic acid	EDC	5	6·0	2	90
NAD⁺	CMC	20	4·7	16	75
Saccharo-1,4-lactone	EDC	6	4·8	16–20	72
Calcium oxalate	EDC	74	4·7	20	67
p-Aminobenzamidine	CMC	2	4·75	5	68
Anthranilic acid	CMC	100	4·9	20	91
α-Lipoic acid	EDC	14	4·7	16	92

pH 11·5 for a few minutes or by treatment with 1M hydroxylamine for 30 minutes. The lability of this bond towards selective chemical cleavage makes it possible to remove an intact ligand–protein conjugate from an adsorbent containing a very tightly bound protein.

The carbodiimide reaction has also been used to prepare an organomercurial-Sepharose which binds sulphydryl proteins.[14] *p*-Chloromercuribenzoate is coupled to aminoethyl-Sepharose in 40% *N*,*N'*-dimethylformamide at pH 4·8 in the presence of EDC. This adsorbent will bind sulphydryl proteins, and effective elution is achieved with 0·5M cysteine.

2. Other Methods for Peptide Bond Formation

Other methods are available for the linkage of carboxyl functions to aminoalkyl-Sepharose and vice versa. Thus, carboxyl groups can be activated by mixed anhydride formation with iso-butyl chloroformate (iso-butyl chlorocarbonate).[87] The dry carboxylic ligand is dissolved in dry dioxan in the presence of a small amount of tri-*N*-butylamine and cooled in ice. Iso-butyl chloroformate is added and the mixture left for 30 minutes, keeping the temperature at 8°C such that the dioxan is just liquid, whence the aminoalkyl-Sepharose is added in a mixture of water, dioxan and NaOH. The pH is maintained at 8·0 for 30 minutes at 8°C and then for 20 hours at 0°C. The gel product can then be thoroughly washed.

$$\text{R—COOH} + \text{Cl—C(=O)—O—iso-Butyl}$$

$$\downarrow$$

$$\text{R—C(=O)—O—C(=O)—O—iso-Butyl} + \text{(matrix)—NH}_2$$

$$\downarrow$$

$$\text{(matrix)—NH—C(=O)—R} + \text{iso-Butanol} + \text{CO}_2$$

Woodwards *K* reagent[88] can be used to effect a similar condensation.

3. Anhydride Reactions

Succinylaminoalkyl-Sepharose can be prepared by treating ω-aminoalkyl-Sepharose with 1% succinic anhydride in aqueous solution, the pH being raised to, and maintained at, 6·0 with 20% NaOH.[14] On completion of the reaction, no further change in pH occurs, and the suspension is left for a further 5 hours at 4°C or for 1 hour at room temperature.

$$\text{(matrix)—NH(CH}_2)_n\text{NH}_2 \xrightarrow[\text{pH 6}]{\text{succinic anhydride}} \text{(matrix)—NH(CH}_2)_n\text{NH—C(=O)—CH}_2\text{CH}_2\text{C(=O)OH}$$

Likewise, the hydrazide derivative of polyacrylamide can be converted into the succinylhydrazide derivative by treatment with succinic anhydride at pH 4.[63]

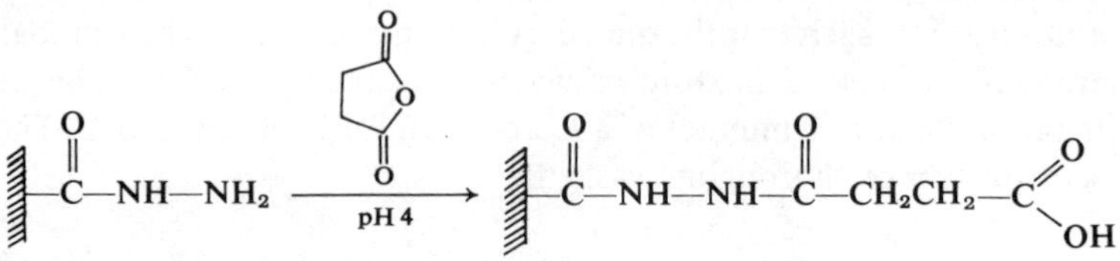

$$\text{P}-\overset{\overset{\displaystyle O}{\|}}{C}-NH-NH_2 \xrightarrow[\text{pH 4}]{\text{succinic anhydride}} \text{P}-\overset{\overset{\displaystyle O}{\|}}{C}-NH-NH-\overset{\overset{\displaystyle O}{\|}}{C}-CH_2CH_2-C(=O)OH$$

Ligands containing primary aliphatic or aromatic amino groups can be condensed with such carboxylated polymers in the presence of the water-soluble carbodiimide, 1-ethyl-3-(3-dimethylaminopropyl) carbodiimide hydrochloride. These reactions have been discussed in Section 1 above.

4. *N*-Substituted Hydroxysuccinimide Reactions

Bromoacetyl derivatives of Sepharose react with ligands containing primary aliphatic or aromatic amines, imidazole or phenolic groups and are readily synthesized from appropriate ω-aminoalkyl-Sepharoses.[14] Thus, bromoacetamidoethyl-Sepharose can be prepared in mild aqueous conditions by reacting aminoethyl-Sepharose with *O*-bromoacetyl-*N*-hydroxysuccinimide (XXI). The hydroxysuccinimide reagent is prepared in situ by *N*,*N*′-dicyclo-

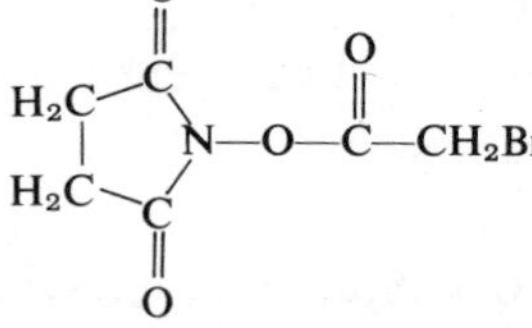

$$H_2C(-C(=O)-)_2N-O-\overset{\overset{\displaystyle O}{\|}}{C}-CH_2Br \quad (H_2C-H_2C \text{ ring})$$

XXI

hexyl carbodiimide coupling of bromoacetic acid to *N*-hydroxysuccinimide in dioxan. After removal of the dicyclohexylurea by filtration, the filtrate is added to a suspension of aminoalkyl-Sepharose at pH 7·5 and left for 30 minutes at 4°C. Alternatively, crystalline bromoacetyl-*N*-hydroxysuccinimide ester can be reacted directly with the substituted gel under the same conditions.

N-Hydroxysuccinimide ester of carboxylalkyl-Sepharose can be prepared in a similar fashion[16] by reacting the agarose derivative for 70 minutes at room temperature with 0·1M *N*,*N*′-dicyclohexyl carbodiimide and 0·1M *N*-hydroxysuccinimide in dioxan. The active *N*-hydroxysuccinimide ester of agarose is more stable at pH 6·3 than at pH 8·6 in aqueous solution, and it can be stored in dioxan for several months with very little loss of coupling potential. Ligands carrying free unprotonated amino groups can be coupled rapidly at 4°C through a stable amide linkage to these activated derivatives

$$\text{—NH(CH}_2)_n\text{NH}_2 \xrightarrow{\text{(succinimido)—O—CO—CH}_2\text{Br}} \text{—NH(CH}_2)_n\text{NH—CO—CH}_2\text{Br}$$

$$\downarrow \text{NH}_2\text{—R}$$

$$\text{—NH(CH}_2)_n\text{NH—CO—CH}_2\text{—NH—R}$$

under mild aqueous conditions and in the pH range 6–9. Of several other functional groups tested, only sulphydryl, and to a lesser extent imidazole groups, reacted with the active ester.[16] This means that it is possible selec-

$$\text{—NH—(CH}_2)_n\text{—COOH}$$

$$\downarrow \text{Dioxan; HO—N(succinimide); } N,N'\text{-Dicyclohexyl carbodiimide}$$

$$\text{—NH—(CH}_2)_n\text{—CO—O—N(succinimide)}$$

$$\downarrow \text{NH}_2\text{—R}$$

$$\text{—NH—(CH}_2)_n\text{—CO—NH—R}$$

tively to immobilize ligands which contain multiple functional groups by their primary amino groups, without employing blocking agents to protect other groups. These procedures are also applicable to the derivatization of porous glass.

5. The General Acyl Azide Procedure

Primary aliphatic amines can be coupled to hydrazide gels (Section B.2) via the acyl azide derivative without intermediate washings or transfers.[63] The hydrazide polymer is cooled to 0°C in 0·25 HCl in an ice bath, then crushed ice is added, followed by the rapid addition of 1M sodium nitrite solution. The mixture is stirred for 90 seconds, whence the amine is added rapidly and left for 0–2 hours (depending on the amine used). Unreacted azide may be reconverted into hydrazide and then to the stable acetyl hydrazide treatment with hydrazine hydrate, followed by sodium acetate and acetic anhydride after suitable washing.[63]

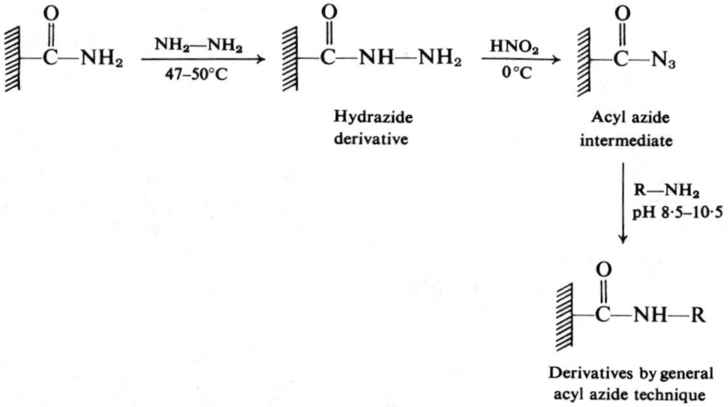

$$\text{—C(=O)—NH}_2 \xrightarrow[47\text{–}50^\circ\text{C}]{\text{NH}_2\text{—NH}_2} \text{—C(=O)—NH—NH}_2 \xrightarrow[0^\circ\text{C}]{\text{HNO}_2} \text{—C(=O)—N}_3$$

Hydrazide derivative

Acyl azide intermediate

$$\text{—C(=O)—N}_3 \xrightarrow[\text{pH } 8{\cdot}5\text{–}10{\cdot}5]{\text{R—NH}_2} \text{—C(=O)—NH—R}$$

Derivatives by general acyl azide technique

Table V.3 lists some derivatives that have been prepared by coupling amines to polyacrylamide hydrazide by the general acylazide procedure.[63]

6. Diazotization Procedures

Diazonium derivatives of agarose[14] or polyacrylamide[63] react with phenols, imidazoles or any other compounds susceptible to electrophilic attack and can be prepared under mild aqueous conditions from *p*-amino-benzamido-alkyl derivatives. The aminoalkyl derivative in an alkaline solution, such as 0·2M sodium borate pH 9·3 or triethylamine, and 40% (v/v) *N*,*N*′-dimethylformamide (DMF) is treated for one hour at room temperature with *p*-nitrobenzoyl azide. On completion of the reaction, the *p*-nitro-benzamido-alkyl gel is washed extensively with 50% DMF, 25% DMF and finally water. The washed derivatized gel is then reduced with 0·1M sodium dithionite for 40 minutes at 40–50°C, and subsequently diazotized by treatment with 0·1M sodium nitrite for 7 minutes at 0°C in 0·5M HCl. These reactions are summarized in Fig. V.11.

Table V.3. Some derivatives prepared by the acyl azide method

Derivative	
Diethylaminoethyl (DEAE-)	$-CH_2-CH_2-N(CH_2-CH_3)_2$
Phosphoethyl	$-CH_2-CH_2-O-P(=O)(OH)-O^-Na^+$
Sulphoethyl	$-CH_2-CH_2-SO_3^-Na^+$
Dihydroxyethylaminopropyl	$-CH_2-CH_2-CH_2-N(CH_2CH_2OH)_2$
3-Picolyl	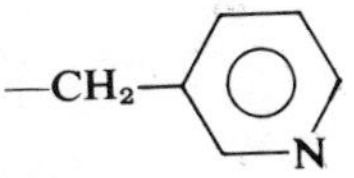 $-CH_2-$(3-pyridyl)
p-Hydroxyphenethyl	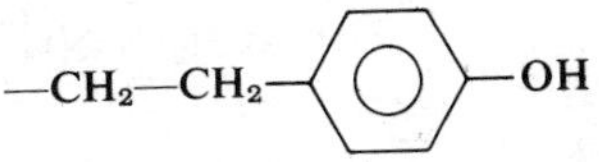 $-CH_2-CH_2-C_6H_4-OH$
Protein	—protein

The diazonium derivative can be used in situ without further washing, by adding the phenol, imidazole or protein to be coupled in a strong buffer such as saturated sodium borate. The pH is adjusted to 8 for histidyl residues and 10 for phenolic groups and maintained for 8 hours at 40°C.

Two important advantages accrue in the use of azo-bonded ligands: first, the bound inhibitor can be rapidly and completely released by reduction of the linkage with 0·1M sodium dithionite in 0·2M sodium borate buffer, pH 9. This permits an estimation of the amount of inhibitor bound to the gel. Secondly, the procedure permits the release of an intact protein-inhibitor conjugate under mild conditions.

The *p*-nitrobenzamido-alkyl derivative of the matrix may also be prepared by treatment of the aminoalkyl gel with *p*-nitrobenzoyl chloride. This procedure has been adopted for the preparation of diazonium derivatives of glass.[93,58] The δ-amino group of silanized glass (Section B.3) was acylated by refluxing for 24 hours in chloroform containing by weight 10% *p*-nitrobenzoyl chloride and 10% triethylamine. The resulting acylated glass beads were thoroughly washed with chloroform whence the diazonium derivatives could be prepared by reduction with dithionite and treatment with nitrous acid.

Ligands containing a diazotizable aromatic amine can be coupled through

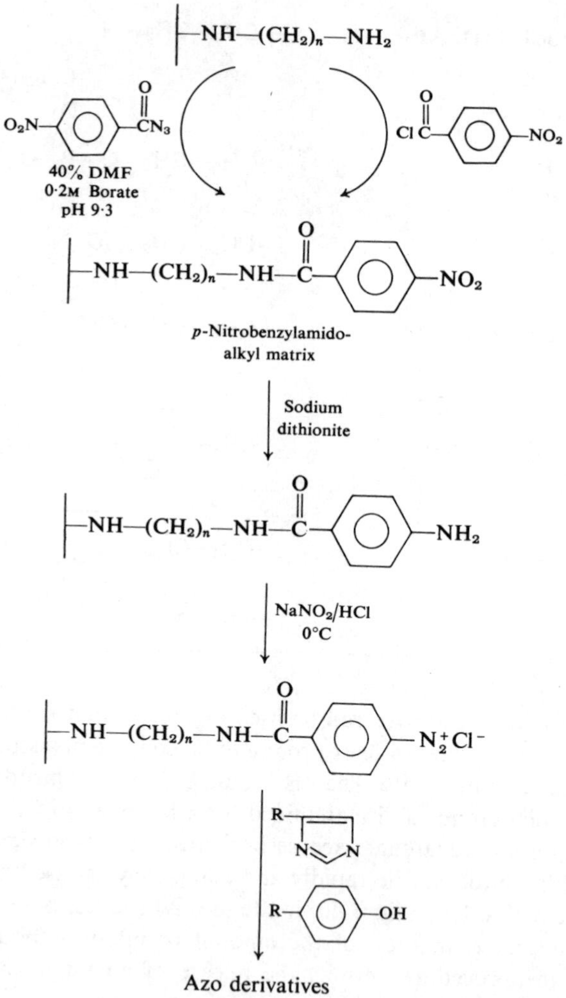

$$\vdash NH-(CH_2)_n-NH_2$$

$O_2N-C_6H_4-C(=O)N_3$ — 40% DMF, 0·2M Borate, pH 9·3

$Cl-C(=O)-C_6H_4-NO_2$

$$\vdash NH-(CH_2)_n-NH-C(=O)-C_6H_4-NO_2$$

p-Nitrobenzylamido-alkyl matrix

↓ Sodium dithionite

$$\vdash NH-(CH_2)_n-NH-C(=O)-C_6H_4-NH_2$$

↓ $NaNO_2/HCl$, 0°C

$$\vdash NH-(CH_2)_n-NH-C(=O)-C_6H_4-N_2^+Cl^-$$

↓ R-imidazole; R-C_6H_4-OH

Azo derivatives

Figure V.11. The preparation and properties of diazonium derivatives

azo linkages to tyrosyl derivatives in high yield.[14] The tripeptide, glycylglycyltyrosine, containing a carboxyl-terminal tyrosine residue can be coupled directly to Sepharose by the cyanogen bromide activation procedure. The ligand is diazotized essentially as described above and the entire mixture added to a suspension of tyrosyl-Sepharose in 0·2M Na_2CO_3, pH 9·4, containing crushed ice. The reaction is allowed to proceed for three hours or so whence the product is thoroughly washed. Azo derivatives prepared in this manner have been used to purify Staphylococcal nuclease.[94]

$$\text{—Gly—Gly—NH—CH(COOH)—CH}_2\text{—C}_6\text{H}_4\text{—OH} \xrightarrow{N_2^+\text{—C}_6\text{H}_4\text{—R}} \text{—Gly—Gly—NH—CH(COOH)—CH}_2\text{—C}_6\text{H}_3(\text{OH})\text{—N=N—C}_6\text{H}_4\text{—R}$$

7. Reductive Alkylation

Aliphatic aldehydes and ketones react rapidly and reversibly with amino groups under mild, slightly alkaline, conditions. The adducts, Schiff's bases, so formed can be reduced with mild reducing agents to yield stable alkylamino groups.[95]

$$\text{—NH}_2 + \text{R—CHO} \rightleftharpoons \text{—N=C(H)—R} \xrightarrow{[H]} \text{—NH—CH}_2\text{—R}$$

Ryan and Fottrell[96] have prepared immobilized pyridoxal-5′-phosphate (PLP) by this general procedure. Solid pyridoxal phosphate was added to a suspension of the ω-aminoalkyl-Sepharose and gently stirred for 30 minutes at 20°C. The resulting Schiff's base was reduced with sodium borohydride at 0°C and the whole procedure repeated several times to ensure complete substitution of the free terminal amino groups of the Sepharose.

Periodate oxidation of sugar moieties also generates aldehyde groups which can be coupled by Schiff's base formation to alkylamino-agarose.[97] Subsequent reduction with borohydride in neutral solution generates a stable alkylamino linkage. Periodate oxidation of the ribose moiety of adenosine-5′-monophosphate (AMP), Schiff's base formation with 6-aminohexyl-Sepharose and subsequent reduction with borohydride have been used to prepare immobilized AMP for affinity chromatography.[86] Furthermore, Robberson and Davidson[98] have demonstrated that periodate oxidized RNA or UMP could be coupled to hydrazide-agarose. Likewise, Jackson *et al.*[99] have covalently attached GTP to an insoluble matrix. The procedure involved coupling ε-aminocaproic acid methyl ester to CNBr-activated agarose followed by hydrazinolysis. Periodate oxidized GTP was then reacted with

$$\text{—NH(CH}_2)_n\text{NH}_2$$

$$\downarrow \text{PLP}$$

$$\text{—NH(CH}_2)_n\text{N=CH—}(\text{pyridine ring: } CH_2OPO_3,\ N,\ OH,\ CH_3)$$

$$\downarrow \text{NaBH}_4$$

$$\text{—NH(CH}_2)_n\text{NH—CH}_2\text{—}(\text{pyridine ring: } CH_2OPO_3,\ N,\ OH,\ CH_3)$$

the hydrazide agarose to form the respective hydrazone derivative. A similar procedure has been used by Remy *et al.*[100] for coupling *t*RNA to Sepharose. Periodate oxidized yeast *t*RNA^{Phe} was coupled to freshly prepared hydrazinyl-Sepharose in 0·1M acetate buffer pH 5·0 for one hour at 37°C and 15 hours at room temperature. The gel was washed thoroughly with 1M NaCl and the hydrazone stabilized by reduction with $NaBH_4$ in tris-HCl buffer, pH 8·0.

8. Isothiocyanate Coupling

The thiourea coupling of primary amines to ω-aminoalkyl derivatives via an isothiocyanate intermediate has been used for the derivatization of porous glass. The ω-aminoalkyl-glass is refluxed for four hours in chloroform with a 10% solution of thiophosgene. The glass is decanted and thoroughly washed with chloroform and then dried. The isothiocyanate product can be coupled to the amino ligand or a protein in pH 9–10 buffer for 2–3 hours at room temperature.

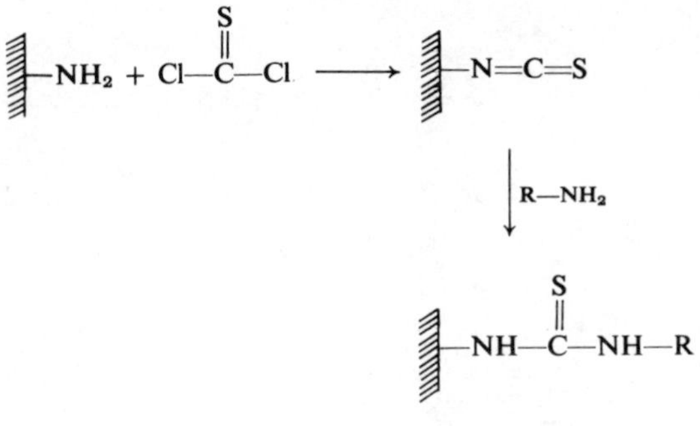

$$\text{—}NH_2 + Cl\text{—}C(=S)\text{—}Cl \longrightarrow \text{—}N{=}C{=}S$$

$$\xrightarrow{R\text{—}NH_2} \text{—}NH\text{—}C(=S)\text{—}NH\text{—}R$$

9. Bifunctional Reagents

Reagents with two reactive groups or bifunctional reagents can be used to cross-link molecules of the same or a different kind.[101] However, the use of potential bifunctional reagents can produce two serious problems: firstly, it is important that both functional groups on the molecule react and that the reaction is not monofunctional. This could, in principle, leave residual active groups which could irreversibly bind the complementary protein on subsequent chromatography. Secondly, a large excess of reactants is required to ensure that the bifunctional reagent does not cross-link the polymer matrix.

The following survey is intended to give some idea of the types of bifunctional reagents available, and to what uses they can be put. These reagents are summarized in Table V.4.

a. Maleimide Derivatives

N-substituted maleimides are specific for sulphydryl groups under mild conditions and exhibit a minimum of side reactions. Likewise, the *N*-substituted-*bis*-maleimides are highly specific (XXII, XXIII) and mild bifunctional reagents. Several *N*-aryl and *N*-alkyl-*bis*-maleimides (XXIV) are

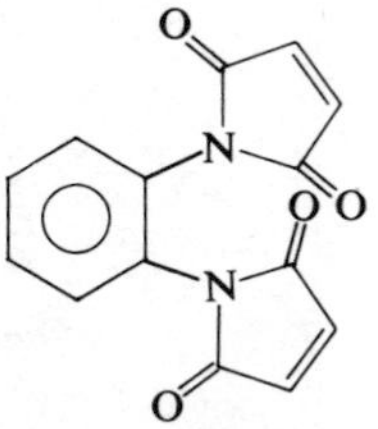

XXII

N,N′-(1,2-Phenylene)-*bis*-maleimide[102]

Table V.4. Bifunctional reagents

		Principal reaction	Reference
Glutaraldehyde	$OHC-(CH_2)_3-CHO$	NH_2—	115, 116
p,p′-Difluoro-*m,m′*-dinitro diphenyl sulphone		NH_2— Phenolic—	117
Hexamethylene diisocyanate	$O{=}C{=}N-(CH_2)_6-N{=}C{=}O$	NH_2—	118
N,N′-(1,3-Phenylene)-*bis*-maleimide		SH—	102
N,N′-Ethylene-*bis*-iodoacetamide	$ICH_2CONH(CH_2)_2NHCOCH_2I$	SH—	106
3,6-*bis*-(Mercurimethyl)-dioxan		SH—	119
bis-Diazobenzidine		Phenolic—	75, 120

Woodward's *K*	$3\text{-}SO_3^-\text{-}C_6H_4$–(isoxazolium ring, N^+–CH_2–CH_3)	—COOH to NH_2	88
bis-Oxiranes	CH_2–CH–R–CH–CH_2 (each CH_2–CH pair bridged by O)	—OH to NH_2	53
Dimethyl adipimidate	$^+NH_2$=C(OCH_3)–$(CH_2)_4$–C(OCH_3)=NH_2^+	NH_2—	121
Dimethyl suberimidate	$^+NH_2$=C(OCH_3)–$(CH_2)_6$–C(OCH_3)=NH_2^+	NH_2—	114
Diethyl malonimidate	$^+NH_2$=C(OCH_3)–CH_2–C(OCH_3)=NH_2	NH_2—	122
Phenol-2,4-disulphonyl-chloride	HO–C_6H_3(2-SO_2Cl)(4-SO_2Cl)	NH_2—	123
Divinylsulphone	CH_2=CH–$S(=O)_2$–CH=CH_2		52

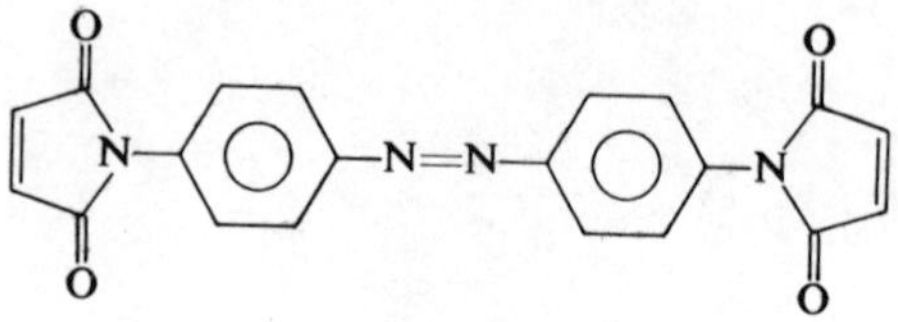

XXIII

Azophenyldimaleimide[103]

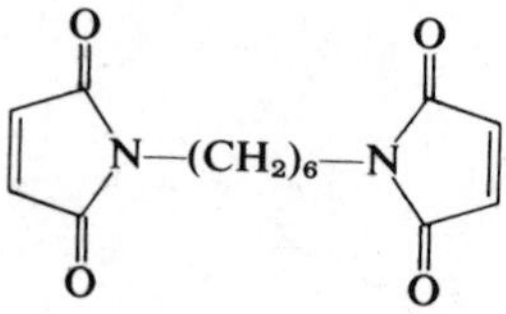

$N—(CH_2)_6—N$

XXIV

N,N'-Hexamethylene-*bis*-maleimide[104]

available.[102] These reagents are all insoluble in water and have generally been added in stoicheiometric amounts as a solid to aqueous solutions (pH 7–8) of the reactants. Azophenyldimaleimide (XXIII) (and its reaction products) is readily cleaved by reduction of the azo groups with dithionite. These reagents have been used for cross-linking protein sulphydryl groups and could be used for linking sulphydryl ligands to thiol matrices.

b. Alkyl Halides

The bifunctional alkyl halides (XXV and XXVI) react primarily with thiol, imidazole and amino groups. At neutral to slightly alkaline pH the reaction with sulphydryl groups is favoured whilst at higher pH values reaction with amino groups is preferred.

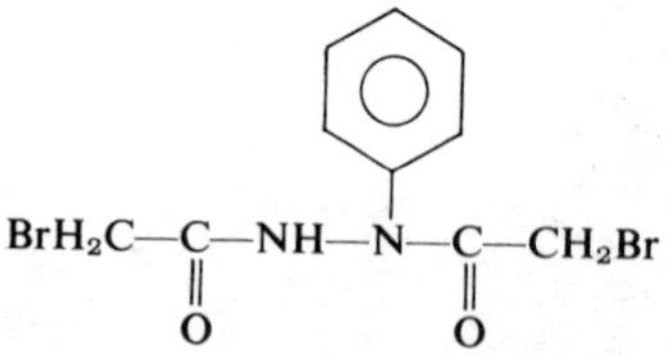

$BrH_2C—C(=O)—NH—N(C_6H_5)—C(=O)—CH_2Br$

XXV

N,N'-Di(bromoacetyl)-phenyl hydrazine[105]

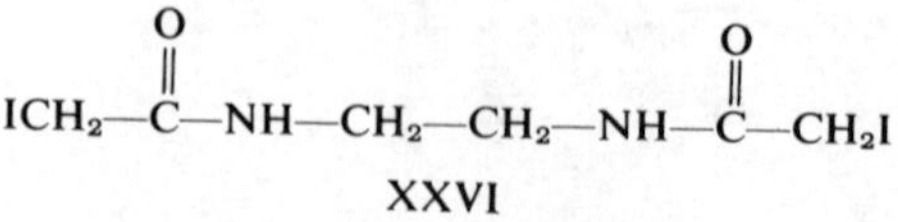

$ICH_2—C(=O)—NH—CH_2—CH_2—NH—C(=O)—CH_2I$

XXVI

N,N'-Ethylene-*bis*-iodoacetamide[106]

c. Aryl Halides

Reagents such as 1,5-difluoro-2,4-dinitrobenzene (XXVII)[107] are insoluble in water and react preferentially with amino groups and tyrosine phenolic

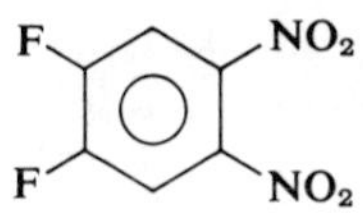

F NO_2
F NO_2
XXVII

groups, but will also react with sulphydryl and imidazole groups. Relatively high pH values are required for a rapid reaction. The reagent is generally added as a concentrated acetone solution to an aqueous solution of the reactants and product formation followed by characteristic spectral changes.[108]

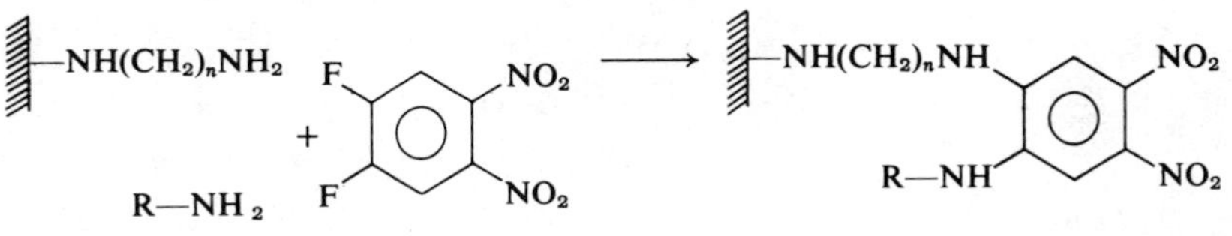

$-NH(CH_2)_nNH_2$ + R—NH_2 + F, F, NO_2, NO_2 ⟶ $-NH(CH_2)_nNH$, R—NH, NO_2, NO_2

d. Isocyanates

Isocyanates react with amines to form substituted ureas, with alcohols to form urethans and with water to give amines and carbon dioxide. At alkaline pH the reaction with amines is preferred. Compound XXVIII, 2,2′-dicarboxy-4,4′-azophenyldiisocyanate[109,110] is water-soluble and has the advantage that the bridge it forms can be readily cleaved by reduction of the azo group by dithionite.

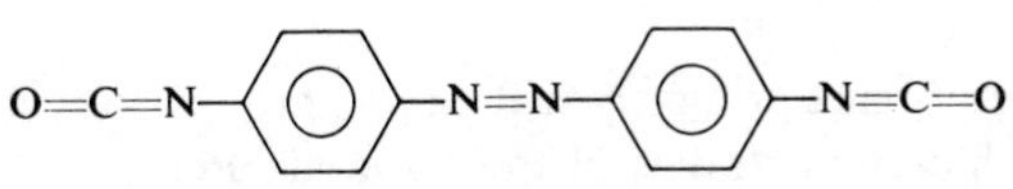

O=C=N— —N=N— —N=C=O

XXVIII

e. Acylating Agents

In principle, any aliphatic or aromatic dicarboxylic acid or disulphonic acid can be activated to provide bifunctional acylating agents capable of reacting under mild conditions. The nitrophenylesters of dicarboxylic acids[111] and the aromatic-*bis*-sulphonyl chlorides (XXIX) are good examples. They are insoluble in water and hydrolyse rapidly. The *bis*-sulphonyl chlorides react with amino groups to form stable sulphonamide linkages which can subsequently be cleaved with HBr in glacial acetic acid.[112]

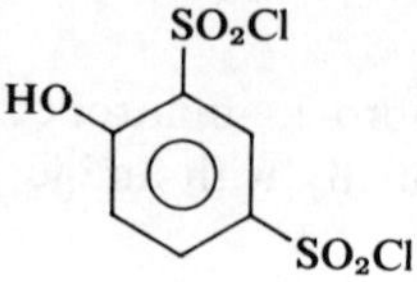

$$\text{HO}-C_6H_3(SO_2Cl)_2$$

XXIX

Phenol-2,4-disulphonyl chloride

f. Imidoesters

The synthesis of a number of bifunctional imidoesters (XXX) has been described.[113] They are soluble in water and react with amino groups under mild conditions and with a high degree of specificity. Dimethylsuberimidate[114] has been used in 0·2M triethanolamine-HCl buffer pH 8·5 for three hours at room temperature. The resulting amidine is stable to acid hydrolysis, but can be cleaved with ammonia.

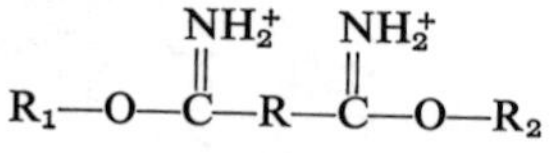

$$R_1-O-\overset{NH_2^+}{\overset{\|}{C}}-R-\overset{NH_2^+}{\overset{\|}{C}}-O-R_2$$

XXX

g. Dialdehydes

Dialdehydes such as glutaraldehyde (XXXI) form Schiff's bases with primary amines which on reduction with borohydride give stable secondary amines under mild conditions.[115]

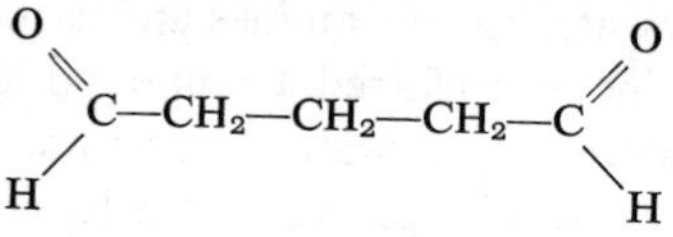

$$OHC-CH_2-CH_2-CH_2-CHO$$

XXXI

h. Vinylsulphones

Vinylsulphones[52] react primarily with amino groups but will, at higher pH values, react with carbohydrates, phenols and alcohols.

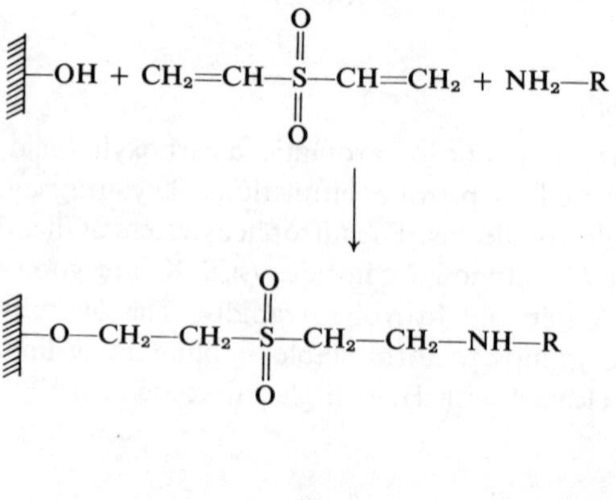

$$\text{[matrix]}-OH + CH_2{=}CH-SO_2-CH{=}CH_2 + NH_2-R$$

$$\downarrow$$

$$\text{[matrix]}-O-CH_2-CH_2-SO_2-CH_2-CH_2-NH-R$$

10. Thiolation Reactions

Thiol groups can be introduced into polymers by reacting ω-amino alkyl derivatives with *N*-acetylhomocysteine thiolactone for 24 hours at 40°C and pH 9·7.[14,124,125] Ligands carrying free carboxyl groups can be condensed with

$$\text{—NH(CH}_2\text{)}_n\text{NH}_2 + \text{H}_2\text{C}\langle\text{CH}_2\text{—S—C=O, CH(NHCOCH}_3\text{)}\rangle \xrightarrow[4^\circ\text{C}]{\text{pH 9·7}} \text{—NH(CH}_2\text{)}_n\text{NH—C(=O)—CH(NHCOCH}_3\text{)—CH}_2\text{—CH}_2\text{—SH}$$

such thiol-agarose derivatives with water-soluble carbodiimides. The resulting thiol ester linkage can be specifically cleaved by a short exposure to alkaline

$$\text{—SH} \xrightarrow[\text{Carbodiimide}]{\text{RCOOH}} \text{—S—C(=O)—R} \quad \text{Thiol ester}$$

$$\text{—SH} \xrightarrow{\text{RX}} \text{—S—R} \quad \text{Thiol ether}$$

pH or 1M neutral hydroxylamine.[14] In contrast, a stable thiol ester linkage is formed on reaction with alkyl halides.

Disulphide bridges can be formed by the oxidative coupling of a sulphydryl ligand to a polythiol matrix, in the presence of alkaline ferricyanide. The linkage is readily cleaved by brief exposure of the disulphide to 0·1M cysteine

$$\text{—SH} + \text{HS—R} \rightleftharpoons \text{—S—S—R}$$

hydrochloride at 25°C. Furthermore, amines can be converted to thiol groups by prior treatment with *N*-acetylhomocysteine thiolactone. Thus, reversible inert supports can be prepared which are of particular value as regenerable matrices for enzyme and proteins.

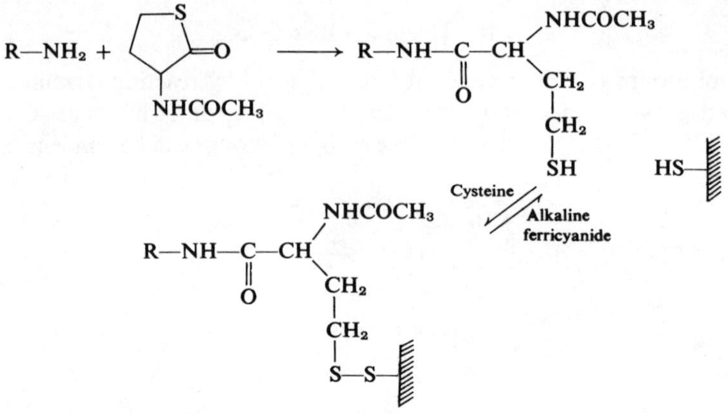

$$R{-}NH_2 + \text{(thiolactone ring: S, =O, NHCOCH}_3\text{)} \longrightarrow R{-}NH{-}\underset{\|}{\overset{}{C}}(=O){-}CH(NHCOCH_3){-}CH_2{-}CH_2{-}SH$$

HS—(matrix)

Cysteine / Alkaline ferricyanide

$$R{-}NH{-}C(=O){-}CH(NHCOCH_3){-}CH_2{-}CH_2{-}S{-}S{-}\text{(matrix)}$$

F. FUNCTIONAL GROUPS ON THE LIGAND

The previous section has suggested methods which should be generally applicable to the attachment of a variety of ligands to derivatized matrices. In some cases, however, it may be necessary to convert a functional group on the ligand into another such that coupling to a spacer arm on the inert support is facilitated. For example, there are few methods available for the linkage of ligands bearing ketone groups to aminoalkyl or carboxyl-alkyl spacer arms. This can, however, be accomplished by reaction of the ketone with *O*-carboxymethoxylamine-hemihydrochloride. For example, 6-keto-oestriol and *O*-carboxymethoxylamine-hemihydrochloride were heated under reflux for two hours in pyridine/water/methanol (1:1:4) and then left overnight at room temperature.[87] The reaction was completed by further refluxing for two hours and the product, Oestriol-6-(*O*-carboxymethyl)oxime extracted with ethyl acetate. The resulting *O*-carboxylmethyl oxime can subsequently be coupled to aminoalkyl-Sepharose by any of the methods described in Section E.

$$R{-}C{=}O + NH_2{-}O{-}CH_2{-}COOH \longrightarrow R{-}C{=}N{-}O{-}CH_2{-}COOH$$

Hydroxylic ligands can be hemisuccinylated with succinic anhydride (Section E.2) to yield carboxylic ligands. The coupling procedures commonly

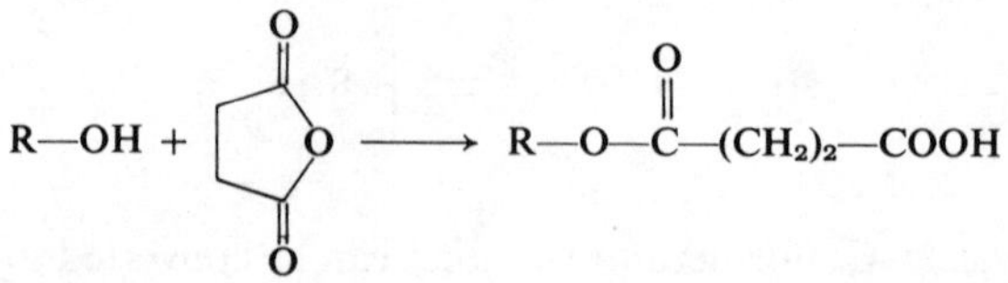

$$R{-}OH + \text{(succinic anhydride)} \longrightarrow R{-}O{-}\overset{\overset{O}{\|}}{C}{-}(CH_2)_2{-}COOH$$

employed for ligands bearing amino, carboxyl, benzenoid or phenolic, imidazole, thiol, hydroxyl, or aldehyde functions are summarized in Schemes I to V.

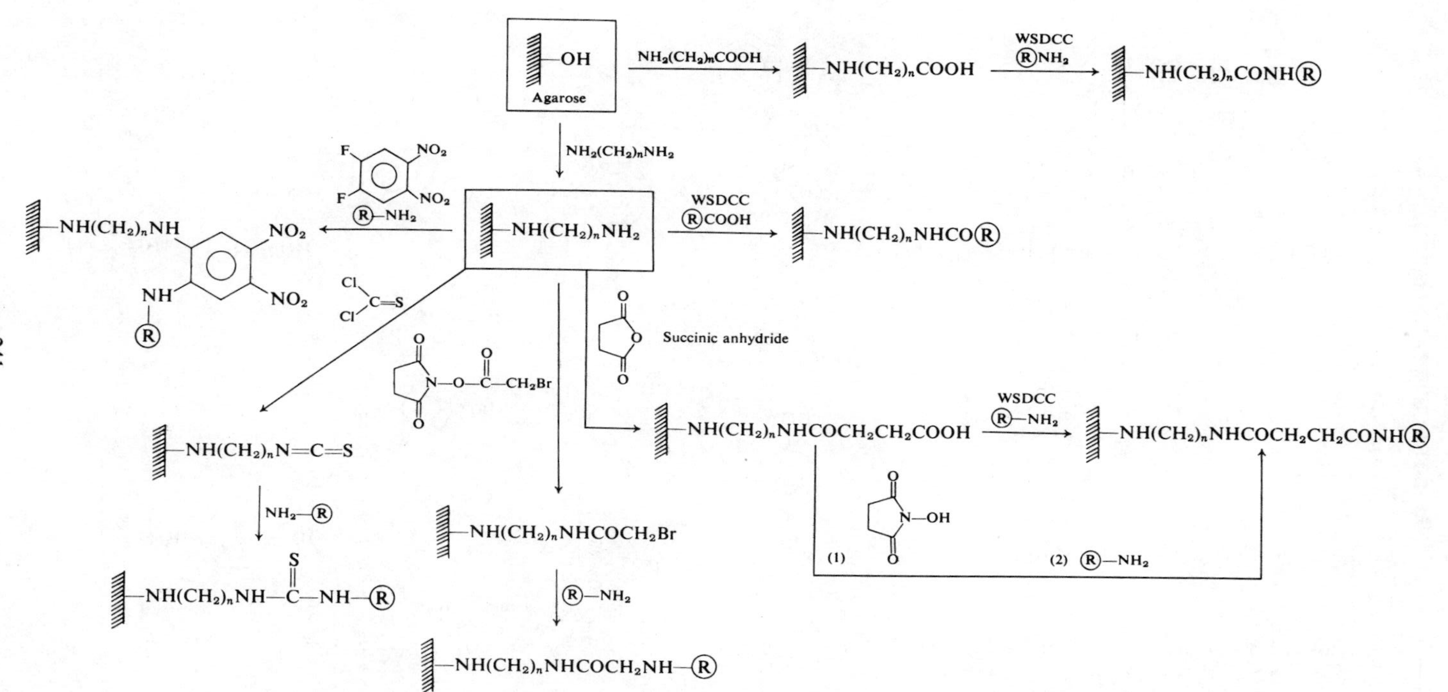

Scheme I. Reactions for coupling amino- and carboxyl-ligands

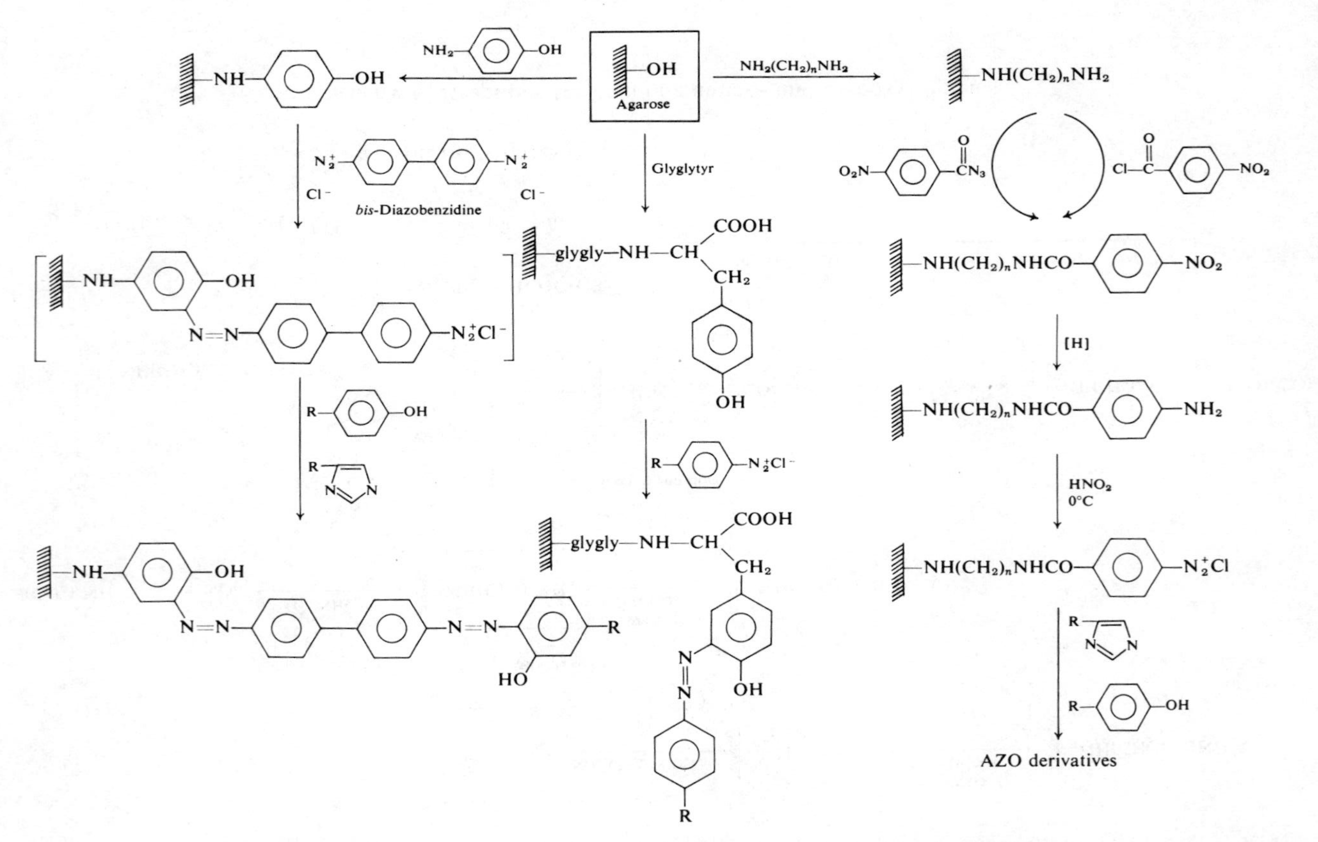

Scheme II. Reactions for coupling benzenoid, phenolic and diazotizable ligands

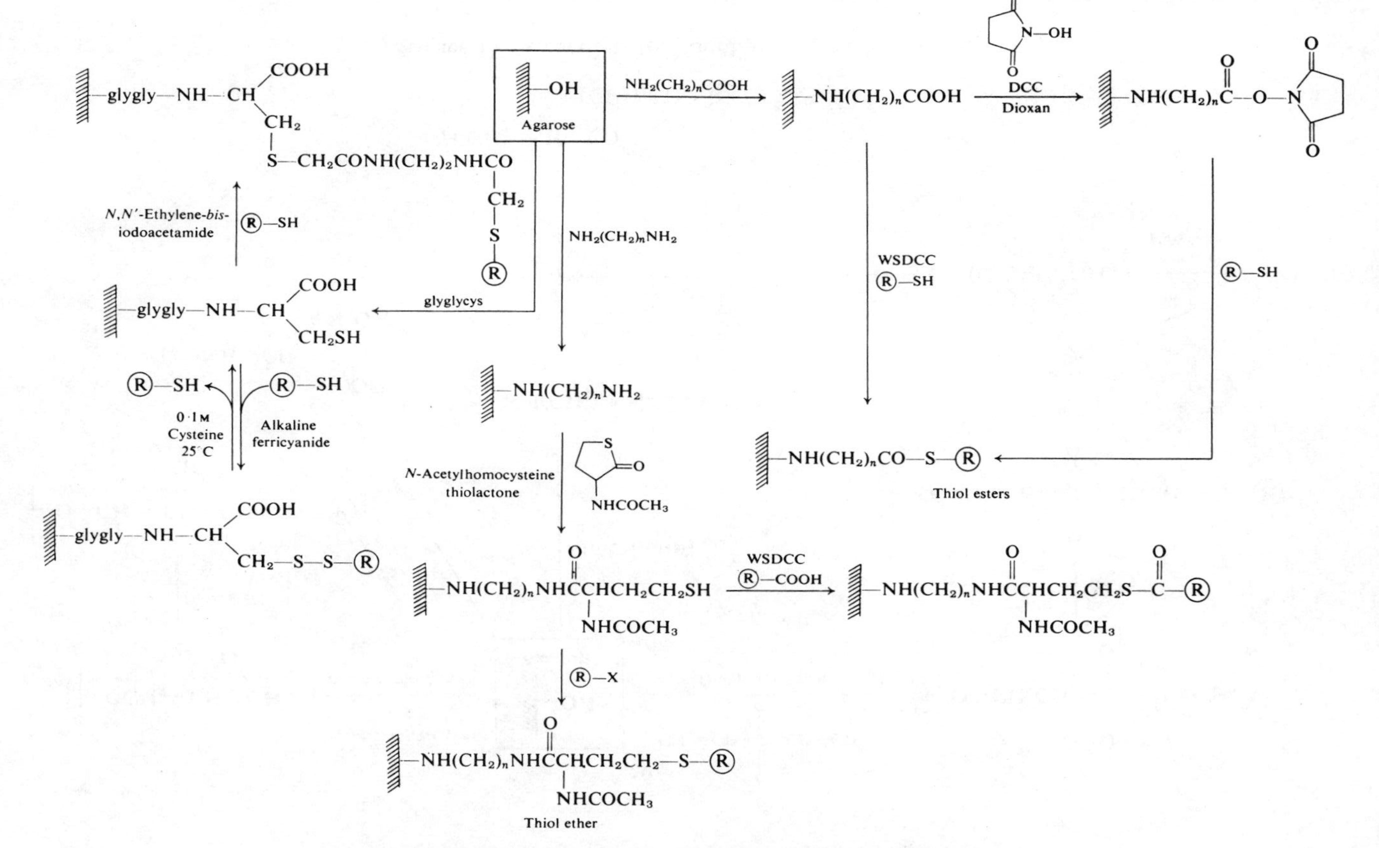

Scheme III. Reactions for coupling thiol ligands

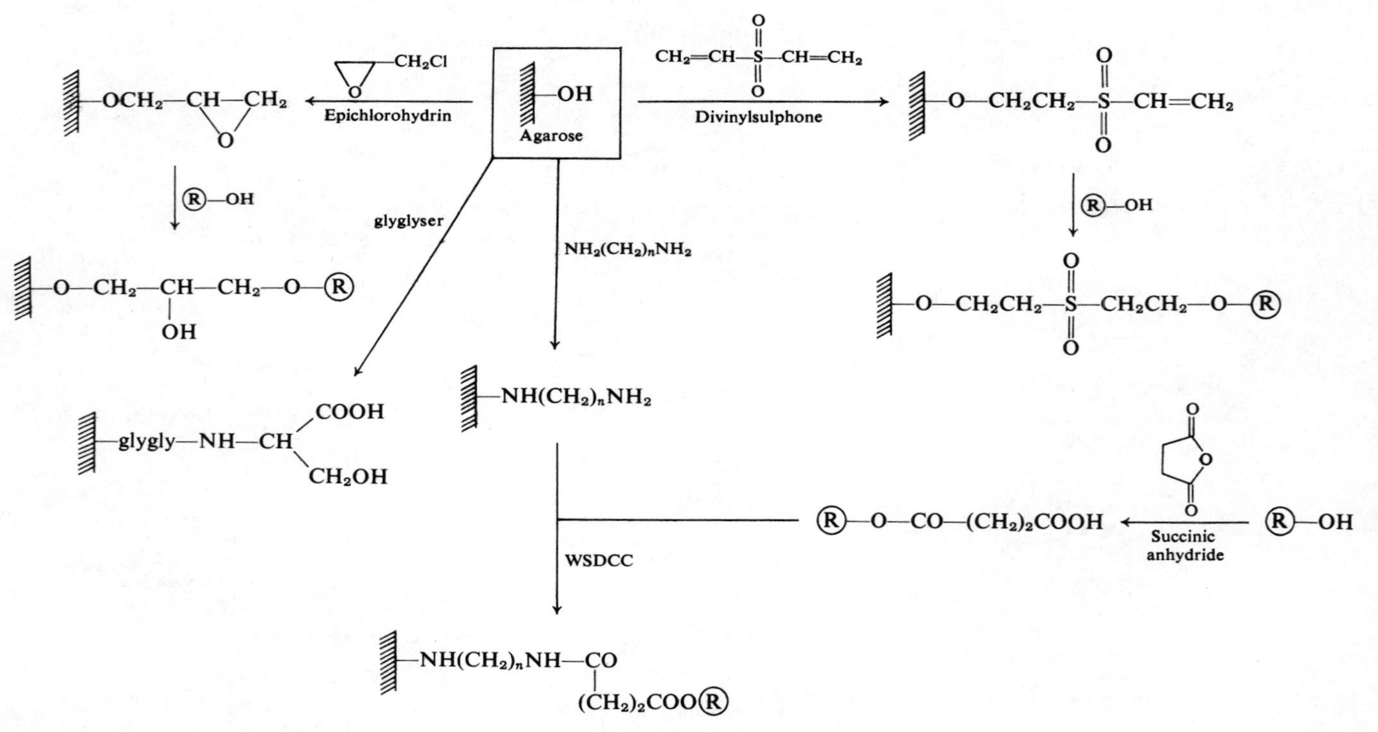

Scheme IV. Reactions for coupling hydroxylic ligands

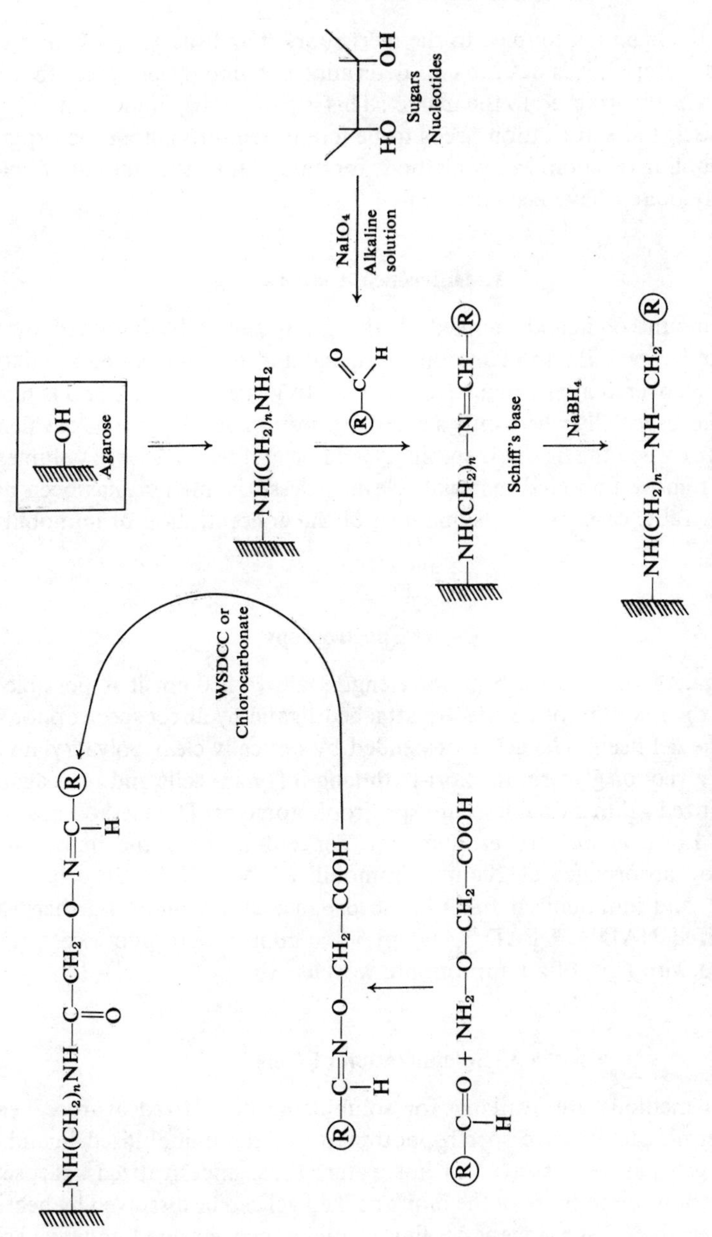

Scheme V. Reactions for coupling aldehydes and ketones

G. METHODS FOR QUANTIZATION OF IMMOBILIZED LIGANDS

When the ligand is coupled to the inert support and the gel washed free of unbound material, it is desirable to determine the amount of ligand that has been covalently attached to the matrix. This is particularly important in cases where the ligand substitution needs to be monitored throughout the progress of the coupling reaction. Many methods for quantitating the amount of ligand covalently bound have been devised.

1. Difference Analysis

The amount of ligand coupled to the gel is generally estimated by the difference between the total amount of ligand added to the coupling mixture and that recovered after exhaustive washing. In general, this method is highly inaccurate, especially where only a small proportion of the ligand is covalently attached or when the ligand is sparingly soluble and requires large volumes of wash to remove unbound material. Nevertheless, the method has been used in innumerable cases to give some idea of the concentration of immobilized ligand.

2. Direct Spectroscopy

For ligands which adsorb at wavelengths above 260 nm it is possible to estimate the quantity of covalently attached ligand by direct spectrophotometry of the gel itself. The gel is suspended by optically clear polyacrylamide, ethylene glycol or glycerol in short-path length (1 mm) cells and read against underivatized gel in a double-beam spectrophotometer. The method has been used by Lowe *et al.*[74] to estimate the concentration of the immobilized NAD^+ by absorbance at 206 nm, immobilized NADH by absorbance at 340 nm[86] and immobilized FAD by absorbance at 450 nm.[92] Furthermore, immobilized NADH,[86] FAD[92] and pyridoxamine[70] were fluorescent when irradiated with light of an appropriate wavelength.

3. Solubilization of Gels

Several methods are available for solubilizing derivatized agarose beads which permit quantitative spectrophotometry of the immobilized ligand.[126] Agarose gels can be dissolved in hot water; thus, underivatized agarose at concentrations up to those of the moist packed gel can be dissolved by heating to 60°C or above. Subsequent cooling of highly concentrated aqueous solutions of agarose generates turbid solutions whereas dilute solutions become viscous but transparent. These effects are independent of pH over the range of 1–14. In contrast, derivatized gels are not solubilized either by this treatment or by warming to 75°C with 8M urea or 6M guanidine–HCl. However, the gels can be rendered soluble by warming at 75°C with 0·1M HCl or in NaOH and, in most cases, by 50% acetic acid. Once solubilized the pH can be adjusted to neutrality without visible precipitation.

The spectra of solubilized agarose are shown in Fig. V.12. Agarose dissolved in water is transparent at wavelengths above 350 nm but absorbs a little between 240 nm and 350 nm. This absorbance is directly proportional to the concentration of agarose and hence easily eliminated with a suitable blank. Treatment of agarose with 0·1N HCl for two hours at 75°C produces a significant absorption peak at 280 nm, probably due to acid-catalysed

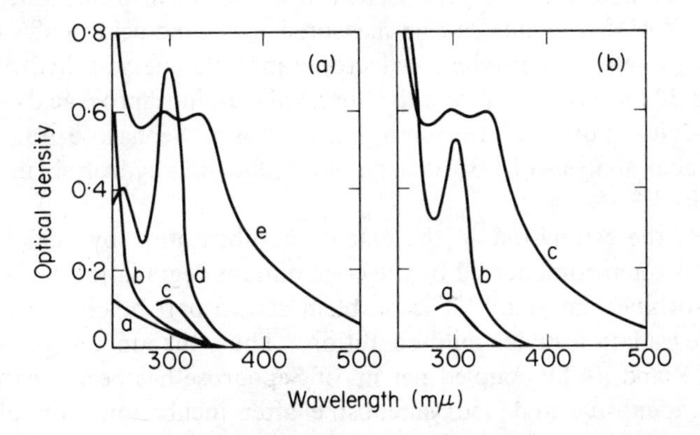

Figure V.12. Spectra of solubilized agarose. A suspension containing the equivalent of 0·2 ml of the agarose in 5·0 ml of the indicated solvent was dissolved by heating at 75°C for 2 hours. (A) H_2O (a), 0·1N NaOH–0·1% $NaBH_4$ (b), 50% HOAc (c), 0·1N HCl (d), N NaOH (diluted to 10·0 ml with N NaOH) (e). (B) 50% HOAc (a), 0·1N HCl (b), N NaOH (diluted to 10 ml with N NaOH prior to recording spectrum) (c). Reproduced with permission from D. Failla and D. V. Santi, *Anal. Biochem.*, **52**, 363 (1973)

furfuraldehyde production.[127,128] In contrast, treatment with 0·1–1M NaOH under the same conditions leads to partial caramelization of the gel beads with a broad absorption in the UV and visible spectra and the appearance of amber-coloured solutions. This effect could be prevented by the addition of 0·1% $NaBH_4$ prior to heating, whence spectra similar to those obtained in water were observed.

This solubilization technique[126] can be used in conjunction with the sodium 2,4,6-trinitrobenzene sulphonate (TNBS) test[14,63] to yield a quantitative assay for amino derivatized agarose. The amino-agarose is reacted with excess TNBS, thoroughly washed to remove picric acid and unreacted TNBS, and then solubilized by warming with 50% acetic acid. The absorbance at 340 nm gives an estimate of the amine content accurate to within 5%.[126]

4. Acid or Enzymic Hydrolysis

More vigorous treatment with acid will hydrolyse the matrix–ligand bond and liberate either free ligand or a degradation product which can be assayed. Thus, 5′-deoxyadenosylcobalamin-agarose was treated with 6N HCl at room temperature for 12 hours and the liberated corrinoids quantitatively estimated by conversion to dicyano derivatives with 0·1M KCN.[129] Galactosyl-pyrophosphate-Sepharose was estimated by phosphate analysis following hydrolysis in 0·5N HCl at 100°C for one hour.[130] Similarly, *N*-acetylglucosamine-Sepharose was hydrolysed in 6N HCl at 110°C and the liberated glucosamine estimated.[130] Many ligands can be measured by amino-acid analysis. Thus, Sepharose *N*-ε-aminocaproyl-β-D-galactopyranosylamine was hydrolysed in 6N HCl for 22 hours at 110°C and the liberated ε-aminocaproic acid estimated by amino-acid analysis.[131] Furthermore, L-tyrosine-Sepharose was assayed by amino-acid analysis of liberated tyrosine following hydrolysis in 6N HCl for 24 hours.[132]

However, the estimation of the extent of substitution by acid-catalysed hydrolysis is often complicated by the concomitant degradation of Sepharose to UV-absorbing material.[126] The problem can in part be circumvented by enzymic degradation under mild conditions. Thus, the amount of *p*-aminophenyl-ATP and *d*ATP coupled per ml of Sepharose has been measured by release of adenosine and deoxyadenosine after incubation with phosphodiesterase and alkaline phosphatase.[76] Similarly, immobilized AMP was determined by enzymic hydrolysis of the terminal phosphate with alkaline phosphatase followed by phosphate analysis.[133]

5. Elemental Analysis

In some cases, elemental analysis can give unequivocal estimates of the ligand concentration. Phosphate analysis has been used by several authors,[74,130,133,134] and sulphanilamide coupled to Sepharose with cyanogen bromide[135] has been determined by sulphur analysis.[136] Furthermore, 3-iodotyrosine-Sepharose[137] has been estimated by a method for protein-bound iodide.[138] Elemental nitrogen analysis can yield misleading values in view of the nitrogen introduced during the cyanogen bromide activation step.

6. Acid–Base Titration

Polar ligands can be estimated by titration of the gel with acid or alkali. Hixson and Nishikawa[139] demonstrated how this method could be applied to thrombin and trypsin adsorbents. Ligands, such as *p*-aminobenzamidine, which carry an amino group were condensed with carboxylated Sepharose with a water-soluble carbodiimide. The amount of ligand substitution was determined by titration. The carboxylate-Sepharose was washed with 50 volumes of 0·5M NaCl and the gel volume determined by settling for one hour or by gently centrifuging the beads in a calibrated cone. The beads were titrated from pH 7 to 3 and the procedure repeated after the amino ligand was coupled. The ligand concentration was deduced by difference. The concentration of *N*-6-(6-aminohexyl)AMP-Sepharose has also been determined by direct titration.[133]

7. Radioactivity

In many cases, radio-labelled ligands have been incorporated into the coupling procedure. For example, aliquots of a gel containing ($\gamma^{32}p$) GTP[99] were resuspended in Bray's solution[140] and counted in a liquid scintillation counter. Likewise, vitamin B_{12}-Sepharose was assayed by (^{57}Co) activity[141] and estradiol-Sepharose by the radioactivity of ^{3}H-estradiol.[14]

8. Special Methods for Sulphydryl Groups

Gel-bound sulphydryl groups can be readily estimated with Ellman's reagent,[142] 5,5′-dithio-*bis*-(2-nitrobenzoic acid), which liberates for each thiol group 1 mole of the strongly coloured thionitrobenzoate anion. The latter is determined by its absorption at 412 nm. This method has been used by

–SH + NO_2–C₆H₃(⁻OOC)–S–S–C₆H₃(COO⁻)–NO_2 $\xrightarrow{pH \geqslant 6{\cdot}8}$

–S–S–C₆H₃(COO⁻)–NO_2 + ⁻S–C₆H₃(COO⁻)–NO_2 + H^+

$\epsilon = 1{\cdot}36 \times 10^4\ M^{-1}\,cm^{-1}$ at pH 8

Cuatrecasas[14] for the determination of SH-groups and Cox and Lowe[92] for the estimation of dihydrolipoic acid covalently attached to Sepharose. Alternatively, 4,4′-dithiodipyridine[143] gives similar results. Sepharose-bound thiols can also be determined by their ability to take up ^{14}C-iodoacetamide when reacted with 0·01M iodoacetamide in 0·1M $NaHCO_3$, pH 8·0, for 15 minutes at room temperature.[14]

-SH + 4,4'-Dithiodipyridine ⟶ -S—S-(pyridyl) + 4-Thiopyridine

4,4'-Dithiodipyridine
$\lambda_{max} = 247$ nm
$\epsilon = 1{\cdot}63 \times 10^4\ M^{-1}\ cm^{-1}$

4-Thiopyridine
$\lambda_{max} = 324$ nm
$\epsilon = 1{\cdot}98 \times 10^4\ M^{-1}\ cm^{-1}$

H. LABORATORY TECHNIQUES

This section describes a few basic practical aspects of techniques commonly used for the preparation and use of affinity adsorbents. For a more exhaustive treatment the reader is referred to Sections B–F of this chapter and the references cited therein.

1. Washing and Storage of Agarose

Agarose must be well washed prior to activation or derivatization to remove bacteriostatic agents and other preservatives present in commercial preparations. The operation is conveniently carried out on a sintered funnel of large porosity. The gel is measured either by sucking moist on the funnel and weighing or by suspending the gel in a measuring cylinder in distilled water and reading the settled volume. The wet weight of the gel is approximately equivalent to its volume. Vigorous or prolonged stirring of agarose should be avoided.

Microbial growth can be greatly reduced by adding an antimicrobial agent such as sodium azide (NaN_3) at a final concentration of 0·02% (w/v) to all buffers and eluants. In the absence of phosphate ions, sugars, proteins or other nutrients, microbial growth is unlikely to occur, and in those cases where it may be detrimental to add a microbial growth inhibitor, flushing with degassed distilled water is recommended.

2. Cyanogen Bromide Activation of Agarose

Cyanogen bromide (CNBr) is available commercially as a white crystalline solid of melting point 51·3°C, boiling point 61·3°C and density 1·9–2·0 g/ml. Some batches of CNBr have a distinct yellow colour and rapidly become orange thereafter. It is important that such materials are destroyed because of the risk of explosion.

The solid very readily sublimes to a highly toxic vapour and consequently all manipulations should be performed in a well-ventilated hood. Cyanogen bromide can be purified by sublimation. The pure solid is sparingly soluble in water (25 mg/ml) and is rapidly hydrolysed in aqueous media to mixtures of hydrocyanous (HOCN), hydrocyanic (HCN), hypobromous (HOBr) and hydrobromic (HBr) acids. The potentially dangerous cyanides can be oxidized with nitrites or complexed with alkaline ferrous sulphate to produce the harmless and easily disposable ferrocyanides. Cyanogen bromide is very soluble in dimethylsulphoxide and acetone, although in the latter solvent bromoform can be formed under some circumstances.

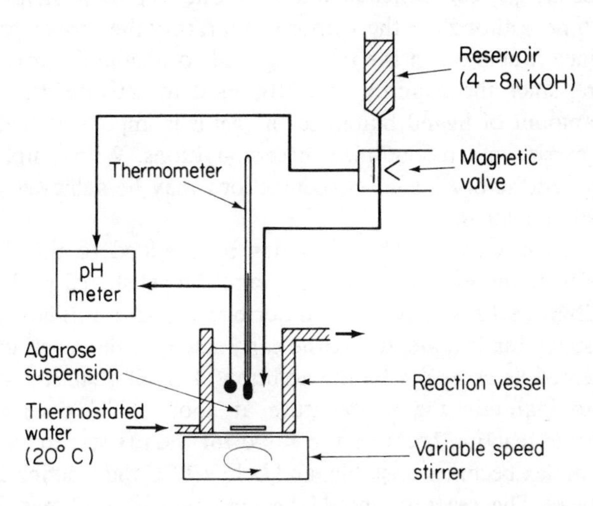

Figure V.13. Recommended equipment for the cyanogen bromide activation of agarose

The most convenient apparatus for the cyanogen bromide activation of agarose is depicted in Fig. V.13. It comprises a radiometer pH meter, an autotitrator with magnetic valve (radiometer TT3 and MM1) and a thermostatically-controlled reaction vessel. The magnetic valve allows a reservoir containing 4N or 8N KOH to empty at a controlled rate into the reaction vessel. The autotitrator is present to titrate upscale with a null point at pH 11·0, such that when the pH of the reaction vessel falls below 11·0, the magnetic valve is opened and strong alkali from the reservoir is injected into the reaction mix. When the pH rises above 11·0 the magnetic valve is closed and should thus be maintained constant at $11{\cdot}0 \pm 0{\cdot}1$. The outlet of the delivery tube from the magnetic valve is a glass tube of 1 mm internal diameter such as a 100 μl disposable pipette and ensures that a convenient volume of agarose (20 ml) can be activated in the optimum time of 8–10 minutes.

The thermostatted reaction vessel is fitted with a magnetic stirrer and charged with 20 ml of washed agarose and an equal volume of distilled water. A preweighed quantity of CNBr is added as a finely divided powder. Commercial cyanogen bromide has a tendency to sublime and coalesce into large crystalline aggregates on storage and consequently prior to addition to agarose it must be finely ground with a mortar and pestle. It is important that the grinding and all other operations with the cyanogen bromide be performed in a well-ventilated hood to avoid the toxic vapours given off by the solid. The powdered material can be weighed into a ground-glass stoppered bottle and transferred straight into the reaction vessel. The cyanogen bromide (2–5 g per 20 ml gel) can either be added in one step or in small quantities (0·5 g) at a time, although in the author's laboratory the former procedure is preferred since it cuts down the time required to handle the toxic material. Furthermore, since the quantity of CNBr used to activate the gel directly affects the amount of ligand bound to the gel it is important that adequate studies are carried out to optimize these conditions. With simple aliphatic amines very satisfactory ligand concentrations may be achieved with 2–5 g CNBr per 20 ml agarose.

As soon as the cyanogen bromide has been added to the agarose the temperature is maintained at about 20°C and the gel stirred to disperse any clumps of CNBr that may have formed during the addition. The thermostatic control is usually inadequate during the early stages of the reaction and must be supplemented as necessary by the addition of small quantities of ice. It is important to maintain the temperature at about 20°C since above this temperature the reaction becomes too rapid for the pH stat to cope and the activated complex becomes unstable, and below 20°C the reaction is too slow to be practical. The reaction should be complete in 8–12 minutes and is indicated by the cessation of proton release. A large amount of crushed ice is then rapidly added to the suspension and the entire reaction mixture transferred quickly to a coarse disc Buchner funnel. The gel is thoroughly washed under suction with a large volume of ice-cold 0·1 M sodium bicarbonate buffer pH 9·0 until all traces of CNBr, cyanide, etc., are eliminated. For 20 ml of activated agarose 1–5 litres of cold buffer are passed through the gel in the space of a few minutes. The final washings are removed under suction and the moist cake of gel transferred to an ice-cold solution of the ligand in the same buffer used to wash the gel and in a volume equal to that of the packed agarose. The entire procedure of washing, transferring the gel cake to the ligand and mixing should be performed as rapidly as possible at a low temperature since the activated agarose is unstable. The gel is gently stirred at 0–4°C on a magnetic stirrer for 16–20 hours to ensure complete loss of reactive groups. The substituted gel is finally washed with large volumes of water, 2 M KCl and appropriate buffers until it is established that ligand is no longer being released.

3. The Preparation of ω-Aminoalkyl Agarose

Aliphatic α,ω-diamines can be directly coupled to Sepharose by the cyanogen bromide procedure described above. A 1:1 (v/v) Sepharose 4B–water suspension is activated with 250 mg CNBr per ml of Sepharose and washed on a sintered funnel. The washed moist gel is added to a solution of 1,6-diaminohexane (2 mM per ml Sepharose) in 0·1M sodium bicarbonate buffer pH 10·0 and allowed to react for 16–20 hours at 4°C. The substituted gel is thoroughly washed with water and 2M KCl to ensure complete removal of the non-covalently bound diamine. This treatment can introduce up to 12 μM of 6-aminohexyl-groups per ml of agarose.

4. The Sodium 2,4,6-Trinitrobenzene Sulphonate Colour Test

This simple qualitative colour test, first introduced by Inman and Dintzis,[63] is a useful way of following the course of substitution in agarose and polyacrylamide gels. A small quantity of the derivatized gel (0·2–0·5 ml in distilled water) is added to a solution of saturated sodium borate (1 ml) and

$$\text{—NH(CH}_2)_6\text{NH}_2 \xrightarrow[\text{pH} > 9]{\text{TNBS}} \text{—NH(CH}_2)_6\text{NH—C}_6\text{H}_2(\text{NO}_2)_3$$

three drops of a 3% aqueous solution of sodium 2,4,6-trinitrobenzene sulphonate (TNBS) added. The colour is allowed to develop for two hours at room temperature. Table V.5 illustrates the range of colour products that can be formed with this reagent. The progress of substitution of the amino-derivatives by carboxylic ligands and of hydrazide derivatives by amino

Table V.5. Colours produced by the TNBS test

Derivative	Colour
Unsubstituted agarose or polyacrylamide	Pale yellow
Carboxylic and bromoacetyl	Yellow
Primary aliphatic amines	Orange
Primary aromatic amines	Red-orange
Unsubstituted hydrazides	Deep red

ligands can be estimated by the relative colour intensity of the gels. Furthermore, the test can be quantitated by washing the gels to remove picric acid and unreacted TNBS, and solubilizing by warming with 50% acetic acid.[126] The absorbance at 340 nm gives an estimate of the amine content of the gel.

5. Water-soluble Carbodiimide Coupling

The coupling of L-threonine via its α-carboxyl function to 6-aminohexyl-Sepharose will be taken to illustrate the general procedure although the precise conditions employed will be to a certain extent dictated by the system under study. To 5 ml of washed 6-aminohexyl-Sepharose containing approximately 2 μmoles of amino functions per ml were added 0·1 mM L-threonine and the pH of the mixture adjusted to 4·7 with 0·1N HCl. The water-soluble carbodiimide, 1-ethyl-3-(3-dimethylaminopropyl) carbodiimide hydrochloride (0·25 mM) in 0·5 ml water was added dropwise over a 5-minute period and the pH maintained at 4·7 for 20 hours at room temperature by continuous titration with 0·1N HCl on a radiometer autotitrator. The product was washed successively with water, 2M KCl, water and buffer prior to use.

Other conditions for water-soluble carbodiimides are outlined in Table V.2. In some circumstances, where the ligand is sparingly soluble in aqueous media, the coupling reaction can be performed in aqueous ethanol, *N,N'*-dimethylformamide or pyridine.

6. The Preparation of Derivatives for Diazonium Coupling

6-Aminohexyl-Sepharose prepared as described above is suspended in 4 volumes of 40% (v/v) redistilled *N,N'*-dimethylformamide containing 0·2M sodium borate buffer pH 9·3 and 0·07M *p*-nitrobenzoyl azide dissolved in pure *N,N'*-dimethylformamide. The resulting mixture is stirred at room temperature for 1–2 hours. The resulting *p*-nitrobenzamidohexyl-Sepharose is extensively washed with 50% (v/v) *N,N'*-dimethylformamide, resuspended in 1–2 volumes of 0·5M sodium bicarbonate buffer pH 8·5 containing 0·1M sodium dithionite and incubated at 38°C for 1 hour. The product is thoroughly washed with distilled water and suspended in ice-cold water for 5 minutes to lower the temperature of the gel. The resulting gel is sucked moist and diazotized by addition of 5 volumes of ice-cold 0·5M HCl containing 0·1M sodium nitrite and stirred at 0°C for 7–8 minutes. The resulting diazotized gel is rapidly washed with ice-cold 1% urea or sulphamic acid to remove excess nitrous acid and finally again with ice-cold distilled water. The moist gel is then added to the phenolic- or imidazole-bearing ligand in 0·2M sodium borate buffer pH 8·3 and the suspension stirred overnight at 0–4°C. The final adsorbent is washed with appropriate buffers.

7. Chromatographic Procedures

The practical aspects of column chromatographic technique are adequately covered in several monographs on gel chromatography[144,145] and will not be described here. However, in the authors' experience rapid assessment of potential affinity adsorbents can be made on miniature columns. Micro-columns (5 mm × 20 mm) prepared from disposable-type pipettes are packed with 0·5 ml of the adsorbent and equilibrated with an appropriate buffer. Enzyme samples (50 μl) are applied to the top of the moist column, allowed to run into the bed and washed in with a small volume of the equilibrating buffer. Non-adsorbed protein is washed off with 6 ml of the equilibration buffer and enzyme eluted with a linear gradient of KCl (0–1M; 20 ml total volume) in the same buffer. Fractions (1·6 ml) are collected at a flow rate of 8–10 ml/hour and assayed for protein, KCl (conductivity) and enzyme activity. Up to 10 miniature columns can be run simultaneously and the whole procedure completed within a few hours.

REFERENCES

1. Arond, L. H., and Frank, H. P. (1954): *J. Phys. Chem.*, **58**, 953.
2. Ricketts, C. R. (1961): *Progr. Org. Chem.*, **5**, 73.
3. Determann, H. (1968): in *Gel Chromatography*, Springer-Verlag, New York, p. 15.
4. Fischer, L. (1969): in *An Introduction to Gel Chromatography* (Eds. T. S. Work and E. Work), North-Holland, Amsterdam, p. 181.
5. Cruft, H. J. (1961): *Biochim. Biophys. Acta*, **54**, 609.
6. Konigsberg, W., Weber, K., Notani, G., and Zinder, N. (1966): *J. Biol. Chem.*, **241**, 2579.
7. Determann, H. (1968): in *Gel Chromatography*. Springer-Verlag, New York, p. 28.
8. Araki, C. (1956): *Bull. Chem. Soc. Japan*, **29**, 543.
9. Russel, B., Mead, T. H., and Polson, A. (1964): *Biochim. Biophys. Acta*, **86**, 169.
10. Hjerten, S. (1961): *Biochim. Biophys. Acta*, **53**, 514.
11. Fish, W. W. (1966): *J. Biol. Chem.*, **244**, 4985.
12. Gwynne, J. T., and Tanford, C. (1970): *J. Biol. Chem.*, **245**, 3269.
13. Fischer, L. (1969): in *An Introduction to Gel Chromatography* (Eds. T. S. Work and E. Work), North-Holland, Amsterdam, p. 193.
14. Cuatrecasas, P. (1970): *J. Biol. Chem.*, **245**, 3059.
15. Larsson, P.-O., and Mosbach, K. (1971): *Biotechnol. Bioeng.*, **13**, 393.
16. Cuatrecasas, P., and Parikh, I. (1972): *Biochemistry*, **11**, 2291.
17. Hjerten, S., and Mosbach, R. (1962): *Anal. Biochem.*, **3**, 109.
18. Haller, W. (1965): *Nature* (*London*), **206**, 693.
19. Levin, Y., Pecht, M., Goldstein, L., and Katchalski, E. (1964): *Biochemistry*, **3**, 1913.
20. Calam, D., and Thomas, H. (1972): *Biochim. Biophys. Acta*, **276**, 378.

21. Epton, R., McLaren, J., and Thomas, T. (1972): *Carbohydrate Res.*, **22**, 301.
22. Barker, S. A., Somers, P., Epton, R., and McLaren, J. (1970): *Carbohydrate Res.*, **14**, 287.
23. Manecke, G. (1962): *Pure Appl. Chem.*, **4**, 507.
24. Manecke, G., and Singer, S. (1960): *Makromol. Chem.*, **39**, 13.
25. Filippusson, H., and Hornby, W. (1970): *Biochem. J.*, **120**, 215.
26. Bar-Eli, A., and Katchalski, E. (1963): *J. Biol. Chem.*, **238**, 1690.
27. Cebra, J., Givol, D., Silman, H., and Katchalski, E. (1961): *J. Biol. Chem.*, **236**, 1720.
28. Patty, F. A. (1949): in *Industrial Hygiene and Toxicology*, Vol. 2, Interscience, New York, p. 634.
29. Axen, R., Porath, J., and Ernback, S. (1967): *Nature* (*London*), **214**, 1302.
30. Porath, J., Axen, R., and Ernback, S. (1967): *Nature* (*London*), **215**, 1491.
31. Axen, R., and Ernback, S. (1971): *Eur. J. Biochem.*, **18**, 351.
32. Bartling, G., Brown, H., Forrester, L., Koes, M., Mather, A., and Stasiw, R. (1972): *Biotechnol. Bioeng.*, **14**, 1039.
33. Addor, R. W. (1963): *J. Org. Chem.*, **29**, 738.
34. Hedayatullah, M. (1967): *Bull. Soc. Chim.* (*France*), **75**, 416.
35. Kagedal, L., and Akerstrom, S. (1971): *Acta Chem. Scand.*, **25**, 1855.
36. Kagedal, L., and Akerstrom, S. (1970): *Acta Chem. Scand.*, **24**, 1601.
37. Ahrgren, L., Kagedal, L., and Akerstrom, S. (1972): *Acta Chem. Scand.*, **26**, 285.
38. Svensson, B. (1973): *FEBS Lett.*, **29**, 167.
39. Axen, R., and Vretblad, P. (1971): *Acta Chem. Scand.*, **25**, 2711.
40. Surinov, B., and Manoilov, S. (1966): *Biokhimiya*, **31**, 387.
41. Kay, G., and Crook, E. (1967): *Nature* (*London*), **216**, 514.
42. Cuatrecasas, P., Wilchek, M., and Anfinsen, C. B. (1968): *Proc. Nat. Acad. Sci. U.S.A.*, **61**, 636.
43. Dean, P. D. G., and Lowe, C. R. (1972): *Biochem. J.*, **127**, 11P.
44. Sanderson, C. J., and Wilson, D. V. (1971): *Immunol.*, **20**, 1061.
45. Head, F. S. H. (1963): *J. Textile Inst.*, **44**, T209.
46. Porath, J., and Fornstedt, N. (1970): *J. Chromatog.*, **51**, 479.
47. Mitz, M., and Summaria, L. (1961): *Nature* (*London*), **189**, 576.
48. Jagendorf, A., Patchornik, A., and Sela, M. (1963): *Biochim. Biophys. Acta*, **78**, 516.
49. Sato, T., Mori, T., Tosa, T., and Chibata, I. (1971): *Arch. Biochem. Biophys.*, **147**, 788.
50. Campbell, D., Leuscher, E., and Lerman, L. (1951): *Proc. Nat. Acad. Sci. U.S.A.*, **37**, 575.
51. Axen, R., and Porath, J. (1966): *Nature* (*London*), **210**, 367.
52. Porath, J., and Sundberg, L. (1972): *Nature New Biol.*, **238**, 261.
53. Porath, J., and Sundberg, L. (1970): *Protides Biol. Fluids*, **18**, 401.
54. Avrameas, S., and Ternynck, T. (1969): *Immunochemistry*, **6**, 53.
55. Baum, G., Ward, F. B., and Weetall, H. H. (1972): *Biochim. Biophys. Acta*, **268**, 411.
56. Weetall, H. H. (1971): *Nature* (*London*), **232**, 473.
57. Weetall, H. H., and Baum, G. (1970): *Biotechnol. Bioeng.*, **12**, 399.
58. Weibel, M. K., Weetall, H. H., and Bright, H. J. (1971): *Biochem. Biophys. Res. Commun.*, **44**, 347.
59. Hornby, W. E., Inman, D. J., and McDonald, A. (1972): *FEBS Lett.*, **23**, 114.
60. Inman, D. J., and Hornby, W. E. (1972): *Biochem. J.*, **129**, 255.
61. Hornby, W. E., and Filippusson, H. (1970): *Biochim. Biophys. Acta*, **220**, 343.

62. Sundaram, P., and Hornby, W. (1970): *FEBS Lett.*, **10**, 325.
63. Inman, J. K., and Dintzis, H. M. (1969): *Biochemistry*, **8**, 4074.
64. Weston, P. D., and Avrameas, S. (1971): *Biochem. Biophys. Res. Commun.*, **45**, 1574.
65. Grossman, S., Trop, M., Yaroni, S., and Wilchek, M. (1972): *Biochim. Biophys. Acta*, **289**, 77.
66. Kristiansen, T., Einarsson, M., Sundberg, L., and Porath, J. (1970): *FEBS Lett.*, **7**, 294.
67. O'Carra, P., and Barry, S. (1972): *FEBS Lett.*, **21**, 281.
68. Schmer, G. (1972): *Hoppe-Seyler's Z. Physiol. Chem.*, **353**, 810.
69. Kaufman, B. T., and Pierce, J. V. (1971): *Biochem. Biophys. Res. Commun.*, **44**, 608.
70. Collier, R., and Kohlhaw, G. (1971): *Anal. Biochem.*, **42**, 48.
71. Steers, E., Cuatrecasas, P., and Pollard, H. B. (1971): *J. Biol. Chem.*, **246**, 196.
72. Harris, R. G., Rowe, J. J. M., Stewart, P. S., and Williams, D. C. (1973): *FEBS Lett.*, **29**, 189.
73. Lefkowitz, R. J., Haber, E., and O'Hara, D. (1972): *Proc. Nat. Acad. Sci. U.S.A.*, **69**, 2828.
74. Lowe, C. R., Harvey, M. J., Craven, D. B., and Dean, P. D. G. (1973): *Biochem. J.*, **133**, 499.
75. Lowe, C. R., and Dean, P. D. G. (1971): *FEBS Lett.*, **14**, 313.
76. Berglund, O., and Eckstein, F. (1972): *Eur. J. Biochem.*, **28**, 492.
77. Mapes, C. A., and Sweeley, C. C. (1972): *FEBS Lett.*, **25**, 279.
78. Claeyssens, M., Kersters-Hilderson, H., Van Wauwe, J. P., and De Bruyne, C. K. (1970): *FEBS Lett.*, **11**, 336.
79. Frischauf, A. M., and Eckstein, F. (1973): *Eur. J. Biochem.*, **32**, 479.
80. Hofstee, B. H. J. (1973): *Anal. Biochem.*, **52**, 430.
81. Er-El, Z., Zaidenzaig, Y., and Shaltiel, S. (1972): *Biochem. Biophys. Res. Commun.*, **49**, 383.
82. Sica, V., Nola, E., Parikh, I., Puca, G. A., and Cuatrecasas, P. (1973): *Nature New Biol.*, **244**, 36.
83. Hoare, D. G., and Koshland, D. E. (1966): *J. Amer. Chem. Soc.*, **88**, 2057.
84. Hoare, D. G., and Koshland, D. E. (1967): *J. Biol. Chem.*, **242**, 2447.
85. Gutteridge, S., and Robb, D. A. (1973): *Biochem. Soc. Trans.*, **1**, 519.
86. Lowe, C. R., and Dean, P. D. G.: unpublished observations.
87. Dean, P. D. G., Rowe, P. H., and Exley, D. (1972): *Steroids Lipids Res.*, **3**, 82.
88. Patel, R. P., and Price, S. (1967): *Biopolymers*, **5**, 583.
89. Whiteley, J. M., Jackson, R. C., Mell, G. F., Drais, J. H., and Huennekens, F. M. (1972): *Arch. Biochem. Biophys.*, **150**, 15.
90. Salter, D. N., Ford, J. E., Scott, K. J., and Andrews, P. (1972): *FEBS Lett.*, **20**, 301.
91. Marcus, S. L., and Balbinder, E. (1972): *Anal. Biochem.*, **48**, 448.
92. Cox, E. A., and Lowe, C. R.: unpublished observations.
93. Weetall, H. H. (1969): *Science*, **166**, 615.
94. Cuatrecasas, P., Taniuchi, H., and Anfinsen, C. B. (1968): *Brookhaven Symposia in Biology*, **21**, 172.
95. Means, G. E., and Feeney, R. E. (1968): *Biochemistry*, **7**, 2192.
96. Ryan, E., and Fottrell, P. F. (1972): *FEBS Lett.*, **23**, 73.
97. Gilham, P. T. (1971): *Methods Enzymol.*, **21**, 191.
98. Robberson, D. L., and Davidson, N. (1972): *Biochemistry*, **11**, 553.
99. Jackson, R. J., Wolcott, R. M., and Shiota, T. (1973): *Biochem. Biophys. Res. Commun.*, **51**, 428.

100. Remy, P., Birmele, C., and Ebel, J. P. (1972): *FEBS Lett.*, **27**, 134.
101. Wold, F. (1967): *Methods Enzymol.*, **11**, 617.
102. Moore, J. E., and Ward, W. H. (1956): *J. Amer. Chem. Soc.*, **78**, 2414.
103. Fasold, H., Groschel-Stewart, U., and Turba, F. (1963): *Biochem. Z.*, **337**, 425.
104. Kovacic, P., and Hein, R. W. (1959): *J. Amer. Chem. Soc.*, **81**, 1187.
105. Gundlach, G. (1965): *Habilitationsschrift*, Thesis, Wurzburg, p. 44.
106. Ozawa, H. (1967): *Biochem. J. (Tokyo)*, **62**, 531.
107. Zahn, H., and Stuerle, H. (1958): *Biochem. Z.*, **331**, 29.
108. Zahn, H., and Meienhofer, J. (1958): *Makromol. Chem.*, **26**, 126.
109. Fasold, H. (1965): *Biochem. Z.*, **342**, 288.
110. Fasold, H. (1965): *Biochem. Z.*, **342**, 295.
111. Zahn, H., and Schade, F. (1963): *Angew. Chem.*, **75**, 377.
112. Herzig, D. J., Rees, A. W., and Day, R. A. (1964): *Biopolymers*, **2**, 349.
113. Fasold, H. (1964): *Biochem. Z.*, **339**, 482.
114. Davies, G. E., and Stark, G. R. (1970): *Proc. Nat. Acad. Sci. U.S.A.*, **66**, 651.
115. Richards, F. M., and Knowles, J. R. (1968): *J. Mol. Biol.*, **37**, 231.
116. Quiocho, F. A., and Richards, F. M. (1966): *Biochemistry*, **5**, 4062.
117. Wold, F. (1961): *J. Biol. Chem.*, **236**, 106.
118. Ozawa, H. (1967): *J. Biochem. (Tokyo)*, **62**, 419.
119. Kay, C. M., and Edsall, J. T. (1956): *Arch. Biochem. Biophys.*, **65**, 354.
120. Silman, H. I., Albu-Weissenberg, M., and Katchalski, E. (1966): *Biopolymers*, **4**, 441.
121. Hartman, F. C., and Wold, F. (1967): *Biochemistry*, **6**, 2439.
122. Dutton, A., Adams, M., and Singer, S. J. (1966): *Biochem. Biophys. Res. Commun.*, **23**, 730.
123. Moore, G. L., and Day, R. A. (1968): *Science*, **159**, 210.
124. Benesch, R., and Benesch, R. E. (1956): *J. Amer. Chem. Soc.*, **78**, 1597.
125. Benesch, R., and Benesch, R. E. (1958): *Proc. Nat. Acad. Sci. U.S.A.*, **44**, 848.
126. Failla, D., and Santi, D. V. (1973): *Anal. Biochem.*, **52**, 363.
127. Haworth, N., and Jones, W. (1944): *J. Chem. Soc.*, **81**, 667.
128. Jencks, W. P. (1959): *J. Amer. Chem. Soc.*, **81**, 475.
129. Yamada, R. H., and Hogenkamp, H. P. C. (1972): *J. Biol. Chem.*, **247**, 6266.
130. Barker, R., Olsen, K. W., Shaper, J. H., and Hill, R. L. (1972): *J. Biol. Chem.*, **247**, 7135.
131. Gordon, J. A., Blumberg, S., Lis, H., and Sharon, N. (1972): *FEBS Lett.*, **24**, 193.
132. Chan, W. W. C., and Takahashi, M. (1969): *Biochem. Biophys. Res. Commun.*, **37**, 272.
133. Craven, D. B., Harvey, M. J., Lowe, C. R., and Dean, P. D. G. (1973): *Eur. J. Biochem.*, in preparation.
134. Frischauf, A. M., and Eckstein, F. (1973): *Eur. J. Biochem.*, **32**, 479.
135. Falkbring, S. O., Gothe, P. O., Nyman, P. O., Sundberg, L., and Porath, J. (1972): *FEBS Lett.*, **24**, 229.
136. Gustavsson, L. (1970): *Talanta*, **4**, 227.
137. Poillon, W. N. (1971): *Biochem. Biophys. Res. Commun.*, **44**, 647.
138. Kingsley, G. R., and Schaffert, R. R. (1958): *Standard Methods in Clin. Chem.*, **2**, 147.
139. Hixson, H. F., and Nishikawa, A. H. (1973): *Arch. Biochem. Biophys.*, **154**, 501.

140. Bray, G. A. (1960): *Anal. Biochem.*, **1**, 279.
141. Allen, R. H., and Majerus, P. W. (1972): *J. Biol. Chem.*, **247**, 7695.
142. Ellman, G. L. (1959): *Arch. Biochem. Biophys.*, **82**, 70.
143. Grassetti, D. R., and Murray, J. F. (1967): *Arch. Biochem. Biophys.*, **119**, 41.
144. Fischer, L. (1969): *An Introduction to Gel Chromatography* (Eds. T. S. Work and E. Work), North-Holland Publishing, Amsterdam.
145. Determann, H. (1968): *Gel Chromatography*, Springer-Verlag, New York.

Index